The Earth:
Past, Present and Future

An Introduction to Geology

The Earth:
Past, Present and Future

An Introduction to Geology

MICHAEL BRADSHAW
Principal Lecturer in Geography and Geology
College of St. Mark and St. John, Plymouth

HODDER AND STOUGHTON
LONDON SYDNEY AUCKLAND TORONTO

Cover Genesis: Artist's impression of the surface of the Earth as it began to crust over. Reproduced from 'Story of the Earth' by permission of the Director, Institute of Geological Sciences: NERC Copyright reserved.

British Library Cataloguing in Publication Data

Bradshaw, Michael John
The earth, past, present and future.
  1. Geology
  I. Title
  550   QE28

ISBN 0 340 23948 4

First printed 1980
Third impression 1987

Printed for
Hodder and Stoughton Educational,
a division of Hodder and Stoughton Ltd.,
Mill Road, Dunton Green, Sevenoaks, Kent
by Colorcraft Ltd.
Hong Kong

Preface

In planning a Third Edition of *A New Geology* (first published in 1968, Second Edition – completely revised – 1973) I found that the extent of revision required called for a completely new approach – hence a new title and format. This is mainly a reflection of changes in the subject, which were only partly covered in the 1973 edition of *A New Geology*, and which have since crystallised and been incorporated into school syllabuses. This new book, *The Earth: Past, Present and Future*, is based on the ideas emanating from plate theory, together with the implications of the 'revolution' in geology which this theory has accomplished in the last ten years. These implications are investigated in relation to studies of volcanic and earthquake activity, rock formation and the historical evolution of the earth's surface features and the successions of rocks. The opportunity is also taken to include new illustrations and pupil investigations, to examine the features of planets studied by American space shots, and to highlight some of the environmental and resource issues which must impinge increasingly on the thinking of all geologists in the 1980s.

The book is aimed at the same groups as *A New Geology* – those who require an introduction to geology for general interest, those who are studying geology for 16+ examinations or those who are studying the subject for the first time in the Sixth Form.

I am indebted to my colleague, Dr Richard Mayhew, for close discussion concerning the initial outline, and also for detailed comments on most of the chapters. Roger Stone, the consulting editor, has also made many suggestions for improving the text. Photographs are acknowledged in the text, but a particular word of thanks must go to Wil Dooley of the United States Geological Survey for his continuing assistance. Mr John Saunders of Exeter University provided the photographs of minerals and fossils. [USGS – United States Geological Survey; NASA – National Aeronautics and Space Administration; USDASCS – United States Department of Agriculture Soil Conservation Service.]

Michael Bradshaw
1979

Contents

Part I
Geology, the earth science

1

The earth – a unique planet?

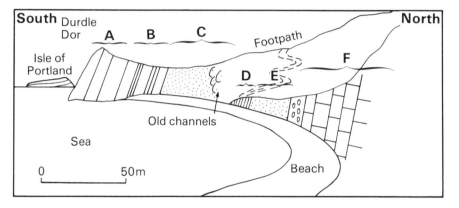

Fig. 1.1 The rocks at Man o' War Bay, near Lulworth, Dorset. They are almost vertical, but were formed as horizontal layers in an ancient sea. An amateur geologist could put the following account together after examining the rocks and extracting some of the fossils they contain with simple equipment. Rock **A** is a hard limestone, containing fossils of marine snails and giant ammonites, which can be seen clearly in the quarries on the Isle of Portland. Although the ammonites have become extinct, they most resembled squids, which are marine animals: taken with the other evidence, this rock was probably formed on the sea floor. Rock **B** is a group of repeated alternations of muddy limestone and limy mudstones. These have been found to contain insect and tree trunk fossils, and were probably formed in a coastal lagoon. Rock **C** is a coarse sand with blackened pieces of plant remains and is crossed by lines suggesting the cross-sections of former stream channels. These channels have their bases on the southern side, showing that the oldest rocks in this sequence (those at the bottom of the pile) are those at **A**. The presence of former stream channels and plant remains suggests that Rock **C** was formed by streams, probably flowing across a coastal plain. So far, the rocks **A**–**C** demonstrate a retreat of sea, exposing more and more land. Rock **D** is a fine mudstone which contains small ammonite fossils, quite different from those in Rock **A**, but also probably of marine habitat. This suggests that the sea advanced over the older rocks at this stage. Rock **E** is a green sandstone containing marine bivalve fossils like cockles. Rock **F** is the very distinctive white chalk. At the base is a band of marine sea urchin fossils. Rocks **D**–**F** thus confirm a pattern of seas extending over the area, and possibly becoming deeper: there was little sediment from the land reaching the area by the time that the chalk was formed.

Questions

We are all conscious that the area in which we live has a foundation of hills and valleys, flat areas or mountains. Most of us realise that the soils in our gardens and the raw materials used in local factories come from resources stored in the earth for millions of years. The fabric of our lives – building materials, metals and plastics and even our food – can all be traced back to the rocks, minerals, and the chemical nutrients supplied by them. Most of us are curious concerning the origins of such resources, and the amounts which still remain for man's use.

When we look at the earth as a whole, we see arrangements of continents and oceans, mountain ranges and ocean deeps, which pose further questions concerning their origins and significance. We also ask about the total quantities of oil, tin and iron in the earth's rocks. How unique is the earth in comparison with other planets in the solar system and beyond? We live in an age where the scientific fiction of one generation becomes the experience of the next, and we follow the expansion of our knowledge concerning the planets with a variety of emotions ranging from awe to indifference. But we must grasp the importance of that knowledge and understanding if we are not to leave insurmountable problems for our children.

Questions about our local environment, and about the earth as a planet, need to be investigated. We can all go to examine a rock exposure in a cliff or quarry; some of us can return to a laboratory and study thin sections of the rocks under a microscope; but few can travel to the moon to collect rock fragments. Geology, the study of the earth, can be carried out at several levels, and at each bring a satisfying sense of achievement and knowledge which may be useful in answering a variety of these questions.

The local environment and the amateur geologist

Collecting evidence about the local geology can be carried out at the simplest level by observation (i.e. using our eyes) and recording this information. Such information may be of different types, but it will include data concerning local rocks where they have a surface outcrop (e.g. quarry, cliff-face, river bank), local relief features such as cliffs or valleys in relation to rock-types, local processes like streams and waves and local uses of geological materials (today and in the past in buildings). This approach may lead to a greater depth of understanding of the geological history and evolution of the local environment: it begins with the collection of evidence and proceeds to interpret this to explain the situations observed (Fig. 1.1).

The professional geologist

The professional geologist may well have before him a particular problem, which he has to solve by using a variety of techniques, justified in terms of costs to the overall project. He might be able to use simple observation and recording techniques similar to those used by the amateur geologist, but he might also use techniques requiring costly equipment and special skills. His problem may be to locate a possible source of oil, to delimit the extent of an underground ore body, or to give advice to civil engineers on the angle of slope adopted for a road cutting. He may have to carry out laboratory tests on materials, instigate magnetic or seismic investigations of subsurface features, or commission aerial photography of various types to assist him in collecting sufficient information on which his conclusions may be based (Fig. 1.2).

Both the amateur and professional geologists adopt similar approaches, in that they rely on the collection of information to judge the best course of action. The amateur may be more general and wider-ranging in his approach to what is evidence,

whereas the professional may have a good idea of the precise type of data he needs for a particular job. The two approaches go together for scientists.

The important point to bear in mind at this stage is that all geologists rely on measured and precisely-described information about the earth, so that, although interpretations may vary over time, the observed information can be used again in a new context.

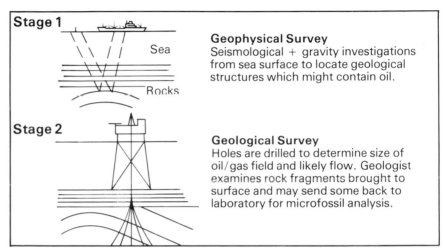

Fig. 1.2 The conclusions concerning the presence or absence of oil/gas can be made only after the use of costly equipment – ship, seismological equipment, gravimeter, drilling rig, laboratory and high-powered microscopes.

A farm in Kentucky with severe soil erosion following removal of woodland and use for growing row crops. A knowledge of geological processes could have prevented such devastation. (USDASCS)

Basic earth facts

Some of the basic facts concerning the earth as a planet can be summarised simply.

The earth's shape

The earth is a sphere, slightly flattened at the poles (Fig. 1.3). It rotates on its axis once every twenty-four hours, and although the earth appears to be a solid ball of rock, it behaves over a period of time as if its interior were molten, and is slightly deformed for this reason.

The earth's surface

The most basic distinction is between land and sea (Fig. 1.4). The highest mountains and deepest ocean trenches (Fig. 1.5) cover small areas compared with the lower-lying lands and ocean-floor plains.

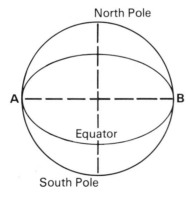

Fig. 1.3   The earth's bulge. The flattening at the poles is scarcely enough to notice. The equatorial diameter (**AB**) is 12 756 km; the line of the earth's axis from north to south poles is 12 713 km long.

Continent	Area (km²)	Ocean	Area (km²)
Eurasia	54 200 000	Pacific	180 500 000
Africa	29 800 000	Atlantic	92 200 000
North America	24 200 000	Indian	75 000 000
South America	18 000 000	Arctic	14 000 000
Antarctica	13 100 000		
Australasia	9 000 000		
Total land	148 300 000	Total ocean	361 700 000

Fig. 1.4   The areas of the continents and oceans.

Earth composition

The earth is surrounded by the ocean waters which cover over two-thirds of its surface, and the gaseous atmosphere. In addition, the great importance of living things on the earth's surface makes a fourth realm in addition to the solid earth, ocean and atmosphere – the biosphere. Each has a distinctive chemical composition (Fig. 1.6).

Earth forces

There are two major earth forces which affect geological processes and aid research into the past history of the planet. The earth is a relatively large and dense body, and so has a gravitational pull which, along with the moderate temperatures at the surface, enables it to retain an atmosphere (in contrast to the moon). Gravitational pull also results in water and ice flowing downhill in contact with the rocks at the earth's surface. The earth's magnetic field approximates to that which would be produced by a permanent bar magnet aligned roughly along the axis of rotation (Fig. 1.7). Aspects of this field millions of years ago are recorded in the rocks and help us to work out past events.

Earth history

Geological studies have shown that the earth is extremely old, and the samples of moonrock have confirmed slightly older ages for the major planets. We are speaking in millions of years. It is difficult for us to think in such terms, and for the moment it will be best to discuss the past in terms of the main periods of geological time (Fig. 1.8).

Earth in the solar system

The earth is one of the smaller planets of the solar system, circling round the vast nuclear furnace of the sun. The solar system, however, is situated towards the edge of a swirling galaxy known as the Milky Way, and there are millions of other galaxies in the universe. Astronomers have been observing the universe for centuries, and, as their instruments have improved, they have found more and more stars at greater and greater distances from the earth, many of the most distant having been discovered by detecting the radio waves they emit.

Such considerations make us feel insignificant, but present information suggests that our planet is unique in supporting such a complex variety of living forms – at least within the solar system.

Although no other planet has been detected outside the solar system, it is at least possible that there are many other planets which support life amongst what must be millions in the universe.

This unique character of the earth is a result of several factors acting together.

(1) The planet is the right size and density to maintain an atmosphere which will let through the sun's rays, but will also allow some of their energy to escape so that a balance of heat is maintained.

(2) The atmosphere acts as a protective layer, filtering the ultraviolet rays which would otherwise burn up living forms at the surface.

(3) The earth is at a distance from the sun – varying from 145–152 million km – which keeps the surface temperatures at levels suitable for living forms to exist over the majority of its surface.

(4) Of greatest importance is the fact that water can exist in the liquid form over the majority of the earth's surface, and is circulated to the continental areas by atmospheric movements. The same movements also spread the heat through the atmosphere.

The atmosphere, the oceans and the rocky exterior of the earth contain a wide variety of raw materials which not only support life, but also enable man to have a full and varied existence. Many of the resources we take for granted have been built up by geological processes over hundreds of millions of years.

The plan of this book

Geology, the study of the earth, has now reached the stage where some major general patterns can form the basis for investigations. This helps us to place new information in a particular context, and to see that it has a significance. In Chapter 2 we examine plate theory, the major explanatory pattern to emerge from geological studies in over 100 years. Chapter 3 studies the interaction of major groups of geological processes, summarising them in the geological cycle diagram.

In Part II of the book there is a consideration of the geological processes which can be seen in action today. This is important, since the geologist must make many of his interpretations on the basis of what he can observe at the present time. Chapter 4 is concerned with the action of the atmosphere, winds, streams and ice. Chapter 5 brings together the study of the coastal zone and the deeper ocean areas. Chapter 6 investigates volcanoes and earthquakes – both manifestations of internal earth processes.

Continent	Highest (mountain)	Altitude (m)
Eurasia	Everest	8 848
South America	Aconcagua	6 959
North America	McKinley	6 193
Africa	Kilimanjaro	5 895
Antarctica	Vinson	5 140
Australia	Kosciusko	2 230

Ocean	Deepest (trench)	Depth (m)
Pacific	Philippines	11 516
	Mariana	11 033
	Tonga	10 882
	(many others)	
Atlantic	Puerto Rico	9 200
Indian	Diamantina	8 047

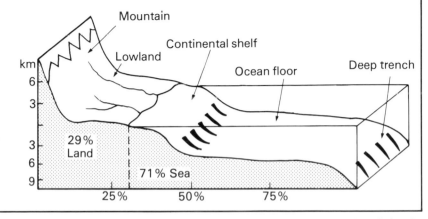

Fig. 1.5 The earth's relief. The relative importance of the different levels in the relief at the earth's surface is shown. Which two levels occupy the greatest area? A large proportion of the earth's surface is just above, or just below, sea level. This means that any movement of the earth's crust which distorts the surface features is liable to alter the extent of water over the continental margins; similarly the freezing or melting of masses of ice can affect the sea level. The numbers along the base of the diagram refer to the proportion of the area of the earth's surface found at the various levels and are expressed as a percentage of the whole.

REALM	Average depth (m)	Average relative density	COMPOSITION (% of atoms)					
			N	H	C	O	Si	Metals
Atmosphere	300 000	0–0·0013	76	2		21		
Oceans	2 630	1·04		66		33		
Biosphere	0·1	1·0	1	60	8	30		
Lithosphere								
Crust	17 000	2·8			3	60	20	16
Mantle	2 883 000	4·5				54	14	29
Core	3 469 000	10·71						100(?)

Fig. 1.6 Earth materials. Compare, for instance, the composition of the ocean waters with that of living organisms; contrast the composition of the atmosphere with that of surface rocks. Why should there be such a high proportion of oxygen in the surface rocks of the earth, and none in the core rocks?

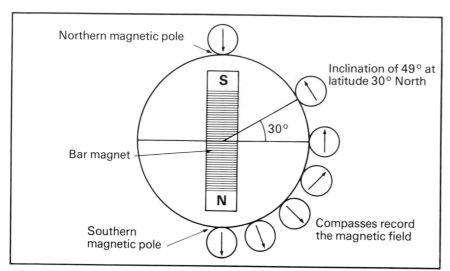

Fig. 1.7 A magnetic field. This shows the result of an experiment you can try yourself, which illustrates the effect of the earth's magnetic field on iron minerals in the rocks. Place a petrie dish over the magnet. The attitude of a mineral is determined by the part in the earth where the rock which contains it solidified.

Part III, Interpreting the Past, takes up each aspect of the evidence available for study by the geologist – the minerals, rocks, fossils and rock structures – and then brings them together in relation to major world structural regions (Chapter 10), and the succession of major changes in the past. These varied aspects also assist us in detailed analysis of earth history (Chapter 12). Finally, the relevance of studies of the solar system planets to our understanding of the earth is discussed.

Throughout the book there are many activities to be carried out, either from the captions of diagrams and photographs, or from the investigations in the text. Geology is most interesting when learnt in the field and in a practical way, so it is important that these should form a part of any course in the subject.

GEOLOGICAL TIME

Millions of years ago	Rocks contain plentiful Fossils		Era	PERIOD/Epoch		Length (Million years)	Characteristic life
						(Million years)	
600			Cainozoic (Recent life)	Quaternary	Holocene Pleistocene	0·01 2	Age of Man
1000				Tertiary	Pliocene Miocene Oligocene Eocene	9 14 15 30	Age of Mammals
	Few signs of life		**70 million years ago**				
2000			Secondary or Mesozoic (Middle life)		Cretaceous Jurassic Triassic	65 45 40	Age of Reptiles, Ammonites
2700	Oldest British rocks		**220 million years ago**				
3000							
3900	Oldest rocks in World		Primary or Palaeozoic (Ancient life)		Permian Carboniferous Devonian Silurian Ordovician Cambrian	50 80 50 40 60 100	Amphibians, Fish, Insects, Coal Forests First Fish, Land Plants Age of Invertebrates (No Land Life)
4000	Moon rocks						
4500	Meteorite						
5000	Origin of Earth?		**600 million years ago**				
			Precambrian			4000	Little evidence of life

Fig. 1.8 This chart shows a time-scale of earth history (**A**), and the main periods of geological time (**B**). What proportion of the earth's time-scale is occupied by rocks containing plentiful evidence of past life?

2
Major earth patterns: plate theory

Geological patterns

Geologists have been able to gather information from a wide section of the earth's surface, and this has enabled them to draw conclusions concerning the formation, distribution and association of rock-types and minerals. It is from this basis of knowledge and understanding that new sources of minerals may be discovered, the historical sequence of geological events leading to the build-up of rocks of different ages may be worked out, and the mechanisms of major earth activity may be understood more fully. Observation has given rise to particular associations of facts, or **patterns**. Such patterns can be generalisations arising from a large number of observations of similar things, or can be used in explanations. Once one is able to understand and explain situations, and the explanations are seen to hold for a variety of cases, then geologists can begin to predict what is going to happen next. Thus, the experience of defining underground bodies of copper ore has led to the formulation of a general pattern, or model, which will probably be repeated again and again. Occasionally a new case will emerge, and this will lead to a modification of the general pattern, or its replacement by a new one. A knowledge of earthquake activity gives rise to a mapped pattern of its distribution around the world, but we do not know enough about earthquakes to enable us to predict precisely when the next one will hit San Francisco. Most patterns in science, and especially in geological science, are partial patterns: we must not expect them to be the final word. Indeed, this is part of the excitement geology holds for many, since each new discovery adds a little more to our understanding and may give rise to new explanations.

The history of geological science is one with many examples of new ideas overturning older explanations. In the early nineteenth century most people believed that the earth had a very short history, measured in terms of a few thousand years, and that the rocks and mountains were produced in a great catastrophe like Noah's flood. This idea gave way to the conflicting idea that the earth had a much, much longer history, measured in hundreds of millions of years. Such a change encouraged explanations of the origin of sedimentary rocks in relation to the slow working of streams, glaciers and the sea at rates we observe today. This meant that there was some point in studying modern processes so that this knowledge could be applied to older rocks formed in similar ways.

A Gemini XI photo of northeast Africa, the Gulf of Aden and southern Arabia, taken from 800 km up. These two sections of the Earth's crust are moving apart as a line of plate divergence is established along the Red Sea. (NASA)

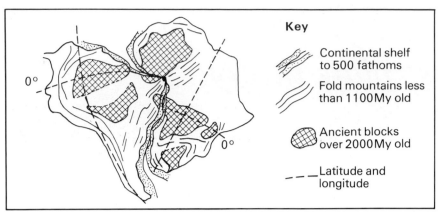

Fig. 2.1 The 'fit' between South America and Africa. How closely do the coastline and the geological features match up across the South Atlantic ocean?

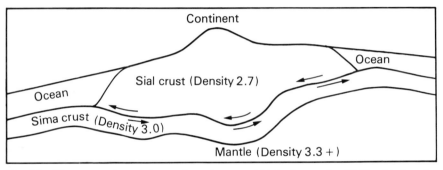

Fig. 2.2 Wegener envisaged the continental masses 'drifting' over the SIMA: this is quite impossible, since the SIMA/SIAL boundary is rigid. His ideas were therefore all dismissed.

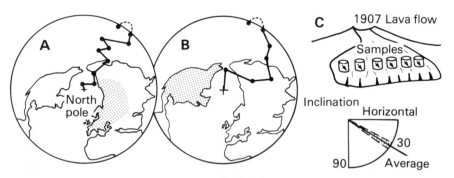

Fig. 2.3 **A, B** Polar wandering. Measurements of 'fossil' magnetism in ancient rocks from a selection of sites across Europe and northern Asia suggest that the position of the magnetic pole has moved over the last 500 million years as shown (**A**). The pattern established in a similar way for North America (**B**) is different: each continent has its own path of polar wandering. This is a convenient way of showing that if the continents wandered (rather than the poles), they each had different paths.
C Palaeomagnetism. The establishment of 'fossil' magnetism is based on the principles shown here for modern lavas. Cores of lava were drilled from a flow which took place in Hawaii in 1907. The direction of magnetic inclination calculated from these agrees with the earth's magnetic field at that date.

The twentieth century has witnessed another major change in our ideas. The major relief features of the earth – continents and ocean basins – were for long regarded as immovable, but it is now agreed that they have changed positions, shapes and sizes over time. This change has resulted from increasing the range of geological observations. By the early years of the twentieth century much of the continental surfaces had been explored, and a lot was known concerning their geology. The shapes of the continents bordering the Atlantic suggested a former close fit (Fig. 2.1), as did other evidence such as fossils in the rocks of Africa and South America and ancient glacial deposits. The mechanism put forward for such 'continental drift' (Fig. 2.2) was, however, unacceptable, and based on a limited knowledge of the earth's interior. Studies of earth magnetism in the 1950s (Fig. 2.3) showed that in the past Eurasia and North America had variable positions with respect to the magnetic poles, and this unexpected information reawakened interest in the idea of mobile continents.

At the same time studies of the relief and geology of the ocean floors had begun. The evidence coming from the 71 per cent of the earth's surface which had previously not been known in any great detail created a revolution in ideas in the 1960s (Fig. 2.4). It was seen that the ocean-floor rocks were moving at slow rates (1–9 cm/year), and that the ocean-floor materials were less than 200 million years old, whilst the ages of some continental rocks went back to 3850 million years. The pattern established by this evidence showed that new ocean-floor rocks were generated along the ocean ridges, and that ocean-floor rocks were lost from the surface by plunging beneath less dense material along the deep ocean trenches. This idea was known as ocean-floor spreading (Fig. 2.5), and was soon combined with information from other sources to produce the all-embracing explanation for the origin of both oceanic and continental relief features known as plate theory. This theory is now used as a basic point of reference in geological studies: observations and new evidence are used to test its validity as an explanation, and to date most have supported the general idea.

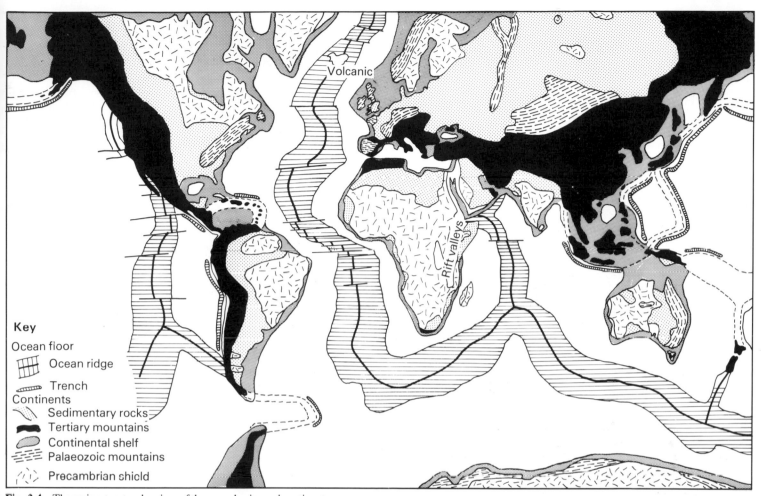

Fig. 2.4 The major structural regions of the ocean basins and continents.

Key

Ocean floor

Ocean ridge

Trench

Continents

Sedimentary rocks

Tertiary mountains

Continental shelf

Palaeozoic mountains

Precambrian shield

Volcanic

Rift valleys

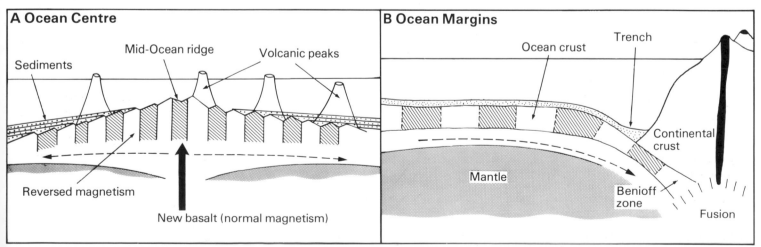

A Ocean Centre

Sediments

Mid-Ocean ridge

Volcanic peaks

Reversed magnetism

New basalt (normal magnetism)

B Ocean Margins

Ocean crust

Trench

Continental crust

Mantle

Benioff zone

Fusion

Fig. 2.5 Spreading sea floors. In the central parts of the oceans (**A**), or at ocean ridges, new molten rock is injected from beneath, pushing aside those already in place. Both the igneous rocks and ocean floor sediments increase in age from the ridges towards the margins of the oceans. The ocean margins (**B**) are often zones of volcanic and earthquake activity – particularly marked in the Pacific where the spreading movements of the ocean floor are most rapid. This diagram shows one interpretation of what is happening.

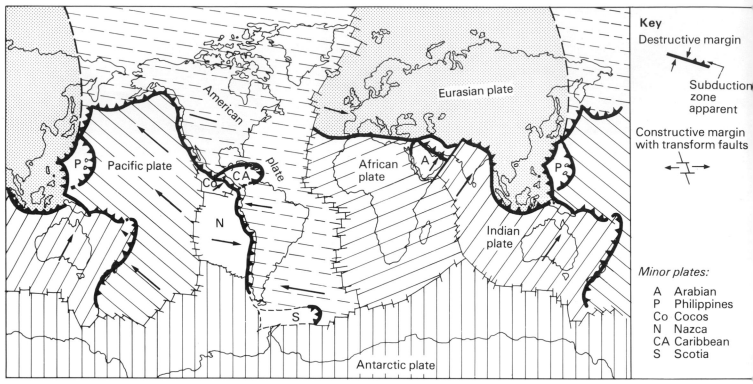

Fig. 2.6 The world distribution of plates and types of plate margin. Six major plates are named on the map and six minor plates in the key. Although this map appears clear-cut, it must be pointed out that not all the details are finally established, and that the polar regions in particular present difficulties of investigation.

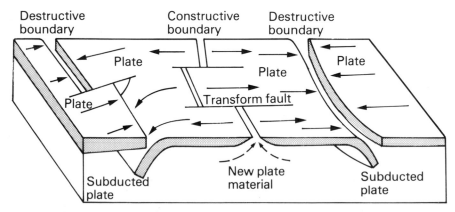

Fig. 2.7 Types of plate boundary.

Plate theory

The idea is basically a simple one. The earth's surface is composed of a series of rigid plates (Fig. 2.6), which are moving constantly against each other. This movement gives rise to major geological events like earthquakes, volcanic activity and mountain building. Each plate has three distinctive types of margin (Fig. 2.7), and the movement of the plates is controlled by rates of gain at the constructive margins and rates of loss at the destructive margins.

A simple initial test can be made using a series of world maps. Plot the distribution of recent earthquakes on one map, and then the distribution of volcanic activity on another. Then plot the location of ocean ridges, trenches and continental fold mountains (Fig. 2.4) on a third, and the plate boundaries on a fourth. Overlay these on a light table, or as transparencies on an overhead projector, and discuss the coincidences and differences. What conclusions do you reach?

Each plate consists of at least the ocean-floor rocks, which are part of a unit with the uppermost mantle rocks having a total thickness of about 100 km, and known as the lithosphere. Beneath the lithosphere is a zone of weakness, also approximately 100 km thick and known as the asthenosphere (Fig. 2.8). A plate may have a continental mass of less dense rock on top of it. It appears that, whilst the ocean-floor plate material has been churned over and over, the relatively light continental rocks have accumulated on the surface like scum on a boiling liquid, so that rocks of much older ages are preserved.

There are more complicated aspects of plate theory, such as the geometry of plate movements on the spherical surface of the earth, but this simple outline is sufficient for us to bear in mind for testing during the course of different aspects of geological studies in this book. At the same time it will be used as the basis for the understanding of a wide range of geological relationships in time and space. Many of these relationships were not understood until the theory was applied to them – a further demonstration of the usefulness of this particular idea.

There are still a number of 'unknowns' associated with plate theory, as with any scientific theory. This is another good reason for people to continue investigating the earth. The mechanism which drives the plates is one of these points at which there is a gap in our knowledge.

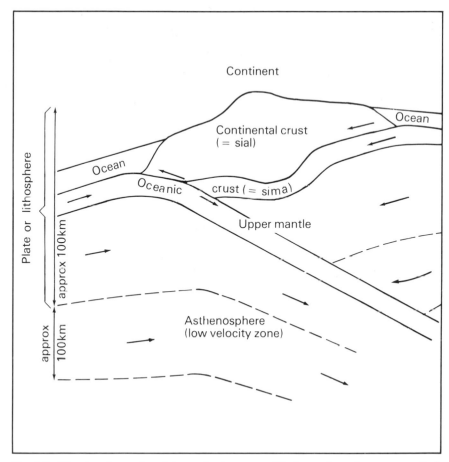

Fig. 2.8 The plate theory view of the mechanism responsible for continental movements. Each plate is 100 km thick and underlain by a weaker zone where the rise in temperature inside the earth overcomes the rise in pressure (the temperature rises sharply just below the surface, whilst the pressure rises more steadily).

3
Major earth patterns: the geological cycle

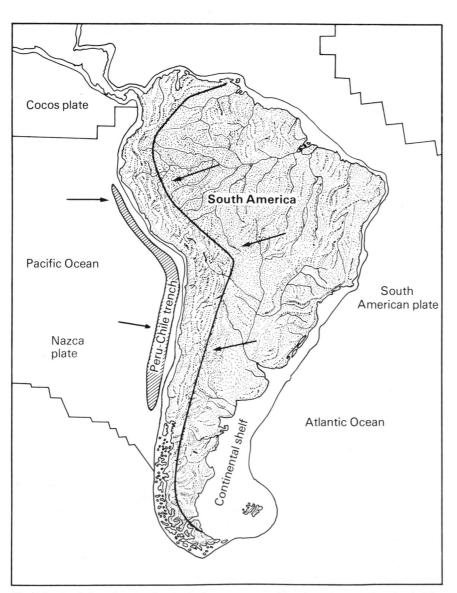

Fig. 3.1 South America, showing the Andean ranges stretching down the western coast, and the relationship to the local plates and plate margins.

Plate theory is one of a number of patterns of activity studied by geologists. Some of these relate to the expression of internal earth energy; others are the result of interactions between solar energy, the atmosphere and the earth's surface. We shall study these patterns in this chapter in relation to the particular example of the development of the Andes mountains in South America.

The Andes mountains

The Andes ranges dominate the western margin of South America, extending from north of the equator to 54 degrees south (Fig. 3.1). They are the second highest ranges in the world after the Himalayas, and provide a good example of the variety of geological processes in action. The history of Andean evolution has been pieced together from a variety of evidence, including the study of surface rock types and the structures affecting them, the relief features, and the subsurface situation revealed by a study of earthquake shock waves (Fig. 3.2). This evidence shows that the Andes have been formed at a convergent plate margin, where the Nazca plate dives beneath the American plate. There has been an unusual thickening of the earth's crustal rocks along this zone, the maximum thickness of 70 km being rivalled only by the Himalayas: elsewhere the crust averages 30–35 km thick.

When the history is unravelled from this evidence (Fig. 3.3) it seems that there have been a series of events over the last 200 million years leading to the formation of the present ranges. Continental movements before this time had resulted in the bringing together of all the continental areas into a single mass, known as Pangaea, and the west coast of South America was a quiet zone of sedimentary rock formation in the sort of situation existing today along the eastern coasts of North America (Fig. 5.16). Somewhere between 250 and 200 million years ago this great continent began to break up and the modern phase of plate tectonics and the evolution of continents and ocean basins started. A convergence zone was established to the west of South America, in which the Nazca plate sank beneath the American. Volcanic activity began when the Nazca plate material, and the sediments on top of it, reached depths of 100–150 km and melted (Fig. 3.3B). The less dense, now mobile, material rose up through the overlying rocks, often melting and absorbing those in the way, until it solidified or reached the surface. A series of volcanic peaks emerged from the ocean as piles of lava were formed: these are igneous rocks, formed by melting and migration in the mobile state. This type of

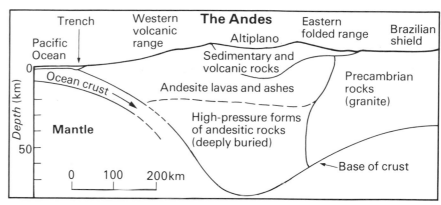

Fig. 3.2 Section across the central Andes, showing the two major ranges and the nature of the crust beneath (over 75 km thick beneath the western ranges).

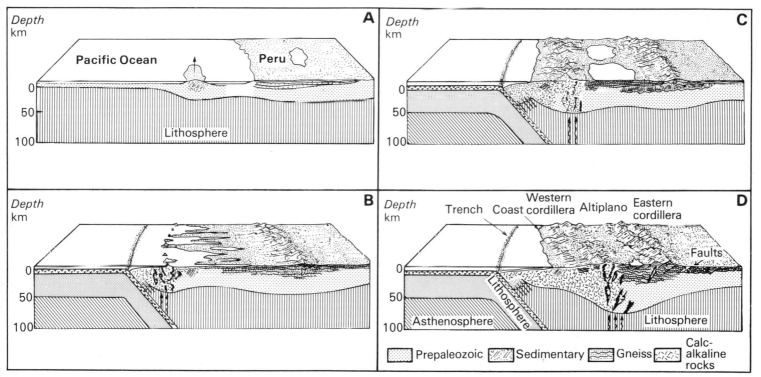

Fig. 3.3 The evolution of the Andes. Late in the Palaeozoic era (**A**), some 250 million years ago, sedimentary deposits blanketed the present coastline, and there were no mountains. In Triassic and Jurassic times (**B**), about 200 million years ago, underthrusting of the lithosphere began, buckling the sedimentary rocks and pushing them upward and eastward. Rising magma from the descending ocean plate formed an arc of volcanoes in the coastal waters of western South America. Some batholiths (bodies of intrusive igneous rock) formed in the sedimentary layers of the eastern ranges. In Cretaceous and early Cainozoic times (**C**), 100–60 million years ago, a second volcanic arc began to form east of the Jurassic arc. Upwelling magma swelled the crust, pushing aside the ancient sedimentary rocks, which crumpled to form the fold mountains of the eastern cordillera. Material eroded from these mountains poured into the altiplano region. Formation of the present volcanic range began 10–15 million years ago, reaching the present structure by Pliocene or Pleistocene time (**D**), one or two million years ago. (*After D. E. James, 1973*)

activity dominated the period from 150 to 50 million years ago, with the centre of volcanic activity moving eastwards and eventually causing the older sedimentary rocks (formed 450–250 million years ago) to be crumpled and thrust into huge slices, leading to their compression horizontally. The combined effect of piling up masses of volcanic lava, and of compressing the ancient sedimentary rocks, led to unusual thickness of low density rocks, and they were uplifted to form the Andean ranges – the Western Cordillera dominated by volcanoes, and the Eastern Cordillera formed largely of the ancient sediments. Such a pile of low density rocks made the earth's crust unstable along the line of deposition, and uplift took place to restore the equilibrium.

As soon as the uplift occurred, the ranges were attacked by streams, the wind and glaciers, and debris from the wearing away of rocks was washed into the sea, or into basins between the mountain ranges. A final phase of volcanic activity began 15 million years ago and resulted in the present relief by building up new peaks on the worn-down remnants of earlier ranges. In the extreme north and south the Andean erosion has been dominated by stream erosion, but in the central area the processes occurring in deserts have affected the relief to a great extent. At the same time the mountain tops in equatorial areas, and a zone increasing in size towards the southern tip of the continent, have been eroded by glaciers and frost action (Fig. 3.4).

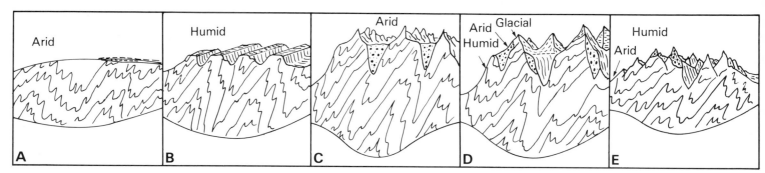

Fig. 3.4   The sequence of events in the development of the landforms of a rising mountain range. This was derived from a study of the central Andes. (*After Garner, 1974*)

Complex folds in Lower Cretaceous calcareous siltstones in the Andes of northern Chile. (USGS)

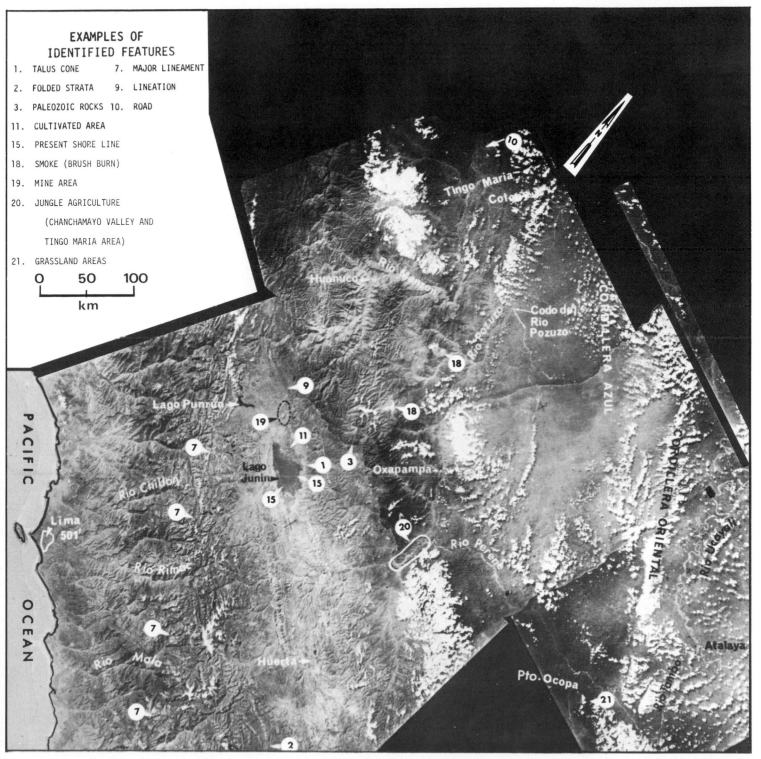

EXAMPLES OF IDENTIFIED FEATURES

1. TALUS CONE
2. FOLDED STRATA
3. PALEOZOIC ROCKS
7. MAJOR LINEAMENT
9. LINEATION
10. ROAD
11. CULTIVATED AREA
15. PRESENT SHORE LINE
18. SMOKE (BRUSH BURN)
19. MINE AREA
20. JUNGLE AGRICULTURE
    (CHANCHAMAYO VALLEY AND
    TINGO MARIA AREA)
21. GRASSLAND AREAS

0 50 100
km

Part of a photomap of Peru, compiled from space images. It shows the coastal and interior cordilleras of the Andes. (USGS)

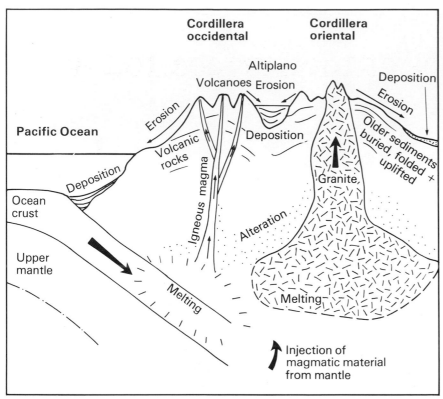

Fig. 3.5   The Andes mountains and aspects of the geological cycle.

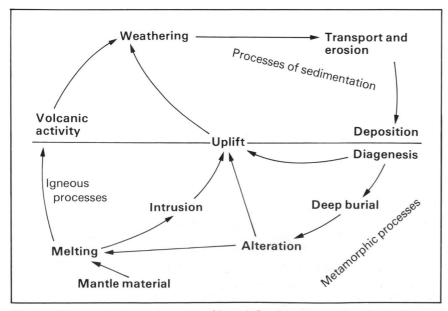

Fig. 3.6   The geological cycle. A summary of the rock-forming processes. Describe the history of a quartz grain from the time of its formation in a granite, through its incorporation in a greywacke to its present outcrop in a gneiss.

The geological cycle

The history of the Andes demonstrates simply how the theory of plate tectonics can be applied, but also introduces a whole range of other processes resulting in the production and wearing away of rocks (Fig. 3.5). These can be summarised more generally (Fig. 3.6), so as to form the basis for more detailed studies in the next sections of this book. Although all are inter-related, it is possible to extract particular aspects.

(1) **The sedimentary cycle** includes weathering and disintegration of rocks in contact with the atmosphere, and the transport and deposition of such debris to form new rocks in layers at the surface, mostly in shallow seas.

(2) **The igneous cycle** involves the melting of rocks at depth in the crust, or just below the base of the crust, and the migration of the molten rock (magma) towards the surface. It may not reach the surface before solidifying, but it may erupt into the atmosphere, to which it supplies quantities of gases. On erosion, rock debris is transferred to sediments, and these may be converted to igneous rocks again on burial, or carried down with a descending plate.

(3) **Metamorphic rocks** are formed by the alteration of other rocks. The changes may be due to crushing near the surface (high pressure, low temperature), to heating in contact with mobile igneous rock (low pressure, high temperature), or to deep burial (high temperature and pressure). The rocks altered in this way do not become mobile (or they would be igneous).

Part II
Geological processes at the surface today

4
Land–atmosphere interactions

Evidence from the present

Direct observation is limited to what is happening now at the earth's surface. Geological science is still only just over 150 years old, and we have to attempt to reconstruct the events of the distant past from our knowledge of present events. 'The present is the key to the past' has become a catch phrase amongst geologists, but we have to be careful about taking its meaning too far. We have already suggested that the continents have changed their positions in the past, and that mountain ranges like the Andes have varied in relief over time. Our knowledge of past Ice Ages shows that climatic zones have also changed, and the changing patterns of life on the earth have had varied interactions with the geological events: the spread of man in particular has produced a new geological agent able to build his own hills and gouge out deep pits. Thus it is true that many things were different in the past.

It is reasonable to assume, however, that the materials from which the rocks and minerals were formed have not changed, and that the physical laws governing the processes acting on the rocks have not changed. Ice, for instance, has always formed when pure water has been cooled to 0°C at present sea-level pressures, and has expanded in the process. Chemical reactions between calcium carbonate and rainwater have given rise to soluble calcium bicarbonate. It would be correct to say that there is little point in pursuing geological (or any

scientific) studies unless we can assume that this is the case. If the temperature at which water freezes changed every day we would be in a terrible mess (think of the implications for car radiators and central heating systems!). Our experience is that this temperature has remained constant for several hundred years, and that when we extend the principle back in time over millions of years we obtain reasonable results.

Our first objective, then, is to study what can be observed today of the processes acting at the earth's surface. The earth's surface is approximately 29 per cent land and 71 per cent ocean, and is covered by a gaseous atmosphere approximately 450 km thick (although 75 per cent of the atmospheric mass and most of the movements causing weather changes at the ground occur in the lowest 15–20 km, or troposphere). The atmospheric movements, driven by energy from the sun, have the effect of transferring moisture from the oceans to the land masses, and of distributing the solar heat more widely across the planet's surface. This results in the **hydrological cycle** (Fig. 4.1), which is an important sector of the geological cycle (Fig. 3.6).

The changes of temperature and the movement of water across the surface of, and into, rocks causes reactions which lead to the breaking up, or disintegration, of surface rocks. This combination of processes is known as **weathering**. Masses of such broken rocks may be moved downslope, possibly lubricated by water, if the angle of slope is too steep for the frictional forces to balance the gravitational forces acting on the mass of rock debris. The flow of water across the land surface in streams gives rise to further movements of rock debris and the wearing away of rocks to form valleys. Some water may move through rocks underground if they are able to transmit it through spaces within them. Some of the moisture is precipitated in cold regions as snow, and accumulates as ice masses like glaciers and ice sheets. This is another aspect of the return of water across the continental surfaces towards the oceans, and the slow, grinding movements of glaciers give rise to distinctive landforms and deposits of debris. Atmospheric movements, or winds, may also cause the movement of surface dust and fine sand.

There is thus a group of processes acting at the earth's surface, where rocks and atmosphere meet, and these give rise to the details of surface relief. The materials which are transported away from their original site are incorporated in a new generation of rocks when they are dropped (in lakes, valley floors or the sea).

The oceans form a blanket over the features of their basins. There is less interaction between the seawater and the underlying rocks than between the atmosphere and the rocks exposed to it. Whilst the main activity at the land surface is wearing away (erosion), the main activity in the ocean is the settling out of rock and organic debris as accumulations on the ocean floor (deposition). At the edge of the ocean and continent the coastal zone witnesses the contact of marine and atmospheric processes with the land.

The other geological events occurring at the earth's surface are those originating within the earth – volcanic eruptions and earthquake shocks. These are surface manifestations of internal earth sectors of the geological cycle (Fig. 3.6).

Weathering: rock disintegration

When rocks are exposed to the atmosphere, changes take place which lead to them disintegrating, being broken into fragments, or being dissolved away. The changes are caused by the variations in atmospheric temperatures, and by chemical reactions – especially in the presence of water. Different minerals and rocks react in different ways, depending on their chemical composition and the number of small fractures which allow the air and water to penetrate their surface.

Changing temperatures may cause the surface minerals to expand and contract at different rates, leading to their separation from one another. This may be a major cause of rock disintegration in desert regions, but even there the chemical changes in the presence of water are thought to have an important effect. Another example of the effects of temperature changes is seen in the fact that forest fires give rise to extensive fragmentation of rocks exposed at the surface, and to the splitting of boulders.

In colder regions at high altitude or latitude the alternations of freezing and thawing result in another type of rock destruction (Fig. 4.2). Water trapped in gaps in the rock, or in the pore spaces within the rock, expands by 10·9 per cent on freezing, and causes pressure to be exerted on the confining rock. This process may result in frost-shattered angular fragments of rock and in irregularly fretted peaks in high regions like the Alps. In dry regions the crystallisation of salts precipitated from solutions moving through the rocks may give rise to expansion pressures, and the rock surface may flake in response to these stresses. The growth of tree roots is another process by which rocks may be broken apart physically.

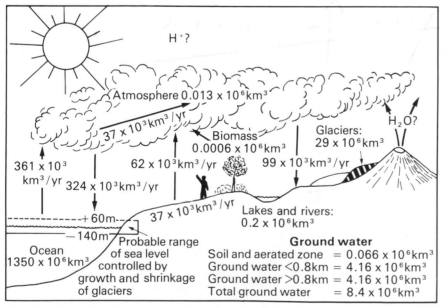

Fig. 4.1 The hydrological cycle – the circulation of water from the oceans into the atmosphere, and then via clouds and precipitation to the continental surface and back to the ocean in glaciers and streams. The amounts of water in various points of storage are given in millions of cubic kilometres ($10^6 km^3$), and the annual exchanges of water amongst these reservoirs are given in thousands of cubic kilometres per year ($10^3 km^3/yr$). (*After Bloom, 1968*)

Chemical weathering is as important as these physical forms caused by temperature changes and expansion due to crystallisation in dry and cold regions: in the warmer and more humid regions it is dominant. A series of chemical processes occur with complex inter-relationships: these include **solution** (the breakdown of minerals in contact with water to give a series of ions in solution); **carbonation** (the reaction of minerals with weak carbonic acid, formed by dissolving carbon dioxide in rain-water and groundwater); **hydration** (adding water to minerals – particularly important in the formation of clay minerals); **chelation** (extracting ions from otherwise insoluble solids by special agents, often provided from the breakdown of plant matter in humus); and **hydrolysis** (a straightforward reaction between water and a mineral). All of these require the presence of water, and most take place at increasing rates as the temperature increases.

Reactions with different minerals vary, due to chemical composition, crystalline structure, size and shape. Rocks produced by igneous activity bring materials from the earth's interior to the surface (Fig. 3.6). Their characteristics vary (Fig. 4.3), but nearly all of their minerals are altered in the process of weathering, as well as being detached from the igneous rock. Quartz is the most resistant of the igneous rock minerals, and gives rise to most of the particles incorporated in sand deposits. The other minerals break down to soluble ions and clay minerals. The chemical formulae of the rock-forming minerals are often complicated, but some of the reactions involved in chemical weathering can be summarised in equations like the following:

$$6H_2O + CO_2 + 2KAlSi_3O_8 \rightarrow$$

rain-water orthoclase feldspar

$$Al_2Si_2O_5(OH)_4 + 4SiO(OH)_2 + K_2CO_3$$

clay mineral silicic 'acid' (in solution)

The products of weathering form the basis from which **soils** are formed, or, when transported to new sites, the materials from which new sedimentary rocks are produced. Physical weathering results in angular fragments or sand grains being produced, but very few silt or clay particles (less than 0·06 mm diameter). Chemical weathering produces clay minerals and soluble ions (Fig. 4.4). The debris remaining on top of the rock following weathering builds up to form a layer known as the **regolith**. This is acted upon further by the weather, water moving through it, accumulating rotted plant matter, and an increasing group of organisms, to form

A dolerite sill, exposed to the atmosphere in Fifeshire, Scotland. What is happening to the rock? (Crown Copyright)

Wastwater, English Lake District. What has caused the formation of the fan-shaped masses of debris to the right of the lake? (Aerofilms)

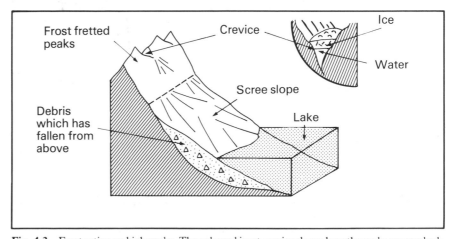

Fig. 4.2 Frost action on high peaks. The enlarged inset crevice shows how the rocks are cracked: the surface water freezes first and when the underlying water nears freezing point it will expand and will either force out the surface ice, or force apart the rocks.

Mineral	Rocks it occurs in	Weathering characteristics
Quartz	Sandstones, granites, many metamorphic rocks	Resistant to chemical weathering: no cleavage to allow water to penetrate. Also resistant to physical weathering: eventually rounded.
Feldspar	Most igneous and metamorphic rocks, some sandstones (arkoses, greywackes)	Almost as hard as quartz, but marked cleavage allowing water penetration and attrition. Potassium feldspars most resistant; sodic/calcic varieties (plagioclase) decompose most easily.
Mica	Granites and other igneous, metamorphic rocks	Easily break into flakes; biotite (dark) soft and attacked by water, but muscovite (light colour) resistant. Alteration to clay minerals.
Olivine	Basalt and other basic igneous rocks	Fractured, easily weathered; alteration begins as lava crystallizes.
Augite (Pyroxene)	Basalt and other basic, intermediate igneous rocks.	Good cleavage, weathered rapidly to clay minerals, soluble ions.
Hornblende (Amphibole)	Igneous and metamorphic rocks	Good cleavage, weathered fairly rapidly to clay minerals, soluble ions.
Carbonates	Limestones, marble	Most soluble common minerals.

Fig. 4.3 Minerals and weathering. Which minerals (a) occur only in igneous/metamorphic rocks; (b) occur only in sedimentary rocks? How is this related to their reactions to weathering at the earth's surface?

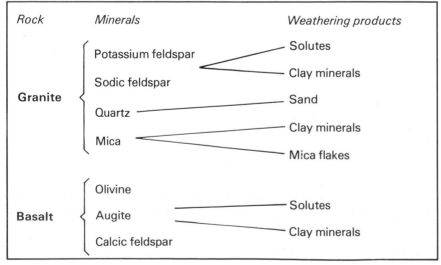

Fig. 4.4 The two main igneous rocks, when weathered, both give rise to sand grains, clay particles and soluble ions – the basic raw materials of sedimentary rocks.

true soils (Fig. 4.5). The greatest thicknesses of regolith (nearly 150 m) occur in the humid tropics on gentle slopes or flat areas, and often consist of deposits of **laterite** or **bauxite** (hydroxides of iron and aluminium).

It is interesting to notice the practical illustrations of weathering processes occurring around us. Quarrymen still heat resistant rocks to crack them apart; we see road surfaces broken and potholed after a severe winter with alternations of freezing and thawing; and the continual fall of frost-shattered debris in mountain regions is a great problem in rail- and road-construction for countries like Switzerland. The weathering processes attacking a mineralised vein where it reaches the surface may lead to the solution and removal of the important minerals at the surface and the enrichment of the lower part of the vein (Fig. 4.6). Later erosion may then expose this richer section.

One word of warning. Similar effects may occur when water, steam and other gases rise through rocks from inside the earth. These processes are known as hydrothermal alteration, and are part of the igneous cycle. They often give rise to clay minerals by reaction with the feldspars. The china clay deposits (kaolinite) near St Austell in Cornwall, and at Lee Moor on south-west Dartmoor, are thought to have been formed by these processes, rather than by weathering.

Mass movements

Weathering may produce so much debris that it becomes unstable and moves downhill under gravity. Such migrations of rock waste may be rapid or slow, and may be 'triggered off' by an earthquake or sudden heavy rainstorm. Many such movements of this type occurred around British coasts during the wet winter of 1976–7, which followed the driest summer on record during which the soil had cracked and been loosened. The combined effects of weathering and mass movements of debris are probably the most widespread method by which the landscape is being worn down.

The slower movements are less dramatic, but affect every slope. **Soil creep** is common in Britain (Fig. 4.7), although the actual movement is imperceptible. It is caused mainly by the heaving of the ground in response to alternations of wetting and drying, freezing and thawing, but the action of organisms in churning over the soil, and the movement of water through the soil, also play their part in moving the particles downslope.

Southwest Butte, Mesa Verde Valley, Colorado. Describe the processes at work in the wearing down of this rock mass. (USDASCS)

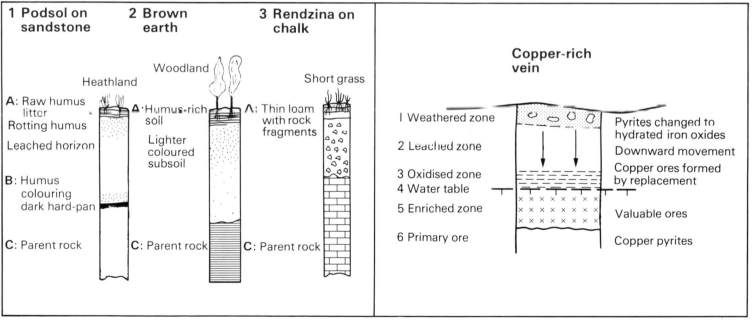

Fig. 4.5 (*left*)   Some British soil profiles. Try and account for the formation of the different soil layers. How far are the different types related to the underlying rock? Podsol is a term originating in Russia, and rendzina is used for thin limestone soils. Find out about other types of profile, including some of those near your home.

Fig. 4.6 (*right*)   Enrichment of mineral ores by weathering. Describe in your own words how the changes take place, using the information on this diagram.

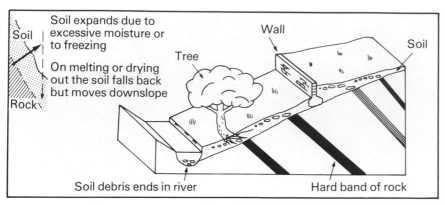

Fig. 4.7   Soil creep. How has the downslope movement of the soil affected the stones in it, the tree trunk, and the level of soil on either side of the wall? Soil heaving is one of the main processes causing this movement in humid temperate regions like the British Isles.

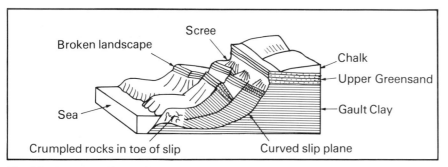

Fig. 4.8   A landslip. Masses of rock have slipped forward along curved planes. This is due to the weak clay becoming saturated and collapsing under the weight of the overlying rigid rocks.

In the cold arctic regions of northern Siberia, Scandinavia and North America the ground below 1 m is frozen permanently (**permafrost**), and only a surface 'active zone' thaws in summer. The melting water cannot percolate downwards through the permafrost, and is concentrated in the surface layer. On slopes this layer may begin to flow beneath the thin cover of vegetation. Such movement is known as **solifluction**. This may produce a separation of coarser and finer rock fragments as stone stripes, or just an irregular movement of boulders down the slope (e.g. the clitters on Dartmoor). On flat ground the movements in the active layer give rise to stone circles and polygons.

The more rapid movements include **rock falls**, **mudflows** and **landslips**. Mountainous regions, and those affected by sudden, heavy storms, are most liable to action of this type, which can often being destruction and tragedy. In Britain landslips are most common where a massive rock which allows water to drain through it lies on top of a weak layer which holds up the passage of water. In 1927 a landslip (cf. Fig. 4.8) carried away the road around the southernmost tip of the Isle of Wight, causing the main road to be re-routed over the hills to the north. The movements continue, and between 1968 and 1970 a further 100 m of road near Blackgang Chine were carried away.

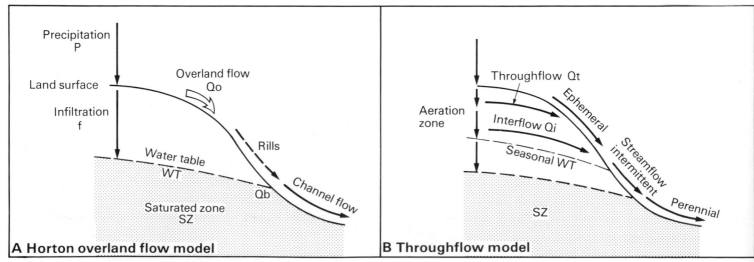

Fig. 4.9   After water hits the ground, it flows to the stream in various ways: some may flow over the surface, whilst some may infiltrate the ground and move through the soil and/or the rock. These diagrams show two ideas of what happens. The Horton model is most appropriate where surface vegetation has been removed, for there is little overland flow in wooded areas.

Streams

Running water plays a dominant part in the wearing down of landscapes by carving out valleys. Most of the world has enough rain to support perennial streams, and even the tiny amount of rain falling in deserts is important, since it falls in heavy showers and flows freely across the surface, unhampered by a covering of vegetation. Water concentrated in streams is effective in erosion and the removal of weathered debris to the sea: the River Mississippi, for instance, carries 2 million tons of silt into the sea every day. The water in streams is provided from rainfall, melting ice and snow, and this may run directly into streams, or pass underground (Fig. 4.9). Once in the stream the water flows through a channel, which it may occasionally overflow (approximately every one to two years in the United Kingdom), carrying mud, sand and silt supplied from downslope movements, and soluble ions from similar sources and from the passage of water through soil and rock.

A landslide near Albany, New York state: the face is 20 m high. Material is uniform fine-grained lake deposit. This landslide dammed the local river. (USDASCS)

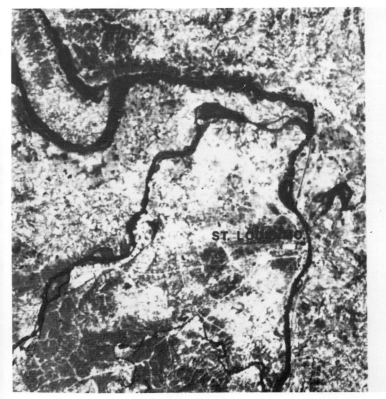

LANDSAT satellite images of the St Louis area, where the rivers Mississippi and Missouri join. That on the left was taken on 2 October 1972; that on the right shows the areas flooded on 31 March 1973. (USGS)

Floods

Flooding rivers are major natural hazards. The flood at Lynmouth in 1952 demonstrates the effects of such rare natural events. During the night of 15–16 August, 1952, a disastrous flood occurred.

'The rain gauge on Longstone Barrow, Exmoor, the tableland above Lynmouth, recorded 230 mm of rain in twenty-four hours. Such rainfall has been exceeded only four times since official records began in Britain 100 years ago. Remember that ordinary steady rain for 24 hours means less than 25 mm in a rain gauge. 230 mm represents over a quarter of a million tonnes of water per square kilometre. Throughout the whole area rainfall was about 95 mm, weighing about 90 million tonnes. The heaviest local fall was probably 100 mm in one hour.

The state of the ground made conditions worse. It was already waterlogged, and just below the surface a layer of rock prevented any appreciable percolation of water. Moreover, the East and West Lyn rivers and their tributaries fall 500 m through funnel-like gorges to Lynmouth in less than 7 km. Down these gorges on that terrible night pounded millions of tonnes of flood water. In dry summer weather the rivers have a depth of a few centimetres; now, at times, a solid wall of water nearly 15 m high raced down to the sea at 30 km per hour. Such a torrent is irresistible to everything except the heaviest and most solidly-based objects. The water gouged out huge rocks and boulders – some weighing 15 tonnes – and carried them to the shore. Telegraph poles and cars followed. Trees, felled by earlier gales, and others washed out by the roots, were swept into the sea. The next morning, a kilometre out to sea, hundreds of trees, presumably weighted down by rocks and soils entangled in their enormous roots, had their upper branches showing above the waves – a fantastic sea forest of stunted trees.

The flood waters dug deep into the earth. Road surfaces were scoured away, and the soft earth of the verges was gouged as by a giant excavator, some gullies being 7 m deep – right down to bare rock. The Lynmouth sewerage system and water mains were wrecked. A vivid illustration of this gouging effect of the flood occurred at a Lynmouth garage. Here petrol tanks were scoured from their foundations and swept away without trace.

When dawn broke the scene on the shore was fantastic. It was littered with the debris of scores of wrecked homes and buildings; smashed cars; telegraph poles; tree trunks, branches and complete trees; the smashed and mangled remains of the undergrowth from the surrounding countryside; some 200 000 m³ of silt, mud, gravel and stones, in some places massed 8 m high; some 40 000 tonnes of rocks and boulders; iron girders and bridges; broken masonry; and the bodies of animals, birds, fishes – and people.'
(From *'The elements of rage'*, by F W Lane)

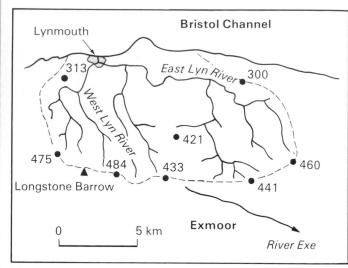

Fig. 1 The Lyn catchment and Lynmouth, north Devon.

Photos taken at the time of the Lynmouth flood disaster in 1952: note the erosion and size of boulders moved by the flood in the village, and the deposits on the shore. (Western Morning News Co. Ltd)

The work carried out by a stream depends on the force of water flowing through its channel. This is related to the amount of water supplied and the overall slope from source to mouth. The work involves wearing away the channel bed (**erosion**), and carrying rock debris, or load, downstream (**transport**). If the stream's energy is lowered, deposition will take place. In most streams in humid areas the amount of water passing a point in a measured time (**discharge**), the stream channel dimensions and the speed of flow (**velocity**) will all increase downstream (Fig. 4.10). As the channel widens and deepens, so the effects of friction of the sides and bed on the flow of water are reduced, so that the water can flow more rapidly despite the slope being lower near the river mouth.

Erosion occurs when the force exerted by the water flow is sufficient to wear away the stream bed or banks and remove more material than is replaced by deposition at that point. The force required will be lower for a stream flowing through regolith than for one flowing across solid rock. In the latter case erosion will be carried out by pebbles being moved across the bed (**corrasion**), or by solution, but few streams in Britain have solid rock beds at present. As the discharge increases, the capacity for carrying solute load increases at a similar rate, but the capacity for a suspended load of fine sand and silt increases much more rapidly than the rate of discharge increase. This means that streams carry out most of their work at the relatively infrequent periods of high stream flow: many streams move half their annual load on two to three days per year. The average flow of most streams fills only one-sixth of their channel, which is able to cope with all but the greatest floods. It is not surprising to discover that boulders in stream beds move very little (Fig. 4.11).

It is necessary to combine the study of an individual river with general statements of the type made at the start of this section. We will take the River Plym as our example. Having learned from this example, you should then examine another near your home and relate information gathered in the field, from maps, and from flow/precipitation data (obtainable from your Water Authority and the Meteorological Office at Bracknell).

The **River Plym** has its source on Dartmoor in Devon, and then flows south-westwards to enter the sea in Plymouth Sound (Fig. 4.12). It begins on granite moorland, an area of heavy rainfall (over 1000 mm per year in the highest parts), and then flows in a deep valley across the slate plateau of

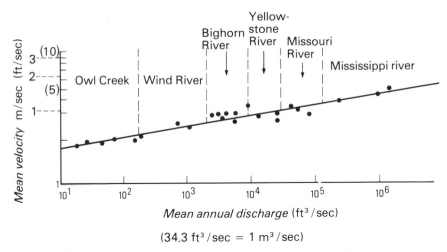

Fig. 4.10 Velocity of streamflow along the Mississippi-Missouri system. Compare the speed of flow in the headwaters of Owl Creek with that near the mouth of the Mississippi. Would you have expected this result?

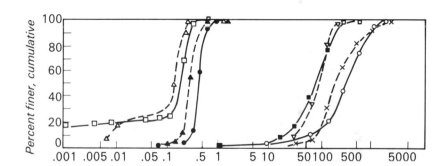

Particle size, millimeters
- Mississippi R. at Mayersville, Miss.
▽ Rock Creek at Roberts, Mont.
× Rock Creek nr. Red Lodge, Mont.
□ Mississippi R. at Head of Passes, La.
△▲ Missouri R. at Omaha, Neb.
○ W. Fork Rock Creek nr. Red Lodge, Mont.
■ Yellowstone R. at Billings, Mont.

Fig. 4.11 Size of bedload in the Mississippi-Missouri system. Use the information from an atlas to plot the distribution of the places mentioned. The valley in Rock Creek and the Yellowstone river is narrow and mountainous; that at Omaha and Mayersville has a wide flood plain, and Head of Passes is near the mouth of the delta. How do these situations relate to the composition of the bedload sizes?

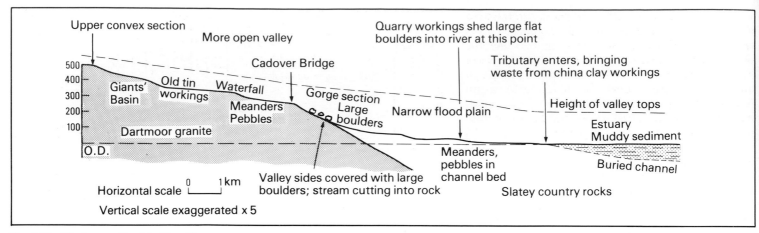

Fig. 4.12   The long profile of the River Plym.

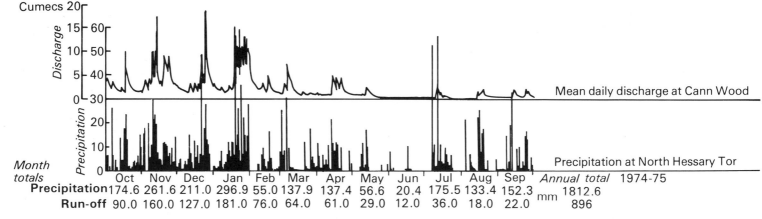

Month totals	Oct	Nov	Dec	Jan	Feb	Mar	Apr	May	Jun	Jul	Aug	Sep	*Annual total* 1974-75
Precipitation	174.6	261.6	211.0	296.9	55.0	137.9	137.4	56.6	20.4	175.5	133.4	152.3	mm 1812.6
Run-off	90.0	160.0	127.0	181.0	76.0	64.0	61.0	29.0	12.0	36.0	18.0	22.0	896

Fig. 4.13   The relationship of inputs of rain (and snow) and outputs of streamflow in the River Plym basin, west Devon. How does the rainfall vary through the water year (October–September), especially between winter and summer? What effect does rainfall have on streamflow in those seasons? What other factors affect the summer pattern? (*Information from South-West Water Authority and Meteorological Office*)

Fig. 4.14 (*right*)   The relationship of channel cross-section to drainage area in the Meavy basin, a tributary of the River Plym. It appears from this that the sites below the Burrator reservoir dam had a smaller cross-section than would have been expected. Water from the reservoir is taken to Plymouth by pipe. (*After K. Gregory*)

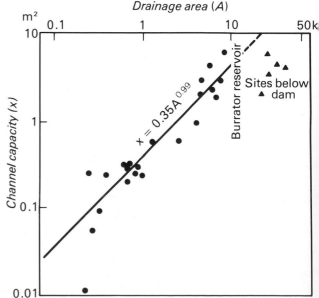

west Devon. All the rocks of its basin are resistant and compact, so that most precipitation flows rapidly to the stream; storage for longer periods occurs in the extensive peat bogs near its source. There is thus a marked contrast between the high and low flow characteristics of the stream (Fig. 4.13): whilst a maximum flow of over 58 m^3/s (cumecs) has been recorded, flows of 0·2 cumecs were common during the dry summer of 1976.

The Plym drainage basin has four major and distinct sections.

(1) The **source area** on Dartmoor lies between 250 and 500 m O.D. It is an open moorland area, where the valley slopes are 5–15 degrees, and where the valleys are generally wide and shallow (maximum relief 100–200 m). The streams have channels up to 4 m wide, mostly meandering across the valley floors, but at times channelled through solid rock with small waterfalls. There is a characteristic rectangular pattern to the drainage (Fig. 4.15), possibly related to the major joint patterns in the granite. The slopes of the valley sides were not formed by stream processes. They are often surmounted by tors and covered by granite blocks (clitter) which were sludged downslope during the very cold conditions experienced by Dartmoor during the Ice Age. The valley floors, on the other hand, are remarkable for the jumble of smaller granite blocks, a result of the active period of surface tin mining (or 'streaming') in the area from AD 900 to 1600. The gravels which had accumulated in the valley floors by solifluction were then excavated for the tin they contained. Man also affected the landforms of this seemingly wild area by destroying the former cover of woodland from the Bronze Age onwards, and this resulted in further movement of soil downslope, exposing summit rocks and burying the valley floor with a metre or so of rubble and soil. More recently the English China Clay Company has not only erected huge mounds of waste from its excavations at Lee Moor, but has diverted the course of the river near Cadover Bridge: the meanders there have thus formed and migrated over the last twenty years.

The major tributary of the Plym is the Meavy, which has been the source of Plymouth's water supply since it was first diverted into a special aqueduct, or leat, by Sir Francis Drake in the 1590s. The building of the Burrator Dam in the 1890s followed winters when the water in the aqueduct froze, and the dam height was raised in 1928. This extraction of water has affected the channel below the dam (Fig. 4.14), a fact which

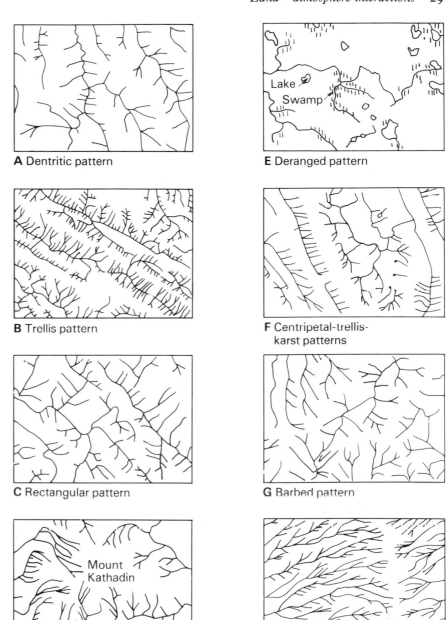

Fig. 4.15 Stream patterns often reflect a relationship with the underlying rocks.

A Dentritic patterns occur where the rocks have a uniform resistance to erosion. **B** Trellis patterns occur with alternations of resistant and less resistant rocks. **C** Rectangular patterns are often related to joint patterns in massive igneous rocks. **D** Radial patterns occur around volcanoes and areas of domed uplift. **E** Deranged patterns are the result of streams becoming re-established on a formerly glaciated terrain. **F** Centripetal patterns may occur with karst landforms. **G** Barbed patterns occur after river capture, or with some joint patterns. **H** Parallel patterns are found on steep slopes with little vegetation cover. (*After Thornbury, 1954*)

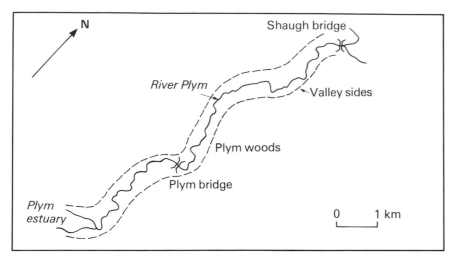

Fig. 4.16 The lower River Plym: the river meanders within the meanders of its valley. The two sets of meanders are unrelated in wavelength, and the river – like many others in Britain – is regarded as underfit.

illustrates the close relationship between stream channels and the water flowing through them.

(2) At the edge of the granite the Plym enters a **steep-sided gorge** between Cadover and Shaugh bridges. The side slopes of the valley are 20–45 degrees, and often reach the channel margins. Large boulders (1–3 m long) were rafted down these slopes during the Ice Age, and many have ended up in the channel. They rest there, often on solid granite, but are not moved by the stream and do not extend below the lowest point of the gorge. This section of the valley is the scene of the most effective stream action when in spate, since there is no flood-plain over which the waters may dissipate their energies: they are confined within the gorge, and the effects of the concentrated erosive power can be seen in the smoothed and fluted granite of the stream bed and the large boulders. The stream long-profile is at its steepest here, dropping 150 m in just over 2 km.

(3) Below Shaugh Bridge, where the Plym and

Studying your local stream

Investigation: measuring geological processes. A number of simple measurements are possible.

(a) **River flow.** You need to measure the cross-profile of the river and calculate its area (roughly depth × width). The velocity can be assessed by timing the passage of a floating object along 50 m of straight river. The amount of water passing in a certain time can be calculated using the following simple formula:

$$\underset{\text{(discharge)}}{Q} = \underset{\text{(velocity)}}{V} \times \text{stream profile area}$$

Carry out these measurements after heavy rain and after a dry period, and compare your results with the local (or school) rainfall records. If you can take a series of readings after a storm you will see how long the river takes to rise and fall. What causes the delay?

(b) **River load.** Remove a bottleful of water on each of the occasions when you measure the river flow, and allow the sediment to settle. The results can be even more worth while if you take bottlefuls from varying measured depths in the stream. The chemists may like to analyse the soluble fraction of the load; the suspended load which has settled out can be dried and weighed, and compared with the volume of water from which it was taken. Calculations can then be made concerning the amount of sediment transported in suspen-

Water-gauging in difficult conditions! The velocity of the stream is measured at several places across the channel to obtain the mean. (USGS)

Meavy have their confluence, there is another **deep valley section**, with relief up to 200 m, but here there is a flood-plain on each side of the channel for most of the way, reaching 75 m wide at times. The stream flows through this portion in a meandering course, although the valley and stream meanders do not coincide (Fig. 4.16). The stream is flowing across the slate rocks surrounding the granite, and the granite boulders extend to 1 km below Shaugh Bridge: they are less than 200 mm diameter. Sand, formed mainly of quartz, with some white feldspar and shiny mica flakes, occurs in the stream bed down to the stream mouth (having been washed down by the tin-streaming processes), but at times there is an addition of slate fragments. This is particularly noticeable for 300 m below a major disused quarry, with slabs up to 600 mm long. The deposits of the flood-plain can be examined in the channel banks, and are formed of pebbles, like those in the present channel, at the base, covered by finer sands and silts.

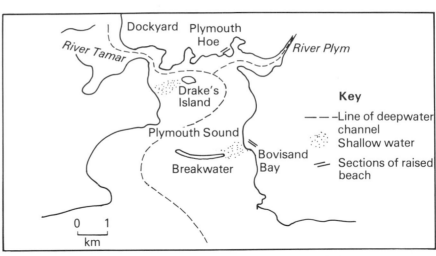

Fig. 4.17 Plymouth Sound: the deepwater channel continues the meandering pattern. It was cut in response to a sea level over 100 m below the present, and this has made it possible for ocean liners and large naval ships to enter the Sound and River Tamar.

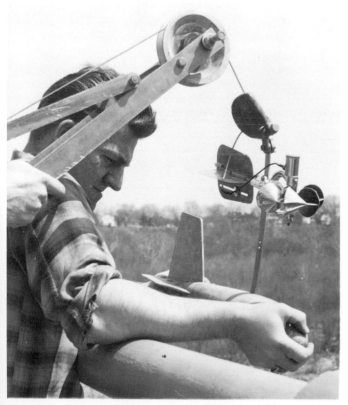

Recording stream-flow using a current meter suspended from a bridge. This method is used in larger streams. (USGS)

sion and solution at the moment when the measurements were taken; an average for the year can be worked out allowing for the differences between periods of high and low water; and the yearly rate of erosion for the whole river basin can be assessed.

(c) Further information on the **rates of weathering and erosion** can be gained by examining fresh road cuttings and ploughed fields after heavy rain. The steepness of the road cuttings often leads to the formation of gullies, and the rate of enlargement can be measured. In many cases the road-constructors leave a bare cutting in soft rock to grade itself and allow the weathering processes to act on it. Once again the rate of change can be measured. You can devise your own means of doing this.

(d) If there is a lake near your home it is worth measuring the **rate of sedimentation** by keeping a close watch on the mouths of the streams emptying into the lake. Examine the type of material brought in and take samples from the bottom at various distances from the shore. Compare the amounts of sediment carried into the lake with that taken out of the lake by the outlet stream.

(e) Has your local **river flooded** over its banks in recent years? Consult the local newspapers for that period to assess the extent of the damage caused. What was the cause of the flooding? What is the river authority doing to cut the risk of future flooding?

All these investigations will help you to increase your understanding of the processes at work and their rate of progress.

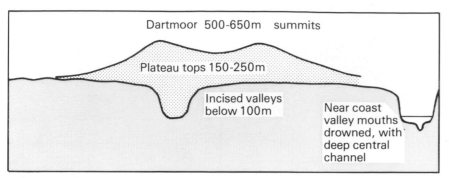

Fig. 4.18 Major aspects of the relief in south-west England. The valleys are cut deeply into the dominant plateaus surrounding the moors, and at times (e.g. River Dart) into the moors as well.

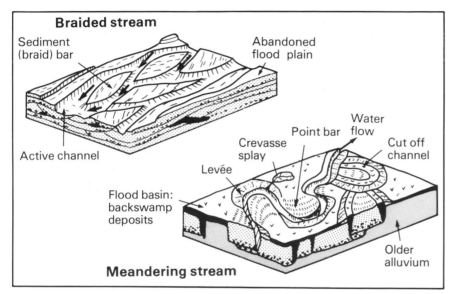

Fig. 4.19 Flood-plain deposits. Compare the variety and types of deposit where the stream is braided or meandering.

(4) The **mouth of the River Plym**, like many streams in south-west England, has been drowned by the rise of sea level at the end of the Ice Age. There is an estuary where banks of light-grey mud (washed down from the Lee Moor china clay workings) are exposed at low tide, and then the open waters of Plymouth Sound. Beneath this the deep channel winds its way towards a former level of the sea over 100 m below the present level (Fig. 4.17). The events which led to the submergence of this portion of the Plym drainage system were only the last in a series of ups and downs of sea level as the ice sheets advanced and retreated during the Ice Age. The features of Plymouth Sound and the lower Plym valley record many of these.

(a) There is a raised beach around the Sound, with portions near the foot of the Hoe and in Bovisand Bay: it is 5–15 m above the present sea level, and there are traces of others at higher levels.

(b) The deep water channel indicates a much lower sea level than at present.

(c) The deeply-cut valleys, like that of the River Plym, suggest that the sea level was higher than the present before the Ice Age (perhaps as much as 100 m), and fell during the times of ice sheet accumulation, when the rivers carved deep valleys to the new 'base-level' of erosion.

When one puts this evidence together, a series of events in the history of the drainage basin emerge (Fig. 4.18), which emphasise the complexity of processes that have given rise to the landforms we observe today.

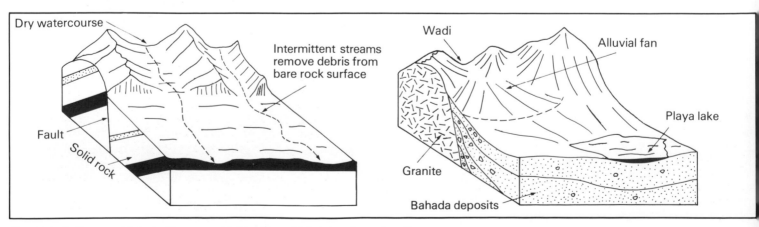

Fig. 4.20 Pediment and bahada. Transport and deposition of debris in a dry region. Can you explain the differences between the two types of terrain?

Apart from explaining the origin of present landscapes and attempting to understand the processes at work in a stream, the study of flowing water also leads to evidence which enables the geologist to interpret the sedimentary rocks of the past. Stream deposits forming today have characteristics which can be traced in ancient sediments.

(1) Most river deposits are associated with the **flood-plain**, having either a narrow, meandering channel, or a broad, braided one (Fig. 4.19).

(2) In areas where streams leave confined valleys and reach an open plain, **alluvial fans** are formed. These are particularly common in arid areas, since the streams dry up as their waters spread out over the flatter land and so deposit their load rapidly (Fig. 4.20), but they do occur in humid regions.

(3) Deltas and estuaries form at the mouths of streams where there is contact with a body of still (fresh or salt) water. **Deltas** form where a vast quantity of sediment is brought down by the river, and estuaries are found where the stream sediment load is too small to build deltaic features. Deltas form as a river's load is checked on entering the sea or a lake. The load is jettisoned, and, if the supply of debris is far too great for the waves or offshore currents to remove it, a delta will build up and extend the land seawards. Many deltaic features are determined by the relationship of the river water to that of the sea or lake into which it flows.

(a) If it enters a freshwater lake, like the Caspian Sea, the river water fans out, dropping its load in a broad arc. Until recently the River Volga delta was advancing 1 km into the Caspian every three years, but has not entered deeper water.

(b) If the river is carrying a great quantity of load, it will form a dense, mud-charged current, which will dive beneath the clear surface waters. The river Rhone enters Lake Geneva heavily charged with rock debris washed from Alpine glaciers, and has carved a deep trench on the lake floor, leaving only a small delta. The sediment accumulates on the lake floor, and the river is clear as it leaves the lake.

(c) When the river enters the sea it is usually less dense than the seawater, and flows over the surface, spreading outwards. The water at the margin of the channel is slowed, and coarse levée banks are built above the flood-plain and sea level. The results are seen best in the Mississippi delta (Fig. 4.21). Its deposits are thousands of metres thick, and have been built up as the sea has risen from its low Ice Age level. They have been likened to a pile of leaves with the veins representing the coarse channel/levée deposits and the leaf flesh the fine swamp deposits

Experiments

Investigation: some laboratory experiments. Many natural processes are hidden from view beneath the ground, or take place on such a large scale that we cannot examine them and make simple measurements. Some can be simulated with simple laboratory equipment.

(a) **Weathering.** Take small, equal-sized pieces of different rocks (e.g. clay, sandstone, chalk, granite). Soak some of each in water, and leave some dry. Then place some in a refrigerator and some in an oven. From time to time take them out and wet the wet ones again. Note the results. You could also include pieces of brick or concrete.

(b) **Porosity and permeability.** You will need a wide glass or plastic tube (diameter 3 cm), with one stoppered outlet; at least 100 cm^3 of glass or plastic beads in each of three sizes (2 mm, 4 mm, 6 mm); a glass beaker; and a measuring cylinder. Place each set of glass beads in the tube in turn, and pour in sufficient water to cover the top level of the beads: measure and record the volume of water used, and this will give you a measurement of the porosity (i.e. the spaces between the beads, which represent rock particles). Then, also in each case, let out the water and record the time taken to empty the tube: this will give you the permeability (i.e. the rate at which a liquid will pass through the sediment). Compare the data for the three sets of beads and draw up your conclusions (e.g. by a graph summary).

(c) A simulation of the behaviour of **density** (i.e. **turbidity**) **currents** can be carried out using the same wide tube. Fill this nearly to the top with water and tilt it at a low angle. Then pour into the top water containing a mixture of sand and mud. Record what happens to this mass and the way it is deposited at the bottom of the tube.

(d) **A delta model.** The equipment needed for this includes a large, round glass trough, a supported metal chute (e.g. a piece of aluminium held up by a retort stand); a rubber tube supplying water to the top of the chute; at least two varieties of sandy sediment, preferably in different colours and textures (stained if not); and a means of emptying the sediment into the chute. The water should flow continuously down the chute and into the glass trough, and some means of emptying (or controlling the overflow) should be provided. The different types of sediment are poured into the chute to build up alternating layers. Turn off the water supply when the sediment reaches the top of the trough; drain out the water and allow the sediments to dry. Cuts can then be made to demonstrate the internal structure of the delta.

(left) Meanders in the river Shenandoah, Virginia. The meander belt is 1–2 km wide, and the stream has cut a valley up to 50 m deep in the underlying shale. (USGS)

(right) The river Ganges plain and the Himalayas, northern India. The river has a braided pattern, related to the monsoonal wet-dry seasonal regime. (NASA)

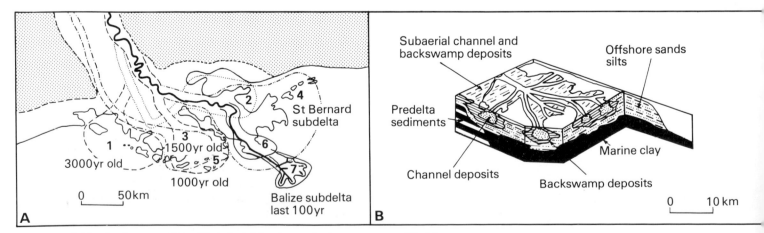

Fig. 4.21   A The development of the Mississippi delta during the last 3000 years. What has happened to the older delta areas? Make your own map of the types of deposit you might expect to find in the area (i.e. fine mud, silt, sand). **B** The features and deposits of a typical delta. Describe the sequence of deposits which would be found if a well was drilled through these deposits in a position just offshore, and compare the sequence found in a river flood-plain (Fig. 4.19).

between. As the channel is built outwards the gradient becomes lower until the river cuts (or crevasses) its banks higher up and takes another course, probably cutting across the lower swamps. Seven deltas have been recognised for the Mississippi over the last 3000 years, and its course has to be artificially regulated to preserve access to the major port of New Orleans.

Estuaries occur at the mouths of rivers bringing restricted quantities of sediment to the sea. They are common, since many rivers have wide openings due to postglacial drowning. The Thames estuary (Fig. 4.22) has two major sections. The inner estuary muds are deposited where the fresh and salt water meet: river flow is checked between Waterloo Bridge and Woolwich, and individual clay particles in the freshwater are able to flocculate (cling together) and sink to the river bed. The deposit of fluid, slimy mud may be dispersed at the next low tide by strong channel currents, but may build up rapidly in sheltered hollows. Dredged tanker channels often have to be emptied continuously, especially in winter when the river brings down the greatest quantities of mud in suspension. In the outer estuary sands cover an older topography of channels cut deeply towards the low sea-level of the late glacial phase. The sands have spread in from the North Sea glacial deposits, and from the deposits of the Rhine, and are arranged in banks separated by tidal scour channels. The deposits shift, filling the tidal channels, which are floored with coarser lag deposits (pebbles, shells, wood, old boots).

The study of these estuarine deposits is important for navigation, and also in connection with such projects as the Thames flood barrage and the reclamation of the Foulness area. It is important to know, for instance, whether the flood barrage will shift the zone of mud formation downstream and cause trouble for the new port developments. The reclamation of the Foulness area would be made more difficult and costly by the variable nature of the underlying, water-laden sediments. It would also restrict the water movements in the estuary and thus affect sediment movements.

Underground water

Much of the rain hitting the earth's surface percolates down into the soil and rock before flowing back into the stream and thence towards the sea (Fig. 4.9). The amount of water moving into the rocks and being stored there depends on two major characteristics of rocks.

Milford Sound, South Island, New Zealand. Notice the pattern of distributaries, the underwater depositional features, and the growth of vegetation on the delta surface. This delta is built from debris brought down a deeply glaciated valley in the Southern Alps. (White Aviation Ltd)

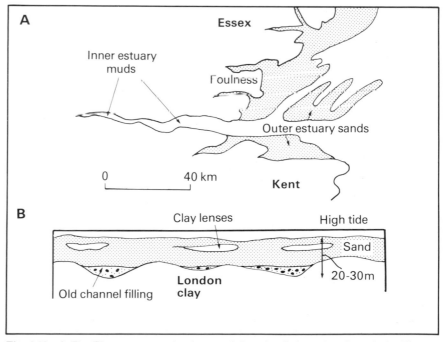

Fig. 4.22 A The Thames estuary: its shape and deposits. **B** A section through the Thames estuary deposits at Foulness. The length of the section is 1 km. If water was extracted from the sands, compaction would take place. Suggest how this might affect the surface of the deposits.

Stream activity and climate

1 Stream processes today. (a) **Humid tropics** are regions of high temperature and fairly constant rainfall inputs (humid temperate regions possess many similar characteristics). Throughflow is more important than surface flow – related closely to the dense vegetation cover and deeply weathered soils. The high degree of chemical weathering in such conditions reduces load to high proportions of solutes and fine materials, so that meandering streams with extensive flood plains result. An integrated, perennial drainage net affects the whole basin, leaving narrow, sharp divides. Q, v, w, d all increase downstream. Stream water contains low concentrations of solutes and suspended load, due to the slow release on slopes stabilised by dense vegetation. Removal of this vegetation, however, leads to rapid movements of materials downslope.

(b) **Arid conditions** are characterised by rare, local inputs of water, most of which runs off rapidly as surface flow, carrying with it the loose, weathered material. A storm may affect only part of a drainage basin, but most storms tend to be localised in the upper regions of basins due to relief. In a former humid region which has become arid close-spaced gullies will tend to form in this upper area, relating to the rapid runoff and increasing discharge rising to a maximum point. Beyond this, however, a combination of evaporation and permeable channel deposits will result in decreasing discharge and deposition, often ending in a mudflow (low water, high sediment). Channel networks are occupied ephemerally, and normally do not flow into the sea.

(c) **A periglacial stream**, like the Colville on Alaska's North Slope, has a very seasonal input of water during the early summer melt and this gives rise to a contrast between high-flow conditions (maintained for only a few days each year), low-flow (during the rest of summer), and the frozen winter situation. Beneath the surface permafrost prevents ground water movement, and throughflow is possible for a short period only. Frost wedging of the ground in fine-grained floodplains gives rise to a particular pattern of meander erosion: large sections collapse along the ice wedge lines.

(d) **Summary.** It would seem that it is not the processes which change, but rather the lengths of time over which they can operate each year. In addition extraneous (i.e.

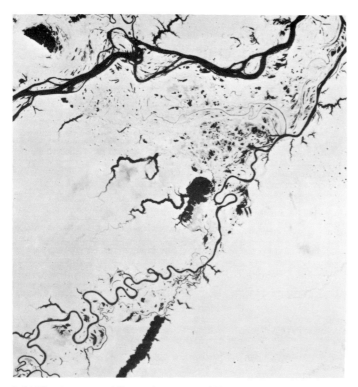

(left) The Amazon and Purus rivers, central Brazil: stream forms in a tropical humid climate. (NASA)

(right) Wadi Hadramaut, southern Arabian peninsula: valley forms in an arid climate. These valleys are filled with sand. (NASA)

non-fluvial) factors, such as vegetation cover, depth of weathering, slope processes, permafrost, and evaporation rates, also give character to the landform development in any area and must therefore be considered together.

2 Changing climates. Any discussion of landform origins is incomplete without acknowledging the fact that climates have changed markedly in the recent geological past. The last two million years have seen a fluctuation of ice sheets and climatic zones; even within the last 10 000 years, since the departure of ice from Europe, there have been major changes which have allowed, for instance, northern Canada to be covered by forest.

(a) **Refuges and relict landforms.** As climatic boundaries changed one area might be invaded by a different climate, whilst another might remain a 'refuge' of its earlier type. Thus in the Amazon basin a few areas have been recognised as refuges of tropical evergreen forest and of humid tropical conditions. In areas overtaken by a different climate there would be landforms which were formed in the earlier climatic phases and would not be related to the present processes: these are 'relicts' – e.g. glacial troughs in Snowdonia, etc.

(b) **Humid-arid-humid alternations.** As an area becomes more arid, the drainage net loses its perennial characteristics; erosion and gullying is concentrated in headwater areas; deposition takes over in the middle course, building this up and filling old valleys; and no water flows to the sea. As valleys fill, planar landforms increase in area until pediments become dominant. On return to humid conditions the stream network has difficulties in re-establishing itself and anastomosing channels are common . . .

(c) **Humid–periglacial–glacial–periglacial–humid alternations.** Think this one through for yourself, referring to British areas.

3 Other factors. Climate is not the only factor we need to consider in attempting to understand the formation of landforms. Sea-level changes may be related to climate, but other important influences include the variable nature of subsurface geology and of tectonic activity on both large and small scales. In fact, there is a growing school of 'structural geomorphologists' as well as of 'climatic geomorphologists'.

In addition the influence of man has been stressed already, and is increasing as his power to modify his environment grows (Fig. 1).

The Tenana river near Delta Junction, central Alaska: stream forms in conditions where summer temperatures exceed 25 degrees C, and winter temperatures are lower than −25 degrees C. (NASA)

Fig. 1 Model based on Piedmont, USA.

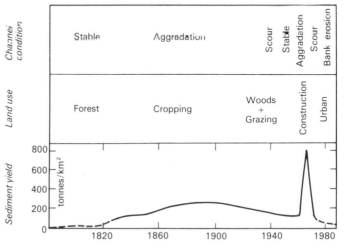

(1) A **porous rock** is one with a large proportion of space not occupied by solid material, and where the grains are not tightly packed or cemented. Such a rock can hold a lot of water (an aquifer), or oil (reservoir rock). Sandstones are often very porous.

(2) A **permeable rock** is one which will allow water to pass through it: permeability is measured by the amount of water which will pass through in a certain time, and may be related to plentiful wide pore spaces (primary or general permeability), or to wide joints or fissures in a largely impermeable rock (secondary or localised permeability). Rocks with the latter characteristic are also termed pervious.

A dune sand, which is very porous, is also permeable; a well-cemented limestone, cut by many joints, is non-porous, but permeable; a clay may be very porous, but the spaces will be too tiny to allow water to pass from one to another, and therefore the rock will be impermeable.

The upper limit of a rock saturated with water is known as the **water table**. This level is more important in rocks with primary permeability (Fig. 4.23). It is more difficult to define in rocks where the water is moving through joints. Above the water table the water percolates downwards after a rainstorm, or may move upwards by capillary action in time of drought. This zone above the water table is one where a lot of solution and replacement of minerals takes place. A fossil shell composed of calcium carbonate may be dissolved away, leaving a hollow mould in the fossil shape, and this may be filled by another mineral (e.g. silica, pyrite) at a later date. Movement below the water table is very slow, and chemical deposition from solution takes place on a large scale, eventually filling the pore spaces and cementing the rock. **Concretions** are formed by the concentration of minerals distributed throughout a rock layer into a single mass. Nodules of flint and pyrite are common concretions in the Chalk, and septarian nodules of impure limestone veined with calcite are found in the London Clay.

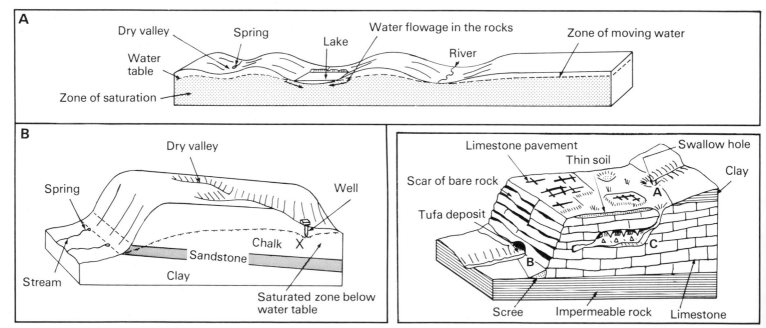

Fig. 4.23 A The water table. How is the water table related to relief? What would be the effect of rainy and dry seasons on its position? Why should a spring rise where it is marked? **B** Springs and wells. Springs and wells in a chalk area (e.g. North Downs). Which of the rocks shown is permeable, and which impermeable? A well sunk into the chalk will create a cone of depression at X.

Fig. 4.24 (*bottom right*) A karst landscape. There is no surface drainage across the limestone: a river disappears underground as it leaves the impermeable clay (**A**), and issues from the base of the limestone cliff (**B**). It flows through the cavern (**C**), which has a series of dripstone formations. What evidence of erosion and deposition can be seen?

A large proportion of the water falling on rocks and unconsolidated deposits with high permeability is moved underground, and such rocks may have little or no surface flow of water. Thus sand dunes absorb most of the rain falling on them, and there are many dry valleys on rocks like limestone (especially Chalk), and on some sandstones. The fact that most water drains underground in limestone areas has given rise to a group of relief features (Fig. 4.24) collectively known as **karst** (named after the area in Yugoslavia where the scenery is common). Features are related to the combined effects of solution and erosion by water. Surface effects include limestone pavement and solution hollows, whilst caverns are hollowed out underground. Such features can be seen in England in the Mendip Hills, the Peak District of Derbyshire and the Malham–Ingleborough area of North Yorkshire (Fig. 4.25). Dripstones are formed where the water percolating through limestone enters the roof of a cavern. The water is a solution of calcium bicarbonate, from which the calcium carbonate crystallises (calcite) and is left behind on the ceiling of the cavern to form a stalactite, part of which drips to the floor to build a stalagmite. The caverns themselves and their connecting galleries are formed by a combination of solution and river erosion. Underground streams are fed by rivers diverted underground through vertical solution pipes known as swallow holes or sinks (dolines in Yugoslavia). Gaping Ghyll, to the south-east of Ingleborough, is 120 m deep and leads down to a cavern. The caverns are enlarged gradually until the roof collapses, leaving an irregular depression or valley.

The Antarctic ice sheet, Marie Byrd land. At the edge of the continent the ice moves through wide gaps between nunatak peaks. (USGS)

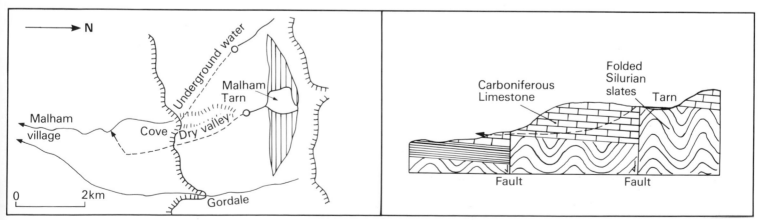

Fig. 4.25 The Malham area, North Yorkshire. Relate the geological structure shown on the section to the landscape and drainage features of the area.

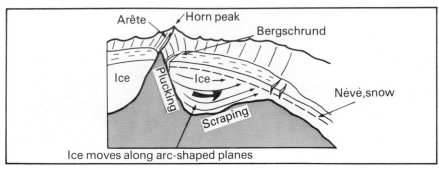

Fig. 4.26 A cirque glacier. The weight of the ice causes the mass to move forward and open up a gap at the back of the amphitheatre. This gap is known as a bergschrund, and it becomes the drain for meltwater, which refreezes around the backwall rocks. When the ice moves forward again large fragments are plucked away with the ice, and these pieces of rock are used to scrape out the lower portions of the hollow. In time two such cirques may be separated from each other only by a steep, narrow, knife-edge ridge (an arête), whilst pyramid-shaped horn peaks (e.g. Snowdon and the Matterhorn) are formed at the junction of the arêtes.

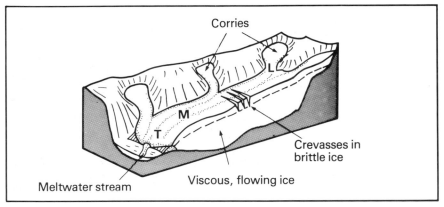

Fig. 4.27 A valley glacier. The glaciated valley shows the contrast between the amount of the valley directly affected by ice and by a stream of running water. **L**, **M** and **T** are three types of moraine on the glacier surface – lateral, medial and terminal.

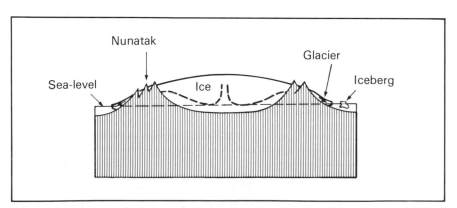

Fig. 4.28 An ice sheet. This shows a generalised section across Greenland, where the weight of 3000 m thickness of ice has caused the central area to be depressed below sea level. As the ice moves towards the margins under pressure it is channelled between the coastal mountains, and the resultant valley glaciers move very rapidly.

Ice

One-tenth of the world's land surface is covered by ice today, and in the recent past great ice sheets extended over a large part of northern Europe and North America. Earlier periods of earth history saw other ice invasions, although their record is less clear. We can therefore study both the areas where ice is active at the present time, and also regions which have just been uncovered after spending many thousands of years beneath ice.

Glacier ice is produced from permanent snow-fields where there is very heavy precipitation. The **snowline** is the line above which snow does not melt completely during the summer: its height depends on latitude, amount of precipitation and position with respect to dominant winds. At the equator the snowline is 6000 m above sea level; at 20 degrees north and south it rises to 6500 m and then it declines to 600 m at 60 degrees north and south. The highest points in the British Isles do not reach the snowline.

The snow layer thickens and is changed by compaction and the addition of meltwater from the surface to the granular **névé**, which increases in density as it is buried deeper in the snow mass. Ice is formed as this hard-packed mass re-crystallises, eliminating the air bubbles from the névé, and changing the colour from white to blue-grey.

A mass of ice and névé in a snowfield high on a mountain slope is unstable, and the increasing weight will eventually force the ice to move down into the valleys. The top ice acts as a brittle solid and fractures easily, but at depth it becomes viscous (it flows stiffly, like pitch). Glaciers are character-ised by deep crevasses in the brittle surface layer, some of them up to 65 m deep. The weight of this thickness of rigid ice changes the properties of the lower layer, which then flows slowly downhill, carrying the surface layer with it. Friction with the rocky sides of a valley will cause the glacier margins to move more slowly than the centre: the movement may be measured by inserting a line of stakes across the glacier between marked valley-side points.

The speed of a glacier is governed by the slope of the valley it occupies, and the rate of snow supply. Thus the Alpine glaciers move up to 80 m per year, some in Greenland have been known to travel 30 m in a day, and those on Antarctica, where very little new snow is provided, scarcely move at all.

Glaciers end at a snout, where the melting processes (ablation) are more powerful than the supply of ice. The position of this snout reflects fluctuations of temperature of precipitation over time. The glaciers descending from Mount Rainier, Washington, USA, had maximum extents around AD 1820 and have since retreated up to 1 km: this suggests that the climate has increased in warmth during that time, or that there has been a drop in precipitation.

Types of ice mass

Glaciation occurs in two main ways. Continental glaciation is the complete covering of a land mass by an ice sheet several thousands of metres thick. The underlying rocks are visible only around the margins. Greenland and Antarctica are still experiencing this today. Mountain glaciation occurs when the ice is confined to the hollows and valleys in high mountain ranges. The valley glaciers may extend to surrounding plains if the climate becomes very cold. This form is characteristic of the Alps, Himalayas and other young fold mountain ranges of the world.

Cirque glaciers are developments from small snowbanks in hollows which face away from the sun's rays and are therefore protected from too much summer melting. Daily thaw and overnight freezing break up the surrounding rocks, and the hollow becomes enlarged. Eventually the combination of ice flow and weathering will give rise to a characterisitc 'arm-chair'-shaped hollow (Fig. 4.26).

Valley glaciers occupy former river valleys or newly formed glacial troughs. They tend to fill the whole valley with ice (Fig. 4.27), with breaks in the valley floor reflected in surface crevasses.

Piedmont glaciers form where several valley glaciers coalesce on reaching a flat plain. Ice movement is slowed as it spreads out, but pressure of ice added from the valley glaciers may give rise to pressure folds in the ice and the debris being carried. The Malaspina glacier of Alaska is the best example today, but in the past Alpine glaciers flowed down their valleys and out onto the surrounding plains of central Switzerland, Bavaria and northern Italy.

Ice caps are local masses of ice which cover an upland area and discharge into valley glaciers. They are found in Iceland, Norway and Spitzbergen.

Ice sheets are the largest form of ice mass, and develop as ice builds up to bury a whole continent without regard for relief features. The ice which covers Greenland (Fig. 4.28) spreads out from the highest points of the flat-topped dome it has built, and then reaches the sea after being extruded through coastal troughs in rapidly-flowing glaciers. At the margin of the major ice sheet, mountain

A cirque in the Swiss Alps. (Swissair)

The Yentna glacier, Alaska—a typical valley glacier. Note the source areas for the glacier ice, the ways in which rocks above glacier level are being worn down, and the surface moraines (black lines on glacier surface). (USGS)

The Malaspina glacier, Alaska—a piedmont glacier. The patterns of moraine are formed where the valley glacier spreads out on a coastal plain and the ice is folded under pressure. (USGS)

The 6 km front of the Columbia glacier, Alaska. This glacier has its terminus in water and discharges icebergs into the shipping lanes used by tankers collecting oil from the Alaskan pipeline port of Valdez. (USGS)

peaks protrude as nunataks. Crevasses in the glacier ice cause it to 'calve' icebergs on reaching the sea. The Antarctic ice sheet is eight times larger, but the thickness of ice is more variable.

Many ice masses are shrinking at the present time, although the largest, Antarctica, is not since it has established its own climate which operates almost independently of the rest of the world and maintains the ice sheet. Elsewhere the ice coverage is much less than it was 150 years ago: paintings of the Rhône glacier in Switzerland show that it was nearly 2 km farther down its valley at that stage. If all the ice in the world now was melted, the sea level would rise 60 m.

Ice transport and erosion

The great volume of ice in a glacier or ice sheet is able to hold a vast quantity of rock fragments of all sizes, and the thickness of deposits in areas which were formerly glaciated testifies to the fact that glacial ice is particularly effective in removing such material from one area to another. Thus much of the soil from northern Canada now resides in the Midwest of the USA.

The material transported comes from rock fragments which fall on the surface from frost-shattered peaks above, from the underlying soil and weathered materials of areas over which the ice passes, and from erosion carried out by the ice itself. It is transported as moraine, sometimes engulfed in the ice (englacial), sometimes in a thick layer at the base, or sometimes in distinctive surface forms (Fig. 4.27).

A glacier can be likened to a giant flexible file as it winds down its valley, smoothing, scratching and polishing the rocks. The larger ice sheets carry out most erosion in the central parts of the areas they cover, from which the loose soils are carried away and zones of weakness etched out to leave a landscape of hummocky bare rock and thousands of hollows filled by lakes after the ice has melted. Northern Canada and Finland are like this, and the 'knob-and-lochan' landscape of the far north-west of Scotland is similar.

Angular rocks held at the sides and base of the ice form deep scratches on bare rock (**striations**), and these may provide an indication of the ice movement direction. Projecting crags of harder rock are smoothed and plucked as the ice passes over them, resulting in **roches moutonnées**; soft rocks in the lee of a hard crag may escape erosion and form a 'tail' (e.g. the Royal Mile slope leading eastwards from Edinburgh Castle).

In upland areas the most dramatic effects of glacial erosion are found. These include **glacial troughs**, with open U-shaped cross section, smooth valley sides from which projecting spurs have been removed, and an uneven floor with rock steps and deep hollows filled by ribbon lakes after deglaciation. Where the ice was restricted to pre-existing valleys a simple stream-like dendritic pattern has resulted (e.g. most Alpine glaciers), but where the ice covered the valley divides new through-routes were carved between one former drainage area and another. This is common in areas like the Highlands of Scotland, where a small ice sheet established a radial pattern of ice flow, thus conflicting with the older valley patterns. This has also led to the radial pattern of glacial troughs in the Lake District of north-west England. Tributary glaciers are smaller and not so thick, and so do not erode their valleys so deeply. This means that when the ice melts the former tributary valleys are left as hanging valleys, and waterfalls commonly occur where they join the main trough (Fig. 4.29).

During the last Ice Age some of the most active glaciers were to be found in areas like western Norway, western Scotland, western Canada and the western coasts of South Island, New Zealand. They carved very deep valleys due to the steepness of the mountain slopes and the heaviness of the local precipitation. When the ice melted the sea rose and drowned the mouths of the troughs carved at that time to form fjords.

Glacial deposits

Glacier ice may deposit its load as it moves across a landscape, but a large proportion of the glacial deposits left by major ice sheets in the past were dumped as the main mass melted. There is often a fairly clear distinction between the major areas of glacial erosion and deposition, although the major areas of erosion, such as the central areas of former ice sheets, and the highland areas, often contain local terminal moraines or patches of glacial melt deposits. Terminal moraines occur, for instance, as dams of the major lakes in northern Italy – Como, Garda and Maggiore – and also around the southern edges of the Great Lakes of North America.

Lowland areas near the margins of former ice sheets often have a covering of **till** (sometimes known as boulder clay). It is typically unsorted, containing a range of fragments from cobbles and boulders to the finest clay. Composition depends on the area over which the ice has passed: in East Anglia it has a major chalky component, whilst in

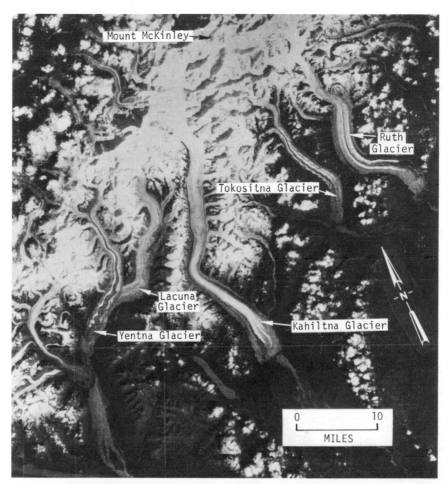

Glaciers and Mount McKinley, Alaska. Ruth and Kahiltna glaciers flow at fairly uniform rates of 10–30 cm per day: their medial moraines are straight. Glaciers like the Tokositna and Yentna (page 41) have wiggly moraines, indicating that there has been a sudden surge of the ice in which movement has been at over 1 m per hour for up to a year after years of stagnation. (USGS)

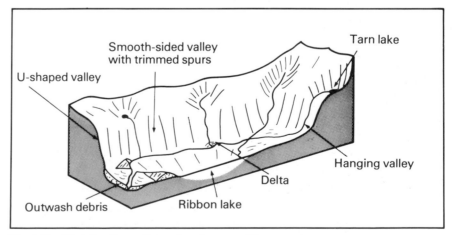

Fig. 4.29 Glacial landforms. Relate these features to Fig. 4.27.

A typical Norwegian fjord. What are the main features of the valley sides, the course of the fjord, and the land on which the fjord-head settlement is placed? (Wideroes Flyveselskap, Oslo)

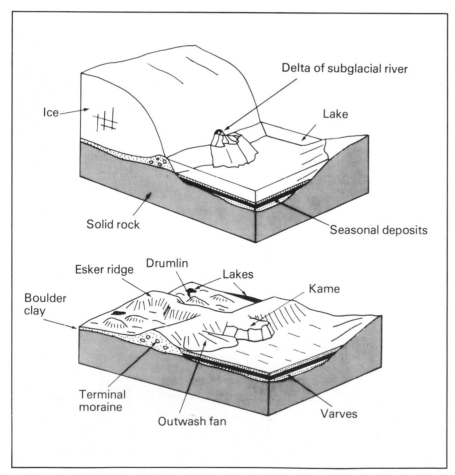

Fig. 4.30 Ice sheet deposits. The diagrams show the formation of distinctive deposits beneath the ice, and in meltwater lakes in front of the ice sheet.

the English Midlands it is often a reddish colour. Till plains are often featureless, or have a low, hummocky relief. Drumlins are elongated, oval mounds with a steeper end facing upstream in the ice flow. They often occur in large swarms in lowland areas, such as central Ireland. Drumlins are often formed of loose till material, but may have a rock core. It is thought that they are formed as the ice smooths out the underlying debris into streamlined forms.

A variety of deposits occurs at the ice margin, composed of **terminal moraines** and deposits formed by meltwater. The marginal moraines are sometimes formed by the accumulation of debris as the ice melts at a point where it stands still for some years; at others they may be bulldozed into a ridge as the ice advances, incorporating a variety of different types of deposit. The meltwater forms include **eskers**, which are sinuous ridges deposited by subglacial streams, and **kames**, formed at the margin of the ice and often in bodies of water (Fig. 4.30). Such marginal lakes are often the site for the deposition of varved clays (see 'Lake deposits', page 48).

Ice may carry large pieces of rock weighing many tonnes from one area to another and then dump them as **erratic blocks**. They may be sufficiently distinctive to allow their source to be located, and this may provide information concerning the direction of ice movement. The shores of Robin Hood's Bay in Yorkshire yield fragments of rock which include the pink granite from Shap in the Lake District, and an igneous rock from Norway: these come from the till which mantles the cliffs. There is a different sort of pink granite on the coast of north Devon, buried beneath marine sands: this is thought to have come from southern Scotland, and to have spent its last journey as part of an iceberg which became stranded on the coast (Fig. 5.7).

Tundra areas

The very cold areas of the world, which are not actually covered by ice, are known as **periglacial** areas. They occur today in areas of tundra climate like Alaska, northern Canada and Siberia, where there is little precipitation, but where the temperatures are below freezing for much of the year, and the bitter winds make it impossible for many plants to grow. Such conditions migrated southwards with the ice sheets in the past, and affected much of western Europe. The freezing of the ground in winter causes it to contract and crack open, and

The snout of the Mammoth glacier, Wyoming. Notice the differences in size between terminal moraine boulders near the ice, and the stream outwash in the foreground. (USGS)

An old terminal moraine carved by fluvio-glacial terraces in Glen Ross, Ross-shire, Scotland. (Crown Copyright)

Ground ice, central Alaska (the white masses). What will happen to this area when the ground ice melts? (USGS)

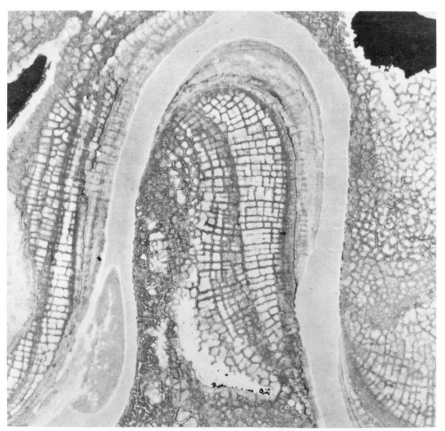

Ice wedge polygons, northern Alaska. These features are common in fine-grained alluvium and affect the detailed form of the stream channel. (USGS)

these cracks become filled with ice wedges: in low-lying areas of fine deposits the whole landscape may be dominated by these ice-wedge polygons. During the short summer the streams receive a single major phase of runoff from the snowmelt, which carries out rapid erosion and transport for a fortnight. The surface soil also melts and becomes very mobile, since the water cannot escape downwards: solifluction on slopes and the heaving of lowland areas gives rise to stone stripes and sorted stone circles and polygons in these areas.

There are major problems in these areas when man attempts to live there: heat from houses will thaw a deeper active zone in the permafrost, and this may lead to subsidence; ploughing of a field will also increase the active zone, since the protective vegetation has been removed; and roads become vulnerable to water seepage which soon refreezes on the surface. These conditions made the construction of the Alaskan pipeline particularly difficult. Crude oil reaches the surface at temperatures of 40–50° C, and is pumped through the pipes at these temperatures to keep it flowing. If the pipes are buried in frozen soils, they have to be lagged or the ice in the soil will melt, causing movement and breaking the pipe. Thus much of the pipe had to be constructed above ground level where permafrost was present.

Wind

Most parts of the world are moist enough to support grasses and other plants which bind together the surface soils. In the British Isles this is true, except for sandy coastal areas and inland areas where the fields are ploughed and left bare during the early months of the year before the crops grow. The processes and results of wind action can be observed in such conditions – as well as attempts to prevent the worst effects of wind erosion.

The deserts and semi-arid parts of the world are particularly vulnerable to wind action, since they do not have a complete vegetation cover over their surface. It is important to notice, however, that grasses have been in existence only since the Miocene period: wind action must have been of greater importance before that time. The most damaging effect of wind action is in the removal of the finest soil particles from an area which has been producing crops. This is happening extensively around the margins of deserts at the present moment, due to over-grazing and over-cultivation of such areas by man. In geological terms, however, wind is a relatively puny agent in comparison with streams and glaciers.

Winds vary considerably in strength from one moment of time to the next, and thus are able to lift and move rock debris only for short periods of time. At wind speeds of 40 km per hour sand grains (0·15–0·30 mm diameter) will be moved, but extremely rare gusts of 150 km per hour are required to roll pebbles along the desert surface. Fine dust (grains of less than 0·06 mm diameter) is moved most easily, and the lightest winds are sufficient to swirl it into the air. Dust storms, caused by temperature differences in the atmosphere, are common in the Sahara, the dust being carried far out over the Atlantic Ocean and northwards across Europe.

Sand grains are seldom raised far into the air, and are usually moved in surface creep and saltation (by being bounced along, usually restricted to the lowest 1 m). Whereas dust particles do not affect the rocks as they waft around them, sand may erode the lower metre or so of a rock mass as it is thrown against it, and will trim the pebbles in its path (Fig. 4.31). The constant collision of sand grains in the air and on the ground makes them rounded and gives them a frosted, 'millet-seed' appearance.

Wind only erodes when it is carrying sand, and then produces only minor effects. This results in a virtual fossilisation of desert landscapes, since wind erosion is so ineffective and weathering in regions with so little water is also very slow. Thus Cleopatra's Needle stood for thousands of years in Egypt without deterioration, but a hundred years on the Thames Embankment has been sufficient to destroy much of the writing.

The wind may lower the landscape by removing large volumes of loose, dry sand or silt and hollowing out large depressions in the process. This is known as deflation, and is common in areas like the 'Dust Bowl' of the central USA or the virgin lands of the USSR, where the land has been ploughed despite being in a semi-arid zone. It is also thought that the large hollows in the Egyptian desert west of Cairo were formed by a similar process: the Qattara depression, 135 m below sea level, is the largest, and has its floor where the water table reaches the surface.

Wind deposits

The most characteristic features of deserts are the widespread areas of bare rock and immovable broken rocks. These remain behind after the fine dust has been removed and the sand grains have

A dust storm blowing fine soil from fields in Colorado. The wind speed on this occasion was 45 km per hour, and the storm lasted nearly 3 hours. (USDSACS)

Wind blowing soil across dry, ploughed fields in early spring in the large grain farm district of central Illinois. (USDASCS)

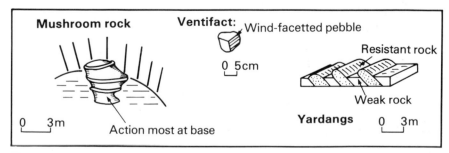

Fig. 4.31 Sand-blast action. Notice the small size of the features produced.

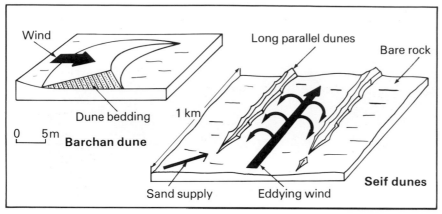

Fig. 4.32 Sand dunes. Make a note of the internal structure of the crescent-shaped barchan dune (dune-bedding).

Sand dunes deposited by the wind in Nevada: the material came from mine tailings at a copper smelter. (USDASCS)

Dune sands. These structures are found in Triassic rocks of Cheshire. How can we tell they were formed in desert conditions? (Crown Copyright)

been rolled into concentrations known as 'sand seas': the latter make up approximately one-eighth of the Sahara. **Sand dunes** are temporary accumulations of wind-blown sand, and may form confused heaps of more regular barchans and seifs (Fig. 4.32). Dunes are also common at the backs of sandy beaches and have resulted from onshore winds blowing across dry, bare sand at low tide. Such dunes tend to move inland unless they are planted with special grasses or trees. Small areas of such coastal dunes are common around British coasts, including the Braunton Burrows in north Devon, and the Culbin Sands in Morayshire.

Dust often settles far from the desert of its origin and becomes anchored by vegetation. The **loess** of northern China covers an area the size of France and consists of a fine-grained, yellowish loam, becoming coarser towards the Gobi desert, its probable source. Similar deposits include the Sudanese 'cotton soils', and those which occur beyond the margins of the former ice sheets in Europe and North America. During the last advance of ice, winds blowing from the ice sheet picked up fine particles from the bare ground in front of the ice and redeposited them to the south to provide some of the best farming soils in the world and leaving gravelly heathland behind.

Lakes

Lakes are water-filled hollows, occurring mainly where the landscape surface has been made irregular by glacial activity, volcanoes and earth movements (Fig. 4.33). Lakes can be viewed as giant settling tanks, into which rivers drop their loads of sediment and emerge as clear streams carrying only the finest matter.

The main process in lakes is therefore deposition, and all lakes are filling gradually. Some have been filled completely, leaving a flat area in the bottom of a valley. It is common to find that deltas form where rivers enter a lake with a good supply of suspended sediment, and some have built out to cut a lake in two by extending their deposits right across, as in the stretch of land dividing Buttermere and Crummock Water in the English Lake District. The coarsest fragments are dropped near the river mouth, but sand and mud may be carried far out into the lake. Organic matter, particularly shells, but also much plant debris, becomes more important towards the central parts. Shallow lakes in cold areas, such as northern Sweden, often have a rather frothy deposit of 'bog iron ore' around their edges.

Varved clays are the result of seasonal deposition in lakes supplied with glacial meltwater. The deposits on the floor of Lake Zurich in Switzerland are carried into the lake by rivers descending from Alpine glaciers. They have an alternation of thin, dark layers of mud rich in organic matter, and thicker, lighter-coloured sands containing calcium carbonate. In winter the rivers carry little glacial debris, but in summer the lighter-coloured layer has a very high proportion of 'rock-flour' transported from melting icefields. The best-known varves are those which were formed in lakes at the margin of the ice sheet retreating northwards across Scandinavia: these have been counted and used to date the various stages of the retreat.

Salt lakes are important areas of chemical precipitation in dry areas, especially where drainage is to an inland basin, and not to the sea. As the water is evaporated from the lake the salts contained in the water become more and more concentrated, and eventually the remaining solution will become saturated with a particular salt, which will be precipitated. The Dead Sea has nearly ten times the concentration of salt in seawater.

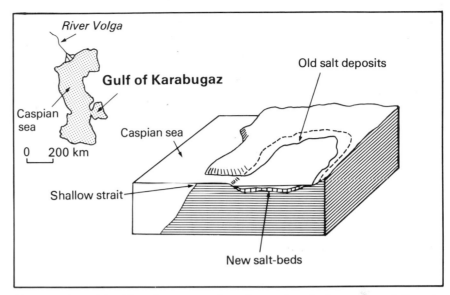

Fig. 4.34 A salt lake. There is a constant flow of water through the narrow strait from the Caspian Sea. The water, only one-third as salty as ocean water, is evaporated in the shallow gulf and becomes nearly ten times as salty as seawater: this leads to the precipitation and deposition of layers of salt on the floor of the gulf. Calcium carbonate and gypsum are overlain by sodium sulphate.

THE ORIGIN OF LAKE BASINS

Origin	Examples	Origin	Examples
Earth Movements Crustal warping	Caspian Sea; Lake Victoria; Lough Neagh	**Deposition** Morainic barrier Irregularities in glacial drift ('kettle holes')	Lake Garda (northern Italy) Glaciated plains, valleys — e.g. Tweed
Backtilting of valley Rift valley Tear fault	Lake Kyoga (upper Nile) Lake Tanzania; Great Basin (USA) Lac de Joux (Jura Mountains)	Ice barrier	Marjelen See (Alps); Glen Roy (Scotland)
Vulcanism Rings of volcanic ash	Maare of Eifel Mountains (Germany)	Landslides, mudflows River deposits — ox-bow lakes Coastal lagoons enclosed by bars, sand dunes	Lower Mississippi valley SW France; Chesil Beach, Dorset
Crater, caldera Lava barrier across valley	Crater Lake, Oregon (USA) Sea of Galilee	**Organic** Coral reef barrier Beaver dams Man-made dams Man's excavations — gravel pits, peat-digging	Pacific Ocean Yellowstone National Park (USA) Kariba (Rhodesia) Norfolk Broads
Erosion Glacially eroded hollows Ribbon lakes Cirque-floor tarns Wind deflation Solution hollows	Canadian Shield; Finland Thirlmere, Lake District Bleawater, Lake District	*Meteorite Impact*	Ashanti Crater (Ghana)

N.B. Many lakes are caused by a combination of factors, e.g. Loch Ness is a glacial hollow along a fault-line; the Sea of Galilee is in a rift valley but is dammed by a lava-flow; the Great Lakes of North America were hollowed out by ice tongues, dammed by the terminal moraines and uplifted at the northern end by isostatic compensation.

Fig. 4.33 The origin of lake basins.

A ribbon lake (Loch Einach) in a glaciated valley in northern Scotland. (Aerofilms)

The type of salt deposited in inland lakes varies according to what has been supplied to the rivers crossing the surrounding rocks, and also to the intensity of evaporation. The Great Salt Lake of Utah is surrounded by layers of carbonate salts, formed when the lake covered a greater area, and sodium chloride is being precipitated on the floor of the present lake. Farther south, in Nevada, there is a series of small playa lakes in which sodium carbonate is being deposited. One of the most interesting of such inland lakes is the Gulf of Karabugaz, beside the Caspian Sea (Fig. 4.34).

Seawater may be trapped in the **coastal lagoon** of an arid area, leading to a consistent pattern of precipitation, since seawater has a relatively constant composition, including 3·5 per cent of dissolved matter, mostly sodium chloride.

(1) When half the seawater has been evaporated, iron oxides and calcium carbonate are precipitated.

Badwater, Death Valley, California. This lake is 90 m below sea-level, and the water contains 5 per cent salts (compare this with seawater—Chapter 5). (USGS)

(2) When 80 per cent of the water has been evaporated, gypsum is precipitated.
(3) When 90 per cent of the water has been evaporated, sodium chloride is precipitated.
(4) Finally, if evaporation continues, the bitterns (magnesium and potassium salts) are precipitated.

Such a sequence is found in the Piano del Sale on the border between Abyssinia and Eritrea. This was once a gulf of the Red Sea, but was cut off by a series of lava flows, and it became a giant evaporating dish in the desert climate. It is lined with gypsum, and rock salt covers the centre of the basin in an area 30 km across. In the very centre, on top of the rock salt, are layers of potash. The original basin was 300 m deep, but that depth of water could have left only a 5 m layer of salt. There must have been a series of invasions by the sea before the pile of lava flows at the entrance to the basin reached its present height.

Water resources

Water is a renewable resource. Although we may use it in large quantities in our homes and industries (Fig.4.35), each shower of rain and snow brings more to the land surface. And yet, we still have to be careful in our use of water.
(1) If we require more water than the normal flow of a river will provide, or if we wish to make provision for the low water season, then we will have to tap stored supplies. River flow fluctuates greatly (Fig. 4.13), but storage of water is possible in a number of ways.
(a) Underground water is stored in pore spaces in aquifers, but requires investment in wells and pumps to gain access to it (Fig. 4.36).
(b) Storage dams can be built across the headwaters of a river system (Fig. 4.12).
(c) Storage reservoirs can be built by the side of the river in its lower course, as in the case of those beside the Thames west of London.

Flush toilet	3 gallons
Take a tub bath	30-40 gallons
Take a shower bath	15-20 gallons
Wash dishes (dish washer)	10 gallons
Washing machine programme	20-30 gallons

Fig. 4.35 (*left*) Domestic uses of water. Calculate how much (a) your home, (b) your town or city, uses in a week and in a year.

Fig. 4.36 (*below*) **A** An artesian basin. The London Basin (a very generalised section). The water table level (X–X) is above the top of the well at Y, where the water will gush out under pressure. **B** West Surrey water supply. Water is pumped from the six wells near Milford to the reservoir at Hydon's Ball, from which it is fed downhill to Godalming and district. The River Wey is also used as a source of water, and wells are dug into the chalk near Guildford.

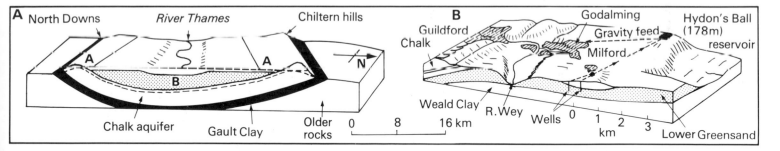

Water reserves

The Grand Canyon, Arizona. The river Colorado obtains over 80 per cent of its water in its mountainous headwaters, and then crosses desert lands, cutting a valley over 1·5 km deep. (USDASCS)

All these methods are costly in terms of capital investment, and the storage dams and reservoirs cover land and influence our use of landscape.

(2) If we return water to the drainage system after use, it should be clean. Either someone farther down the river will be using it again, and will have to pay extra costs of purification, or it will enter the oceans with possible effects on fish stocks and re-cycling (a surface film of oil may reduce evaporation).

(3) There is a limit to the use of water in any area. Whilst water is re-cycled continuously, we cannot persuade the atmosphere to deliver more to a place than it would do naturally.

The **River Colorado** in the United States of America provides a good example of the problem of using water resources to their limits. As man's engineering skills have increased, so the needs for water in the dry western parts of the USA have been met by the taming and use of this river. The waters are used largely for irrigation agriculture, for the domestic and industrial needs of cities like Los Angeles, and for generating hydro-electricity. These uses are not always compatible, since electricity cannot be stored, and peak generating times (and thus flow of water from the dams) coincide with peak morning and evening electricity demands in Los Angeles and Las Vegas. Farmers do not require water at these times, and so nearly every electricity-generating dam has a further storage dam down-stream.

The River Colorado and its tributaries have become a series of lakes and dams. The reservoir lakes, however, lose water by evaporation in the arid climate, and by percolation into the underlying permeable rocks. The reservoirs, like any lake, are also filling with sediment, and are given lives of 75–200 years. The overuse of water on irrigated farms has washed salts back into the river, and by the time it reaches the Mexican border the salt content is too high for the water to be used for further irrigation. When the Mexicans complained, the Americans decided to build a giant nuclear-powered desalination plant on the border!

The first dam was built on the river in 1909, but most have been constructed since the 1930s, when government policies led to the spending of vast sums of money on public works so that people could find some form of employment and so that the economy could be re-stimulated during the great depression. Plans were laid at this stage, based on rainfall measurements in the Colorado basin over the period 1905–20. This happened to be a period of abnormally high rainfall (nearly twice the sub-sequent average), and it is estimated that demands along the river could exceed supply by 1980. Despite this there are plans to use the Colorado water in oil shale mining, and to transport further quantities to the Phoenix–Tucson area.

Reactions to this situation of reaching the limits of water supply from the Colorado are varied. Some look to even more ambitious schemes, such as those which involve transporting water from the melting Alaskan glaciers to a giant storage reservoir in the Rocky Mountain Trench of British Columbia from which the water could be distributed throughout the USA and particularly to the south-west. Others have tried to induce more rain to be precipitated, but they are running into problems associated with increased avalanching after added snows in the mountainous headwaters of the river. Whatever means are used to augment the Colorado supply will be extremely costly and will merely put back the time when the limits of water usage in the area are reached. The best solution might be to encourage water users to be more careful with this resource, and there are signs that habits of conserving supplies – developed during the California drought of 1976–7 – are still being practised.

5
Coasts and ocean basins

The continents have been more fully investigated by geologists than the ocean because they are the home of man, who finds it difficult to endure existence beneath the ocean surface for very long. Much of our knowledge of what happens below the sea level is gathered by remote means – by sending down dredges, drills and cameras – or by using sonar techniques. And yet the oceans cover 71 per cent of the globe at present, and until recently have concealed many features and processes which are now seen as vital for our understanding of what we read in the continental rocks.

Coasts and the offshore zone

Coastal features attract geological studies. This is partly because the rocks and structures are often best exposed in this zone, but also partly because the landforms and deposits forming there today result from the interaction of a wide range of geological processes (Fig. 5.1).

The action of the sea

The sea meets the land and acts on it mainly by the force of surface waves and the changing tidal levels: the combined result is to wear away the rocks and redistribute sediment in the coastal zone. In addition it may dissolve rocks, although it is often saturated with dissolved minerals and cannot take up more in this way. **Waves** are formed by friction as winds blow across the ocean surface, and may grow to 15 m in height; anything over 10 m is unusual even in a strong gale. Wave motion extends to a maximum depth of 100 m, but is very small below 30 m. Wave height depends on the wind strength, depth of water and the distance of water over which the wind is blowing (fetch). The

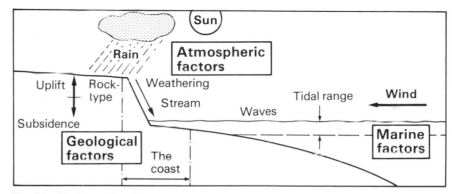

Fig. 5.1 The interaction of forces in the coastal zone.

A coast of bays and headlands. Notice the relationship of the beaches to wave direction. How are the wave directions affected by the headlands? (USGS)

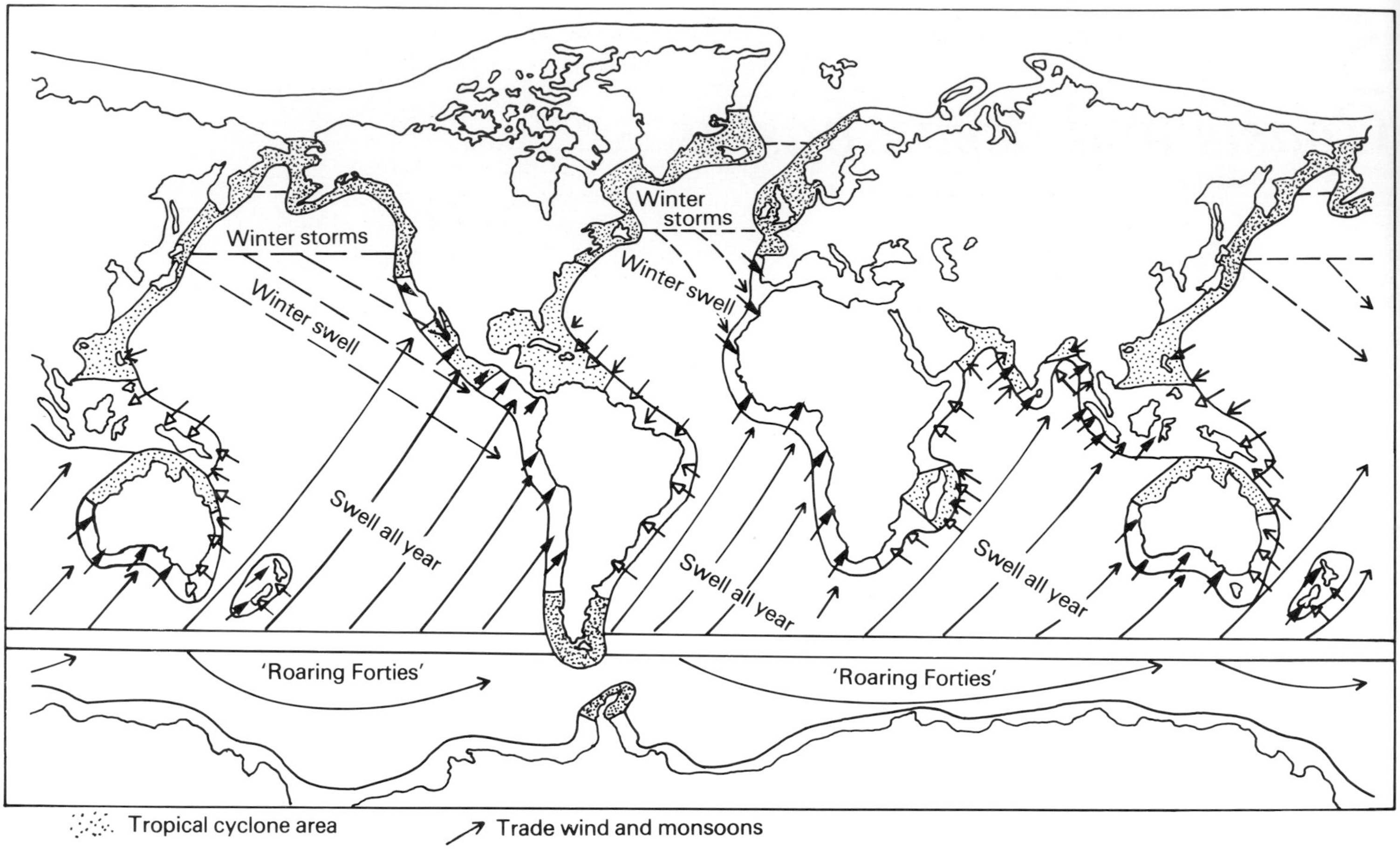

Map	Wave environment	Potential wave energy	Wave characteristics	Directions
	Storm wave (temperate storm belts)	High all year from south, less in summer from north. Swell waves too.	Height often over 5 m. Short waves (period 6-12 s) with shallow wave base	Local storms cause variation. Refraction has less effect near shore. More constant swell wave background
	West coast swell	Medium-to-high, diminishing from south, Little variation.	Low waves, declining in height from centres. Occasionally (W. Mexico) subject to tropical cyclones. Long waves (period 14-16 s) from south	Very consistent, especially from the south. Refraction important due to deep wave base
	East coast swell	Medium-to-low, and less regular. High variation in tropics.	Low waves, and long. Subject to tropical cyclones	Fairly consistent, except in the tropics. Refraction important
	Protected	Very low. Little swell penetrates. Some with monsoons.	Very low waves. If subject to tropical cyclones, extreme contrast occurs	Variable

Fig. 5.2   Major world wave environments. (*After Davies, 1972*)

greatest frequency of large waves is experienced in the midlatitude cyclonic storm belts: in the southern hemisphere the 'roaring forties' zone is virtually uninterrupted by land masses (Fig. 5.2) and has storm conditions throughout the year; in the northern hemisphere the North Atlantic and North Pacific oceans have these conditions in the winter season. Coasts in these latitudes sustain prolonged battering by storm waves, but account for a relatively small proportion of the world's coasts. The British Isles are in this abnormal situation of sustaining such high energy wave attack, especially on their western coasts. It is not surprising therefore that the British Isles is one of the main centres for research into the problem of extracting energy from ocean waves and making it available to mankind. The successful outcome of this research could be of great value in the future as supplies of fossil fuels run short.

Most parts of the world receive waves generated in the storm belts only after they have calmed down and the variety of waves has been sorted into a series of regular, long wavelength waves which approach coasts with moderate energy. These are known as swell waves, and are characterised by low erosive capabilities. Some localities in these zones do receive storm waves at intervals, especially if they lie in the tropical cyclone or hurricane belt, but other parts of the world are in even more protected circumstances, where effective wave action is limited.

Tides are oscillations of ocean water caused by the attraction of the moon and sun. They are scarcely noticed in the open oceans, but when constricted in narrow channels strong local currents may be formed. There are also major differences in the tidal range (i.e. the difference between high and low tide). Areas with tidal ranges over 4 m have macrotidal environments, and those with ranges under 2 m have microtidal environments. This characteristic determines the height range of wave activity, so that it may be concentrated over a narrow range or spead over a greater distance; it also controls the extent of the zone exposed to subaerial weathering at low tide.

The reaction of the land

The waves and tides act on lowlying or mountainous land, weak or resistant rocks, and this interaction gives rise to the detailed coastal features. The waves may approach a coast with a gentle offshore bottom gradient. This will result in the waves being slowed as they approach the coast, whereas their full energy may be brought to bear on rocks in deep water. Headlands extending into deeper water will cause refraction of wave paths so that wave action is concentrated on headlands and the waves spread out in adjacent bays (Fig. 5.3). In addition, waves may approach a coast at right angles, or at an oblique angle. In the latter case material will be moved along the coast (Fig. 5.4).

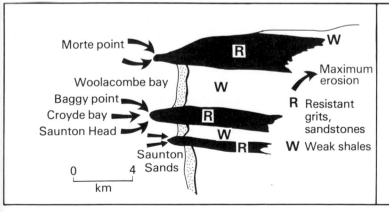

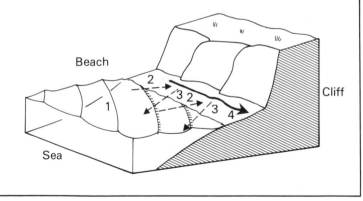

Fig. 5.3 (*left*) Differential marine erosion. The headlands, subject to the greatest wave attack, will eventually be cut back to form a straight coastline.

Fig. 5.4 (*right*) Longshore drift. Longshore drift causes pebbles to be moved along the shore as follows. 1 – the waves meet the shore at an angle; 2 – pebbles are carried up the beach in the direction of wave movement; 3 – they roll back down the steepest slope taken by the undertow; this happens many times and the result – 4 – is that there is an overall movement of rock debris along the beach.

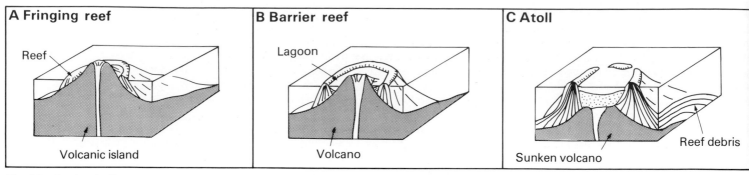

Fig. 5.5 Coral reefs. Charles Darwin's interpretation of coral reef development. Assuming that the reef would begin to develop near the volcanic island, he concluded that the island must have subsided for the reef to become separated from it (**B**), and for atolls (**C**) to be formed. As the island slowly sank, the reefs would be built up to sea level, or would cease to exist.

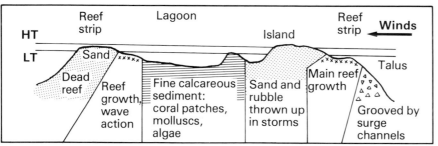

Fig. 5.6 The coral reef environment. Notice how the zone of maximum growth is related to the area of maximum wave action. Draw a sketch-map of a circular coral atoll and plot the different types of sediment you would expect to find.

Helford river, Cornwall—a ria. What effect has a rising sea-level had on the landforms? (Aerofilms)

The effect of climate

Variations of coastal features in different parts of the world may be related to climatic differences. It is only in the warm parts of the globe that corals and other reef-forming organisms can exist, and so biogenic (i.e. formed by living forms) structures are characteristic of tropical coasts. **Coral reefs** are built by lime-secreting organisms with special living condition demands. They need sunlight and warmth, with a temperature of at least 18°C, though they are at their best between 23° and 25°C. The most rapid growth occurs in the agitated, oxygen-rich surface waters. Some reef corals have been found growing as deep as 60 m below the surface, but these are isolated, poor specimens. Muddy water also inhibits growth. Perhaps the strangest feature is that corals grow most strongly in the zone of wave action just above low tide, and are most easily worn away in quiet waters.

The largest reefs are in the Indian and Pacific oceans. This is where Charles Darwin studied them, and where he suggested that the three major types of reef (Fig. 5.5) were three stages in a common series of events. Other scientists have pointed out that the valleys on volcanic islands still remaining above sea level confirm this subsidence theory because they are drowned at their mouths. Darwin's theory did not take into account the sea level changes which occurred during the Ice Age: the maximum drop of 150 m below the present would have killed most of the coral reefs. American geologists like Daly studied the West Indian reefs and were quite sure that this would have had a devastating effect all round the world, planing off the tops of the volcanic islands. When the sea level rose again the islands would be recolonised by reef corals, growing upwards to keep pace with the rising sea level.

It has recently become possible to test these theories, because a series of borings have been made into Pacific atolls. These have shown that whilst many do have solid rock at the height predicted by the glacial control theory of Daly, some borings have penetrated thousands of metres of coral debris before reaching volcanic rock. It also seems that the continental margins have been affected more by the glacial changes of sea level than the islands in the middle of the oceans. As more evidence comes to light we have to admit that both Darwin and Daly could have been right.

Further investigations have also been carried out into the detailed composition and sediments formed on coral reef structures (Fig. 5.6).

It has already been made clear that midlatitude temperate regions on west coasts have storm wave environments and high rates of coastal erosion. Farther towards the poles along the northern shores of North America and Siberia the sea ice produces distinctive beach ridges as it pushes soft sediment forward in the early winter; later melting of ice floes may leave boulders stranded on the beach.

There is also the effect of **past climates**. Coasts in midlatitude areas like the British Isles, and the adjacent offshore areas like the floor of the North Sea, have high proportions of pebbles and boulders due to the fact that they were formerly covered by ice sheets: the rock debris was dumped at the time of melting (Fig. 5.7).

Ups and downs of the land and sea

The features of coasts have been complicated by changes in sea level, and by uplift or subsidence of the land masses. It is not always possible to tell whether it was the sea level or the land which moved, and we often have to talk in terms of relative movements between the sea and land.

We do know, however, that the Pleistocene Ice Age, with its fluctuating ice sheets, experienced rises and falls of sea level. When the ice accumulated, the effect was to slow the return of water to the ocean in the hydrological cycle, and so the ocean level fell all round the world; melting ice caused ocean levels to rise. The different advances and retreats of the ice were associated with a variety of levels, the lowest being over 100 m below the present sea level (Fig. 5.8). When the ice finally melted and the sea rose to the present level it drowned the mouths of the deepened valleys carved down to the lowest base level. The rias of south-west England are not simply drowned, or submerged, features, since they also record evidence of higher sea levels in raised beaches. There may have been some order imposed on these ups and downs of the sea level, since it might have become successively lower with each ice advance. This would suggest that there was an overall fall of sea level occurring at the same time due possibly to deepening of the ocean trenches. Increased plate activity would have made this possible.

World-wide changes of this type are known as **eustatic**. Movements of the land masses are usually

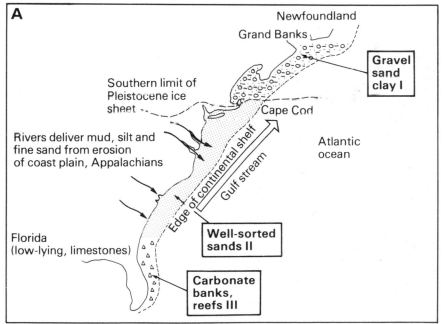

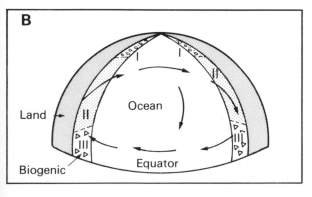

Fig. 5.7 Continental and shelf sediments. How are the sediments off the east coast of North America (**A**) related to the geology of the nearby land, the recent geological history of these areas, and latitude? **B** is a generalised model of the North Atlantic situation, relating sediment to dominant conditions of climate.

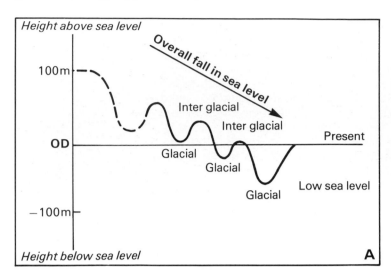

Fig. 5.8   Changing sea levels and the Ice Age. **A** Glacial advances were associated with falls in sea level; retreats with rises. It is probable that there was an overall drop in sea level during this time, so that the last glaciation gave rise to the deep channels at our river mouths. **B** Raised beach deposits at Saunton, north Devon. Relate this sequence of events to **A**. The erratic blocks are granite from southern Scotland. 'Head' is broken, angular rock debris resulting from frost action.

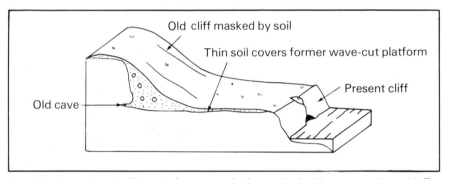

Fig. 5.9   A raised beach. The main features are the flat area backed by a steeper rise or bluff.

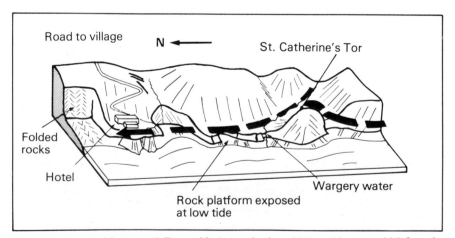

Fig. 5.10   Hartland Quay, north Devon. Marine erosion is rapid along this coast, which faces the full force of the Atlantic rollers. The old valley, marked by the broken arrow, has been cut into and dismembered, and streams are often left hanging above the beach as waterfalls.

much more localised, and certain areas are particularly prone. Those areas which experience frequent earthquakes are areas in which vertical and horizontal movements of the land are taking place, often with fracturing between sections which may be moving at different rates or in different directions. Other areas which once supported a load of ice are now recoiling after being weighed down by the ice: rock platforms and cliffs carved before such recoil may now be 10–30 m above present sea level – a common sight in northern Scotland and around the Baltic Sea. Changes of this type, due to loading or unloading of the earth's crust, are known as **isostatic** (Box 6.1).

The coastal features produced by sea-level changes are distinctive. A relative rise in sea level will lead to a drowning of river mouths, producing rias in upland areas like south-west England and estuaries in lowlying areas. Regions which have been glaciated and have deeper valleys produce fjord coasts, as in western Scotland, Norway and South Island, New Zealand. Further evidence of a recent rise in sea level around British coasts is found in submerged forests and peat below low water mark. In addition, such areas tend to have deeper water offshore, and this means that wave action can be more effective.

A relative fall in sea level will leave an emerged coast, with the typical shoreline features several metres above the present high tide mark as a raised beach (Fig. 5.9). Raised beaches are particularly well-developed around northern Scotland, where they provide flat ground for farming and roads. Even older and higher shorelines have been traced round the margins of the Thames basin, along the South Downs and into the Salisbury Plain, with further evidence in the south-west and Welsh peninsulas.

Cliffs and shore platforms

The ways in which the sea, the land, the atmosphere and changes of sea level relative to the land interact with each other are seen by the study of the major types of coastal landform – cliffs and beaches.

Cliffs and adjacent shore platforms are formed by the erosive action of the waves. This is particularly effective in storm wave environments, where the force of the water and the pebbles it hurls at the cliff base wear away a notch, undermining the rocks above and producing vertical cliff faces and a sloping shore platform beneath (Fig. 5.10). Weak rocks are eroded most rapidly. The coast of Yorkshire between Flamborough Head and the Humber estuary has been pushed back 9 km in 1800 years as

Raised beaches on the Isle of Islay, western Scotland. What features shown here suggest that the sea-level may have fallen three times before reaching its present level? (Aerofilms)

Beachy Head, Sussex. What features of this coast show that the chalk rock is being eroded by the sea, and the cliffs are retreating? (Aerofilms)

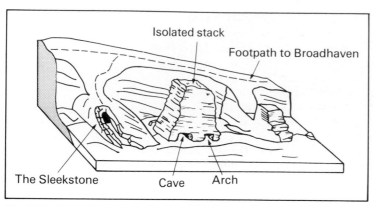

Fig. 5.11   Den's Dor, Pembrokeshire. This feature, just north of Broadhaven, illustrates the undermining effect of the sea. The Sleekstone is an overfold in the Coal Measure rocks.

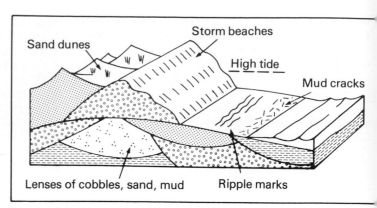

Fig. 5.12   Beach deposits. Notice the type of material included in these deposits and the size and extent of the structures affecting them. Make your own observations of such features when you visit the coast.

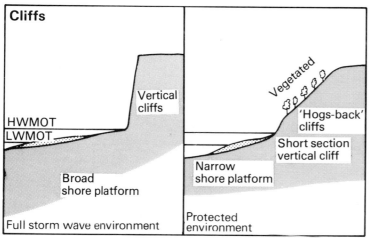

Fig. 5.13   Cliffs and beaches in Cornwall. How are the forms related to the force of the waves on the north and south coasts?

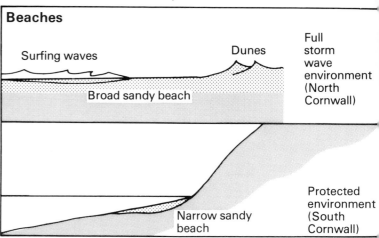

Fig. 5.14   Hurst Castle Spit, Hampshire. Analyse the forces acting to form this feature.

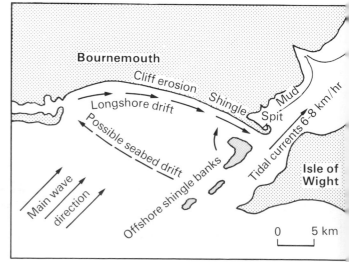

the sea has washed away the soft cliffs of unconsolidated glacial deposits: over thirty villages have disappeared. Even cliffs of resistant rocks may be worn back at such a rate that short rivers emptying into the sea may be continually rejuvenated. Coasts are eroded most rapidly when they face into the full force of storm waves, a fact which is illustrated in the section of coast in north-west Devon and north Cornwall between Hartland Point and Bude (Fig. 5.10). The brown sandstone cliffs north of Sandown, Isle of Wight, also demonstrate the speed of cliff erosion, for a brick boundary wall was overhanging the cliff top by 2 m until the 1964–5 winter, when it crashed down to the beach beneath: the sea had gradually worn away the cliff base until the wall collapsed.

Most erosion is accomplished by the sheer force of the breakers, compressing air in rock fissures, and hurling rock fragments against the cliff base. The force of waves in winter storms in Britain is great enough to shift huge boulders, and the general effect of such action is to undermine the cliffs by a horizontal sawing mechanism. The rock fragments involved are soon worn down to rounded pebbles or cobbles. The sea also supplements the undercut-

ting of the cliffs by working on the weakest places to open up small inlets along faults, joints, bedding planes and weaker rock layers. Small features of cliff scenery, such as caves, arches and stacks (Fig. 5.11) are a sign of erosion taking place.

As erosion proceeds the cliff-line is pushed back and a shore platform is exposed at low tide. If this is in shale or slate rocks, with frequent partings between thin layers, the alternate wetting and drying may lead to a lowering of the surface of this platform. Solution affects some cliffs, especially those formed of limestone, and much of the chalk around Flamborough Head in Yorkshire is pitted with solution hollows.

Thus the erosional force of the sea is concentrated at sea level, between tides, and any wearing away carried out at greater depths is insignificant. Swift currents rushing through narrow straits, as between Dover and Calais, merely prevent deposition taking place, and do not carry out erosion.

Beaches

Beaches are generally formed as banks of sand or pebbles between tides (Fig. 5.12). Debris from cliff erosion is drifted along the coast and built up

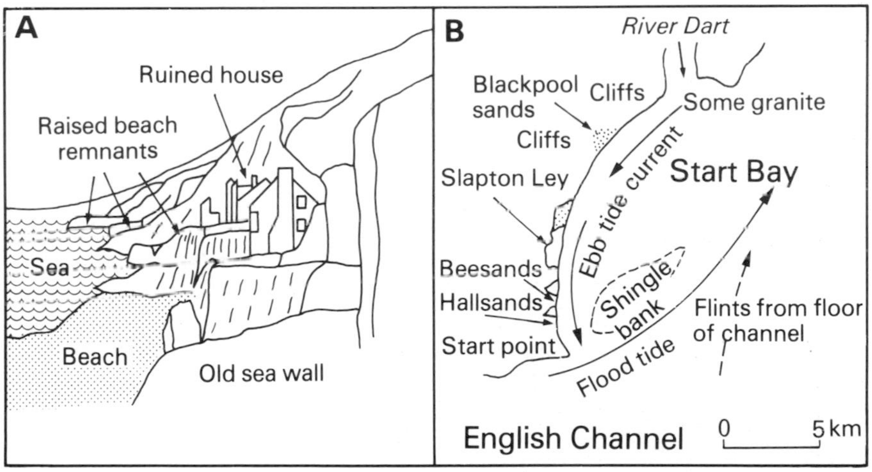

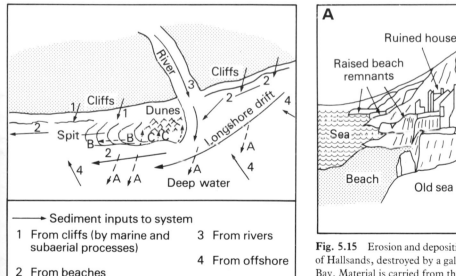

Fig. 5.15 Erosion and deposition on a coast in south Devon. **A** A sketch of the ruined settlement of Hallsands, destroyed by a gale in 1917. **B** The movement of sediment (mainly shingle) in Start Bay. Material is carried from the shingle bank northwards, whence it migrates southwards along the coast. Much of the coast is protected by beaches and small lagoons, such as Slapton Ley. Very small additional inputs come from the River Dart and the floor of the English Channel, but most sediment circulates within the bay. In 1897 650 000 tonnes of shingle were removed from the bank for the use of dockyard extensions at Plymouth. The beach soon disappeared from in front of Hallsands, and within twenty years the village was destroyed. (*After Robinson, 1961*)

Fig. 5.16 The coastal system. Notice the interaction of processes involved in the supply and movement of sediment. The balance between input and output will determine the dominance of erosion or deposition, but both will occur within a few kilometres of each other.

by the waves into ridges of rounded pebbles or flatter stretches of sand or mud. The constituents are controlled largely by the types of rock found locally, and the steepness of the beach is determined by the size of the debris contained and the nature of the dominant type of wave action. The coasts of north Cornwall, exposed to continuous storm wave activity, are characterised by flat surfing beaches and steep cliffs, in contrast to the features of southern Cornwall (Fig. 5.13).

Spits are produced on headlands when pebbles drifted along the coast accumulate in a region of slack water at a turn in the shoreline. Hurst Castle Spit, at the western entrance to the Solent (Fig. 5.14), has been formed in this way. Such features result from a tendency of the sea to even up the coastline: headlands are eroded, bays are cut off and filled in, and estuaries are constricted by spits.

A spit may completely block the entrance to a bay, or shallow water offshore may lead to the waves breaking out to sea and building up a shingle bar. In each case a **lagoon** will be enclosed with water protected from wave action in a situation where the undisturbed waters are warmer than the local seas. They are colonised by a variety of animals and plants, and soon fill with sediment unless scoured regularly by tidal channels. Some turn into swamps and vegetation may take over to convert them into boggy, and later peaty, areas. Slapton Ley in south Devon is a good example of this. In arid regions, where little land-derived sediment reaches such lagoons, they become the site of chemical precipitation and high salinity in contrast to the brackish (i.e. seawater diluted by freshwater) state of the lagoons in humid regions. Continued evaporation will lead to the formation of evaporite salts in the manner described in Chapter 4. Such arid lagoons are known as **sabkhas**.

The coastal sediment budget

Most coasts experience a mixture of erosion and deposition. Thus the Hurst Castle Spit results from erosion at Bournemouth and deposition on the spit, and the rapid erosion along the east coast of Yorkshire is related to the formation of Spurn Head at its southern end. There is often a close relationship between the two, and interruption of such a linkage can lead to disastrous consequences, as in Start Bay, south Devon (Fig. 5.15). This relationship has been summarised as the coastal sediment budget, with inputs from rivers, cliff erosion and offshore deposits balanced by outputs to beach deposits and offshore zones (Fig. 5.16).

Continental margins and offshore sediments

The material which is moved to offshore areas by coastal processes (wave movements, longshore drifting), and by the flow of rivers, may settle to form layers of sediment below wave base. Many coasts are bounded by extensive **continental shelves**, varying from a few to several hundred km in width, but with shallow water up to 100–200 m deep at the most. The average depth of the line where this gives way to deeper water, known as shelf-break, is 130 m.

The nature of the sediments produced on the continental shelf is determined by the material supplied by rivers and the coastal system. The River Amazon carries vast quantities of mud 320 km offshore, but this is exceptional. Variations may also be related to the climatic conditions of the present and immediate past (Fig. 5.16). In general there may be a gradation between those deposits formed near the shore (**littoral**) and finer grades dropped farther out to sea on the shelf surface (**neritic**) (Fig. 5.17).

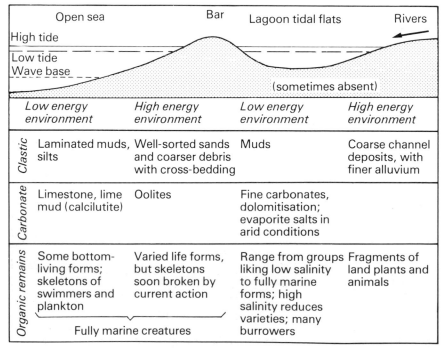

	Open sea	Bar	Lagoon tidal flats	Rivers
	Low energy environment	High energy environment	Low energy environment	High energy environment
Clastic	Laminated muds, silts	Well-sorted sands and coarser debris with cross-bedding	Muds	Coarse channel deposits, with finer alluvium
Carbonate	Limestone, lime mud (calcilutite)	Oolites	Fine carbonates, dolomitisation; evaporite salts in arid conditions	
Organic remains	Some bottom-living forms; skeletons of swimmers and plankton	Varied life forms, but skeletons soon broken by current action	Range from groups liking low salinity to fully marine forms; high salinity reduces varieties; many burrowers	Fragments of land plants and animals
	Fully marine creatures			

Fig. 5.17 Shelf sea sediments. Relate the position of each zone to the energy available (i.e. degree of wave and current action) and to the sediments produced.

Beyond the edge of the continental shelf the **continental slope** falls away to the deep ocean floors (Fig. 5.18). This slope is often cut deeply by **submarine canyons**, which may be incised up to 1000 m into solid rock and sediment, and have steep walls with a branching system of tributaries. Many follow the lines of river systems out to sea (e.g. the River Hudson which enters the Atlantic Ocean at New York, and the River Zaire in central Africa), but some have no such connection. It has been suggested that these canyons have been carved by rivers extending their courses across the continental shelf to a low, Ice Age, sea level. Some, however, extend down to 3000 m below the present sea level, and the sea could not have fallen so far. Recent investigations have shown that seawater charged with suspended sediment stirred up by a storm often rushes down the continental slope, causing damage to cables and even ripping through rocks. These currents or water are known as **turbidity currents**, and may be the main factor at work in the production of submarine canyons. At the foot of the continental slope any sediment which has been moved down the slope or canyons accumulates: the coarser sand or silt settles out first, followed slowly by mud particles. The area of accumulation of such deposits is the **continental rise**.

Coasts and plate theory

Most of this discussion of coastal and offshore geology has been related to local details and the relationships in the situations we can examine on a field visit to the coast. On the world scale other features emerge. The trend of the coast, the degree of land instability (i.e. likelihood of uplift or subsidence), and the width of the continental shelf can all be related to the position of each section of coast with respect to plate margins (Fig. 5.19). Thus the shores of the Atlantic Ocean have wide continental shelves because they form the margins of an opening ocean in which they are moving away from a constructive margin (i.e. trailing edge coasts). Rocky coastlines around the Atlantic often have promontories protruding into the sea at right angles – a sure sign that older structures in the rocks have been truncated by the opening of the ocean. By contrast, the shores of the Pacific Ocean, where destructive plate margins occur close to most coasts, have narrow or absent continental shelves and geological structures which run parallel to the coastline.

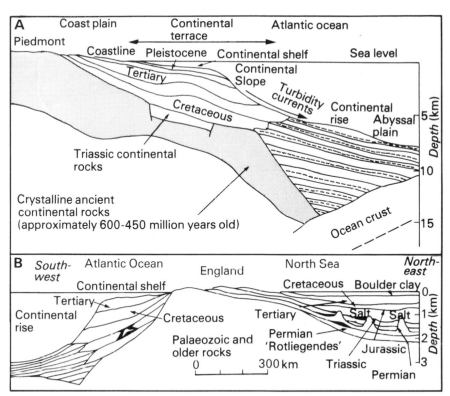

Fig. 5.18 A The continental margin of eastern USA: the construction of the continental terrace and continental rise. **B** Continental margin sediments around the British Isles. Compare the thickness and the completeness of the sedimentary successions of Mesozoic and Cainozoic age on the land and offshore. The North Sea sequence is better known than that off the western coasts due to the recent search for oil and gas (found at x).

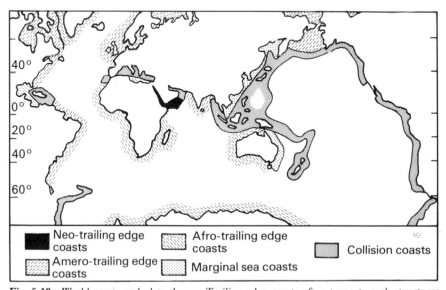

Fig. 5.19 World coasts and plate theory. Trailing edge coasts often truncate rock structures which meet the coast at an angle. Collision coasts are marked by geological structures which run parallel to the coast.

Ocean basins

The continental margins, including the shelf and rise, form the site of distinctive sediment accumulation, having in common the fact that nearly all the material incorporated in new sediments comes directly from erosion of the continents. Beyond these zones lie the deep ocean basins, over 3000 m deep, and forming a vast realm covering over two-thirds of the earth's surface. The dificulty of studying this region in the past means that our knowledge of it lags behind other areas, but many nations are now making intensive efforts to investigate what lies beneath the ocean waters, and are using many new techniques.

Ocean basin relief

The ocean basins have a set of relief features which are quite different from those on the continents (Fig. 2.4): there are no fold mountains or shield areas, and all the rocks and sediments are less than 200 million years old (cf. up to 3850 million years

on the continents). It was a study of these ocean basin features in the 1960s which led to the plate theory which has become a basic part of the geologist's modern understanding of the earth and its evolution. There are four main groups of ocean basin relief forms.

(1) The **ocean ridges** form a system of mountain ranges rising from the ocean floor at 4000–5000 m and extending for 40 000 km through the world's ocean basins. They do not pierce the surface waters, although in Iceland the intensity of volcanic activity has brought many of their features above sea level (Fig. 5.20). The ridges include the mid-Atlantic ridge, which runs parallel to the coasts of the bounding continents, but in other oceans the ridges are off-centre. In the Pacific Ocean, for instance, it is close to the eastern margin, and is partly covered by western California. These submarine mountain ranges are quite different from the continental fold mountains. They are composed of a series of parallel ridges getting higher towards the ridge axis, but culminating in a rift valley (Fig. 5.21). The ridges are produced by molten rock rising to the surface in the rift area, and there solidifying in contact with the cold, deep ocean water (approximately 4°C). When an American expedition descended into the rift in the mid-Atlantic ridge in a bathyscaphe, it was able to take pictures of the bulbous forms taken by the lava: a skin forms quickly on the surface, preventing further flow. Such forms are known as 'pillow lava'. As the new lava solidifies and more begins to rise, the boundary ridges are pushed aside to make way for it; these become fractured to form new ridges. As they get farther from the ridge axis they become lower. Some volcanic cones are associated with the development of the ridge. The Azores islands in the Atlantic are close to the centre, but the islands farthest from the centre are subsiding with the ridge, and may soon disappear.

(2) **Abyssal plains** are the flat ocean floor areas on each side of the ridges. They are flat because they are the site for accumulating sediment. This sediment is extremely fine-grained, and is often known as **ooze**. Much of it is derived from ocean surface organisms (plankton), which produce fine skeletons of calcium carbonate or silica, and shed them on death or during reproduction. The subsequent 'rain' of skeletal remains makes up the greater part of the ocean-floor sediment, accumulating more rapidly than detrital sediments due to distance from the continents. Another important component is fine dust blown over the oceans from the deserts or from major volcanic outbursts.

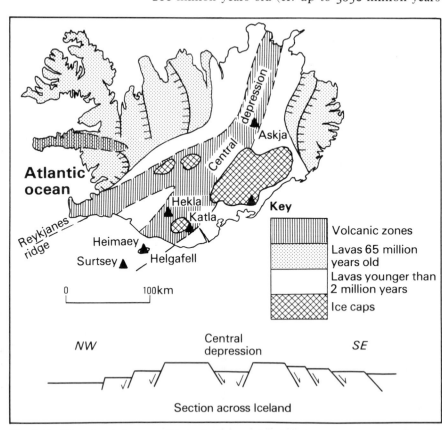

Fig. 5.20   The volcanic features of Iceland. How can these be related to a position on the mid-Atlantic ridge?

The character of oozes depends on the depth of water and the surface climate of the area in which they form (Fig. 5.21). Calcareous oozes are found on the ocean floors down to 3500 m, and are formed of the skeletons of such microscopic animals as the foraminifera (mainly *Globigerina*) and pteropods, mixed with clay. Globigerina ooze, for example, covers over 200 million km² of the ocean floor. Oozes composed of silica dominate the areas between 3800 and 5000 m deep, the silica being less easily dissolved and therefore able to descend farther, although the accumulation rate is slower. Radiolarian oozes cover over 20 million km² of the tropical Pacific Ocean, and diatomaceous oozes are common in polar regions. Red clay occurs at depths beneath 5000 m, and is formed of iron and manganese from volcanic and wind-blown dust, together with nickel possibly from meteoritic particles. Some radiolaria and sharks' teeth are also included in this deposit, which forms at a rate of approximately 1 mm every 1000 years. The fact that the oozes accumulate so slowly is of use in working out some of the historical details of our planet, since so much time is recorded in such a small thickness of sediment. The type of mineral used in the building of tiny skeletons is often diagnostic of ocean surface temperatures when the organisms were alive, and we can therefore assess the climatic conditions of that time. The fossils composing the oozes on the floor of the Atlantic Ocean are no older than the Cretaceous period. The oldest in the Pacific is of Jurassic age.

(3) Although these abyssal plains are normally flat, some relief may be present. **Oceanic rises** may cover hundreds of square kilometres, and occasionally break the surface. The Rockall rise in the North Atlantic is one such feature, being the foundered portion of a continental fragment. **Seamounts** (known as guyots in the Pacific Ocean) are isolated peaks now covered by up to 1000 m of water. They are possibly volcanic in origin, and may have had their topmost points cut off by erosion, or the weight of the rock may have caused the underlying crust to subside. Volcanoes are common in oceanic areas: they have to erupt a vast quantity of material in order to raise their heads above sea level, and then to maintain them there. If they become extinct they sink below the waves. One of the most interesting studies of such features has involved the Hawaiian islands in the central Pacific Ocean and the line of seamounts running north-westwards to the Kamchatka peninsula in eastern Siberia. It seems that there has been a spot in the central Pacific Ocean from which lava has poured for some seventy million years. The movement of sea floor due to plate tectonics has caused each volcanic peak to be separated from the supply of lava after several thousand years, and a new peak has been begun.

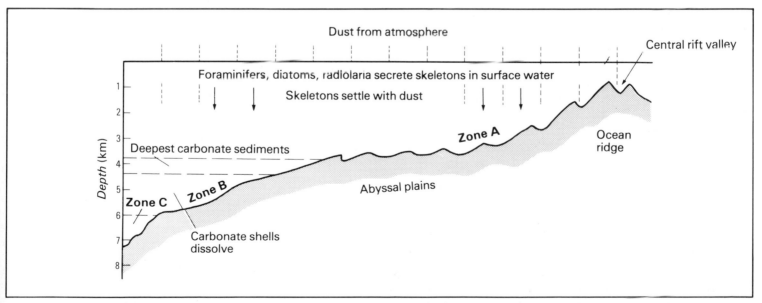

Fig. 5.21 Features of ocean ridges and abyssal plains. Compare the deposits likely to form in zones **A**, **B** and **C**.

Today only the south-easternmost of the Hawaiian islands is active (Fig. 5.22).

(4) **Ocean trenches** are the deepest parts of the oceans, extending below the abyssal plains, often to twice their depth (the deepest trench is over 11 000 m below sea level). These trenches are often thousands of km long, but only 100–200 km across, and are found lying parallel to arc-like chains of volcanic islands (e.g. south of Sumatra–Java; east of Philippines–Japan; south of the Aleutians), or ranges of fold mountains (e.g. west of the South American Andes). They are also associated with major zones of earthquake activity.

Ocean basins and plate tectonics

The study of ocean basins gave rise to the idea of ocean-floor spreading (Fig. 2.5), and then plate theory. Ocean ridges are formed at constructive plate margins where new rock material is added to a plate. Ocean trenches are at destructive margins, where ocean-floor crust dives beneath the margin of another plate. The relatively low ages of ocean-floor rocks compared to those of the continents also

suggests that the ocean floors act as giant conveyor belts moving from constructive to destructive plate margins, re-cycling rock materials as they do so. This view is supported by the history of the Hawaiian islands and adjacent seamounts (Fig. 5.22), which show that the movement of the ocean floor has affected every feature of the ocean basin.

It is now realised that the ocean floor is a basic part of the plate structure, whereas the continents are merely the low density 'scum' which is building up on the surface due to reactions of the ocean rocks with the atmosphere to produce minerals such as clay minerals, carbonates and iron oxides having higher oxygen contents and lower densities.

The study of the ocean basins in this way has also led to an understanding of the sequence of events involved in their history (Fig. 5.23).

Seawater and its movements

The ocean basins are filled with water, the level of which varies with time according to the way in which it is returned from the continents after

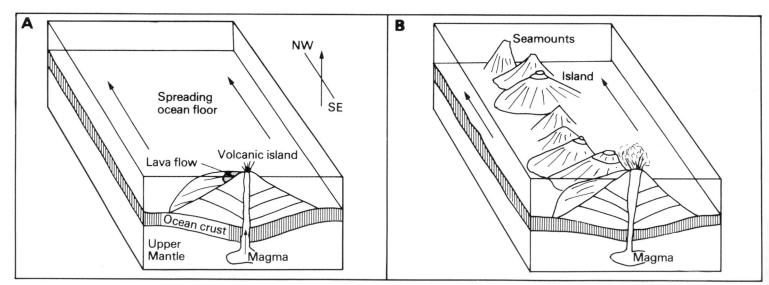

Fig. 5.22   Ocean-floor spreading and the relationship of a 'hot spot' or plume to the formation of volcanic islands and seamounts. The plume is a magma source, which supplies molten rock from a particular point over a period of time. This process can be seen to account for the Hawaiian islands, the Midway group and the Emperor seamounts, which stretch in a chain from Hawaii to the Kamchatka peninsula in Siberia.

A Approximately seventy million years ago: the first volcanic peak is formed. **B** Today: ocean-floor spreading (at 9 cm/yr) has carried a series of volcanic peaks away from the plume, which has then supplied fresh magma to form new peaks. Subsidence of the oceanic crust beneath the weight of rock has caused many former volcanic islands to sink beneath the waves.

passing through the hydrological cycle, and the changes in the ocean basin volume related to continental movements and alterations in the shape of ocean basin features.

The character and movements of seawater also have important geological effects. We all know seawater is salty. The chemicals come from the minerals carried in solution by streams. Almost all the calcium carbonate entering the sea is removed from solution by marine organisms which then use it to build bones and shells. Seawater is thus a solution of very soluble salts which are not used by sea creatures. Sodium chloride is the main one (over 77 per cent), followed by magnesium chloride (11 per cent), and small proportions of other magnesium, calcium and potassium salts (Fig. 7.1).

The proportions of the various minerals change little throughout the world, but the actual concentration of salts (the **salinity**) alters from one sea to another due to different amounts of rainfall, rates of evaporation, supplies of fresh water and communication with the open oceans. Whilst the open oceans have a salinity of 35 parts per 1000, the Red Sea has one of 38·8 and the Baltic Sea of 7·8.

Seawater also contains a certain amount of oxygen, which is vital for marine animals. Plants produce some of this, and can live only in the surface layers where sunlight can penetrate to enable photosynthesis to take place. Thus the oxygen is concentrated near the surface, and the proportion of carbon dioxide increases deeper in the oceans and especially in stagnant basins like the Black Sea (Fig. 5.24) and some deep fjords.

Large-scale movements of ocean waters are known as **currents**, and are caused by differences in the temperature and salinity of the water. Cold water is more dense than warm, and drifts towards the equator from the poles, sinking beneath the surface layers of lighter tropical waters (Fig. 5.25). **Drifts** are surface movements of the water caused by the prevailing winds, and, like the currents, are modified in their courses by the shapes of the continents. The North Atlantic Drift, for instance, is connected closely with the westerly winds of that zone: whilst the winds cause the waters to move towards northern Europe, the relatively warm waters transfer heat to the air above them and this gives the British Isles milder winters than the average for their latitude.

Stage	Process	Sediments	Typical age (million y)	Examples
Embryonic	Uplift and rifting	None	10	East African rift valleys
Young	Rifting with slight lateral movement	Little	25	Gulf of Aden; Red Sea; Gulf of California
Mature	Extensive lateral movement	Moderate: thickest at margins	150	Atlantic Ocean
Declining	Extensive trenches; over-riding of ridges by crustal blocks	Extensive at margins	over 200	Pacific Ocean
Closing	Isolated basins with deformation of sediment and crustal rocks	Thick, deformed	over 200	Black Sea; Caspian Sea; Mediterranean Sea

Fig. 5.23 The evolution of ocean basins: a sequence.

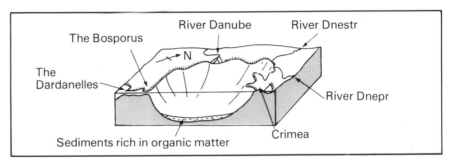

Fig. 5.24 The Black Sea. There is a shallow outlet via the Bosporus, but the deepest parts are 2000 m below the surface. Whilst the surface layers are rich in life the stagnant waters beneath are without oxygen, and undecomposed organic matter accumulates on the sea floor.

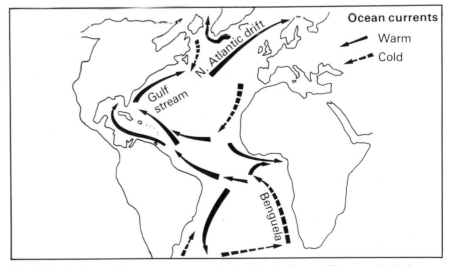

Fig. 5.25 Surface and deep-water currents in the Atlantic Ocean. This generalised diagram shows the main relationships. There is a rapid fall of temperature beneath the surface waters. The figures give the temperature and density of the waters.

6
Volcanoes and earthquakes

The eruption of Cerro Negro, Nicaragua, on 5 February 1971. Notice the importance of volcanic cones and older lava flows in the landscape. What is the relationship of this region to plate margins? (USGS)

Introduction

The processes acting at the earth's surface today are largely a result of the interaction of atmospheric forces powered by solar energy and the exposed rocks. As we saw in our study of coasts, however, there are also other forces at work. The earth's interior is another source of energy which provides the motive power for the large-scale plate movements. This internal energy source is most evident in the short-term and fairly localised events which include volcanoes and earthquakes. Many hundreds of these occur around the world each year, and each one is just a part of the major pattern of earth events – just as a still picture from a moving film merely captures a momentary situation in the overall plot in time and space. Each event provides further evidence for the presence of energy in the earth's interior by virtue of the movement and heat involved. In total these events are a vital sector of the geological cycle (Fig. 3.6), since they cause molten material from the earth's interior to reach the surface and lead to further uplift of sections of the crust.

Volcanoes

Volcanic eruptions

Nearly 700 volcanoes are known to have been active since man began keeping records. This represents a tiny fraction of geological time (and many erupting in the deep oceans have gone unnoticed), and in total the volcanic materials have provided a major proportion of the rocks at the earth's surface. Those who live in a country like the United Kingdom have the feeling that volcanic activity is of rare occurrence, but it has been powerful enough to build the Hawaiian islands from the floor of the Pacific ocean so that Mauna Loa is higher than

Mount Everest, having its base 5000 m below the ocean surface and its top rising to over 4300 m above sea level. Vast plateaus of lava have been formed, and huge quantities of gases (mostly water vapour, carbon and sulphur dioxides) have been poured into the atmosphere, providing the basic materials for maintaining the atmospheric composition and the ocean water store.

One example of a volcanic eruption which has been well documented from its initial eruption to the point when it became dormant is Mount Paracutin in Mexico, which began its life on 20 February 1943. Two weeks of minor tremors came to an end when a crack opened across a ploughed field and vapours and hot stones were ejected. Within twenty-four hours there was a pile of debris 10 m high, and this grew to a cone-shaped feature 18 m high in a week. Lava soon flowed out, inundating local villages, and clouds of ash rose over 600 m into the atmosphere, descending to cover fields and kill trees. By 1946 the cone, composed of lava and ash, was 500 m high, but after that date activity became less intense, and the volcano has been quiet since 1952.

The Hawaiian volcanoes have perhaps been the most studied of all, and their manner of eruption is quite different from Paracutin, with little major explosive activity or ash formation. This has allowed an observatory to function over a long period of time, recording changes and major events (Fig. 6.1). Observations made include regular measurements of temperature and the composition of the gases and molten lava. Lava is probed, cored and drilled, and the nature of the materials is examined. After one 1965 eruption a number of lakes of molten lava were left, and their progress has been monitored. By October 1967 the surface crust was 13 m thick on the 100 m deep Makaopuhi pit, and it is estimated that the entire lava lake will have solidified by the year AD 2000. Such phenomena have allowed scientists to study the sequence of events in the cooling and crystallisation of this lava (Fig. 6.2).

Neither Paracutin, nor Hawaii, has experienced the most explosive types of volcanic activity. The most famous example of an extreme volcanic explosion is that of Mount Pelée on the island of Martinique in the West Indies. In 1902 a volcano which had not erupted in living memory burst open and cleared the top of its vent. Over the next sixteen months a steep-sided dome of lava grew 300 m high, composed of very viscous rock which was virtually solid as it reached the surface – although the outer skin was often broken by internal explosions. Then

An earlier eruption of Cerro Negro (November 1968), seen from above. The photo was taken during a major surge of ash and gases, which included volcanic 'bombs' over 30 cm long; an earlier surge has moved downwind and become whiter as the bombs and larger ash particles have fallen out; between the two surges a quieter interval is marked by the emission of steam and gases. Note the shape of the ash cone around the vent, and the relation of this to the lava flows. (USGS)

The summit of Mauna Loa, Hawaii. Mauna Kea is in the distance beneath the clouds. Compare the features of this volcano with those of Cerro Negro (pages 68–69). (USGS)

	Temperature	Sample from	Characteristics
(1)	1170°C	Melt	Olivine crystals in light-brown glass with tiny augite and feldspar crystals.
(2)	1130°C	Melt	Olivine crystals, small augite, feldspar lath-like crystals in glass.
(3)	1075°C	Partly molten crust	Augite, feldspar laths surrounded by darker glass (increasing in iron, titanium).
(4)	1065°C	Partly molten crust	As above, plus blade-like ilmenite (iron-rich) crystals.
(5)	1020°C	Partly molten crust	Interlocking network of augite, feldspar, magnetite (iron oxide), ilmenite; glass in interstices.
(6)	865°–920°C	Solid crust (just after solidification)	Larger olivine crystals, with augite and feldspar. Some patches of clear silica glass between. No iron minerals.
(7)	725°–795°C	Solid crust	Olivine affected by rising gases, giving red haematite rims.

Fig. 6.2 The process of crystallisation of lava, as studied in the Makaopuhi pit on Hawaii. Samples were taken from areas of molten, partly molten and solid sections.

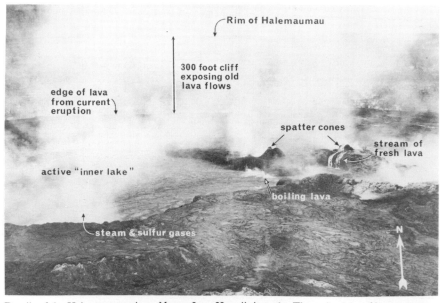

Details of the Halemaumau pit on Mauna Loa, Hawaii, in 1967. The active lake of lava contrasts with the explosive activity of Cerro Negro. (USGS)

quite suddenly one side of the peak opened and black smoke gushed out. This fiery cloud (**nuée ardente**) was composed of droplets of molten lava, ashes and huge boulders of rock lubricated and carried along by expanding gases. It rushed 8 km downhill at speeds of over 300 km per hour, and inundated the town of St Pierre, where heavy statues were carried several metres by its force, and glass was melted. Twenty-eight thousand people were killed as the gases overwhelmed them at temperatures of over 700°C, and then the whole town was buried under ash and lava. The only people to witness this and live were aboard departing ships, for the water in the harbour boiled and many of the boats at anchor were destroyed; one man survived in the depths of the town gaol.

Similar and even greater explosions have been experienced in what are fortunately less populous parts of the world. The uninhabited island of Krakatoa in the East Indies blew away the upper part of its volcanic peak in 1883 with an explosion that was heard nearly 5000 km away. Incandescent clouds of gas and molten lava billowed up to 80 km high, and were carried around the world by the winds of the upper atmosphere, providing brilliant sunsets for the next three years. In 1955–6 Mount Bezimiannyi in the Kamchatka peninsula of eastern Siberia exploded with great violence and ejected so much rock that it has been calculated Paris would have been completely covered to a depth of 16 m. A little farther to the east in the Aleutian islands, Mount Katmai was decapitated by an explosion in 1912, when the mingled gases and lava formed a great frothy mass of pumice, which filled a valley 20 km long and 5 km wide, now known as the Valley of 10 000 Smokes.

Volcanoes, eruptions and rock types

The world-wide study of volcanic activity has demonstrated a range of landforms produced by eruptions of different characteristics, and associated with varied types of lava (Fig. 6.3). Volcanoes like the Hawaiian islands are composed of layer after layer of black **basalt** lava made up of the minerals olivine, augite and feldspar. This lava reaches the surface in a very liquid state at high temperatures and flows easily and for some distance before solidifying. It does not block up the vents, but spreads out to form thin layers. At times the lava withdraws down the vent, causing surface collapse, but generally the activity is quiet and continuous with gases breaking the surface of the lava lakes in spectacular incandescent fountains. This type of

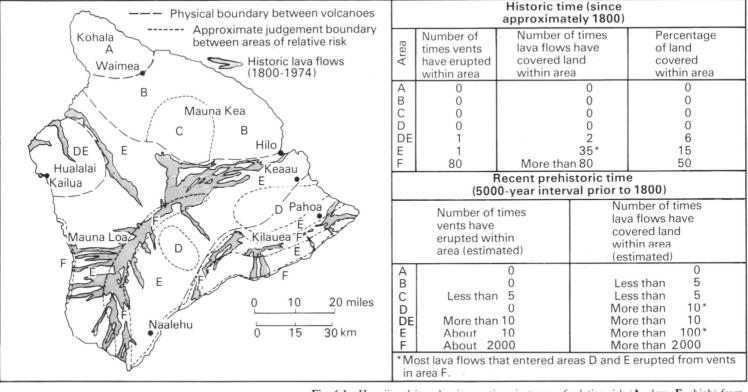

	Historic time (since approximately 1800)		
Area	Number of times vents have erupted within area	Number of times lava flows have covered land within area	Percentage of land covered within area
A	0	0	0
B	0	0	0
C	0	0	0
D	0	0	0
DE	1	2	6
E	1	35*	15
F	80	More than 80	50

	Recent prehistoric time (5000-year interval prior to 1800)	
	Number of times vents have erupted within area (estimated)	Number of times lava flows have covered land within area (estimated)
A	0	0
B	0	Less than 5
C	Less than 5	Less than 5
D	0	More than 10*
DE	More than 10	More than 10
E	About 10	More than 100*
F	About 2000	More than 2000

*Most lava flows that entered areas D and E erupted from vents in area F.

Fig. 6.1 Hawaii and its volcanic eruptions in terms of relative risk (**A** = low, **F** = high) from renewed activity. The table shows the number of eruptions recorded in each zone.

activity is characteristic of many oceanic volcanoes away from the constructive or destructive margins of plates. The lava cools in contact with the atmosphere and may have a smooth, rope-like surface ('pahoehoe'), or a broken, tumbled mass of cindery blocks ('aa') due to gases escaping from the still-molten interior of the flow (Fig. 6.4).

Many of the volcanoes associated with young fold mountains and island arcs have what is regarded as the typical volcanic cone shape with smooth, concave slopes rising to a central peak. These volcanoes are often built up by several distinct processes. At times there will be an explosive eruption and the molten lava will be blown into millions of tiny fragments which cool in the air then rain down to form layers of ash. Ash cones may be formed, like that of Paracutin. Such activity may be followed by liquid lava eruptions. These often break through the ash cone to spread out in a flat sheet. Later eruptions will cover these rocks with alternate layers of ash and lava (Fig. 6.5), gradually accumulating above a single vent to form high peaks like Fujiyama (Japan). The rocks erupted from such volcanoes include the ash layers, thinning away from the vent,

Pahoehoe lava on the slopes of Mauna Loa, Hawaii. (USGS)

Type of volcanic activity	Landforms produced	Rock-type formed	World position	Examples
Hawaiian: lava wells up in collapsed craters, fissures continuously; fine fountains of incandescent liquid lava	Shield volcanoes: broad flat dome built of lava sheets	Olivine basalt lava	Ocean areas over plumes. East African rift valley	Hawaiian islands Africa
Strombolian: regular, closely-spaced small explosions with ash, volcanic bombs falling locally	Conical cinder cones	Ash deposits; tuff (pyroclastic)	Island arcs Cordilleran fold mountains	Stromboli (Italy) Porcaya (Guatemala) Mount Erebus (Antarctica)
Vulcanian: rarer eruptions of ash, bombs, thrown further (1 km)	Composite volcanoes	Andesite lava		Vulcano (Italy), Irazu (Costa Rica)
Plinian: highly explosive with large quantities of gas and steam	Calderas			Vesuvius (Italy), Krakatoa (Java) Bezymianny
Pelean: nuée ardente, with long intervals between. Maximum intensity eruptions	Steep spines, domes Calderas	Ignimbrite Rhyolite lava	Island arcs	Mont Pelée (Martinique) Soufrière (St Vincent)
Solfataric: geysers, hot springs, fumaroles — steam and gases	Vent with mineralised terraces	Calcareous tufa and other mineral encrustations	Most volcanic areas	North Island, New Zealand, California, Iceland
Fissure eruptions: emission of lava along fissure, rather than vent	Lava plateaus	Basalt lava	Continental margins	Antrim (Northern Ireland), S.E. Brazil, Deccan (India), Columbia-Snake area (USA)

Fig. 6.3   The varied types of volcanic eruption and general relationships with the landforms and rock-types produced.

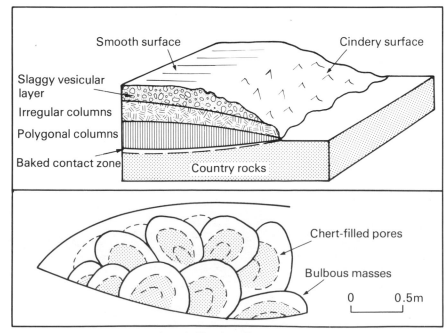

Fig. 6.4   Lava flow structures (**A**) and pillow lava (**B**).

and also **andesite** lava: this is lighter in colour than basalt, and its main minerals are hornblende and feldspar. The types of lava occurring in the oceanic and fold mountain volcanoes are formed in different ways. In the Hawaiian islands it comes from beneath the oceanic crust, and moves upwards through rocks of similar composition: if rocks from the vent walls are incorporated in the mobile rock, they do not alter its composition. In fold mountains, like the Andes, the rock material is melted from the upper part of the oceanic crust as it dives beneath the American plate, and then has to make its way to the surface through a pile of continental rocks which will alter its composition if they are mixed with it. Local melting of rocks in the continental crust may also occur, and this gives rise to lava material which melts at lower temperatures and flows very slowly (i.e. is viscous). The rocks produced at volcanoes erupting this last type of lava are **rhyolites**, which are mixtures of feldspar and quartz with a light colour. They give rise to small, dome-shaped masses, and may be associated with the most extreme forms of volcanic violence.

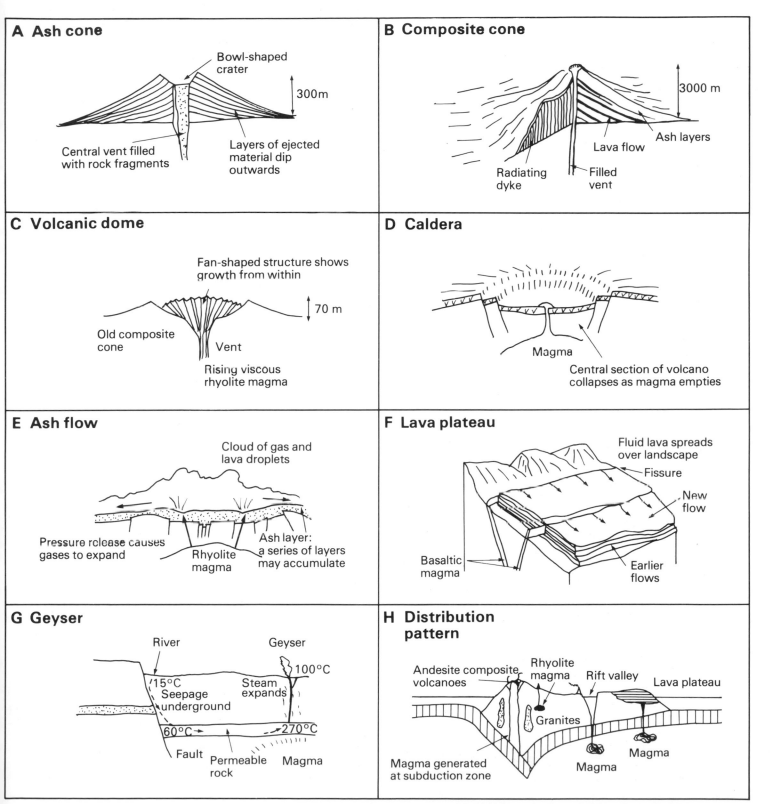

A Ash cone

Bowl-shaped crater

300 m

Central vent filled with rock fragments

Layers of ejected material dip outwards

B Composite cone

3000 m

Ash layers

Lava flow

Radiating dyke

Filled vent

C Volcanic dome

Fan-shaped structure shows growth from within

70 m

Old composite cone

Vent

Rising viscous rhyolite magma

D Caldera

Magma

Central section of volcano collapses as magma empties

E Ash flow

Cloud of gas and lava droplets

Pressure release causes gases to expand

Rhyolite magma

Ash layer: a series of layers may accumulate

F Lava plateau

Fluid lava spreads over landscape

Fissure

New flow

Basaltic magma

Earlier flows

G Geyser

River

Geyser

100°C

15°C

Steam expands

Seepage underground

60°C

270°C

Fault

Permeable rock

Magma

H Distribution pattern

Andesite composite volcanoes

Rhyolite magma

Rift valley

Lava plateau

Granites

Magma generated at subduction zone

Magma

Magma

Fig. 6.5 The variety of continental volcanic activity. Compare this variety with the volcanic features produced in the ocean basins. Apart from the rift valley situation, nearly all volcanic activity is marginal to the continents.

Another major source of volcanic rocks and landforms occurs along rift valleys, as in eastern Africa, and on the margins of opening oceans. At times volcanoes form along such rifts in the crust, but a long fissure may open, pouring mobile lava across the landscape and burying it. Such lava plateaus occur notably on the coast of south-eastern Brazil and in the Deccan plateau of India. They are formed of numerous flows of basalt, averaging 10–20 m thick and added over thousands of years. Some flaws may develop soils and have plants and trees growing on them before the following eruption takes place.

Volcanoes erupt ashes and volcanic bombs as well as molten lava. Each results in distinctive deposits and rock layers. Ashes and volcanic bombs are associated with more explosive types of activity, giving rise to ash cones round the vent (Fig. 6.5). The cone may be formed of ash, volcanic bombs which have formed from clotted lava, and pieces of country rock torn away from the sides of the vent by the explosion. The largest fragments fall back to earth first and nearest the vent. The deposits formed in this way are known as **pyroclastic** (i.e. fire-fragments). The main varieties are **tuffs**, in which the particles are sand, silt or clay grades, and **agglomerates**, in which the bulk of fragments are larger. Tuffs are divided into groups according to their composition and the proportions of crystalline or glassy lava and country rock fragments. Welded tuffs and pumice result from gas-rich nuée ardente explosions, and may resemble lava flows since they form layers 10–13 m thick.

Lava flows are formed by the solidification of molten rock material issuing on the earth's surface. If the eruption takes place under water in the depths of the oceans, the gases cannot escape due to water pressure, and salts may be absorbed from the water to produce sodium-rich rocks. The tumbled masses of lava which have solidified under such circumstances are pillow lavas, like those of the ocean ridges. When we examine a lava flow erupted on land in detail we find that the top layer has a rough, slaggy nature with many cavities, or **vesicles**, where gases were trapped by the cooling processes. In older lavas these have often been filled by percolating waters and the minerals they carry in solution, like calcite. Thus the dark-coloured lavas may be spotted with small, white, almond-shaped **amygdales**. The lower parts of lava flows often solidify in columns as the lava contracts, with cracks forming at right angles to the cooling surface and so isolating the columnar blocks. The Giant's Causeway in Northern Ireland is a lava flow which has been partly decapitated by marine erosion to expose these columns. Another effect of hot molten lava is to bake the underlying rocks as it passes over them. The movement of the lava flow may also cause the crystals forming in it to be dragged out so that their longest dimension lies in the direction of movement, and this often leads to **flow-banding structures**.

One of the most important characteristics of rocks which solidify from lava flows results from the fact that they cool rapidly in contact with the atmosphere and as they lose their gas component to the atmosphere. The crystals which form are usually so small that a high-powered microscope has to be used for examining them. At times the cooling is so rapid that crystals do not have time to form and a glass is produced. The lavas which were erupted recently on Tristan da Cunha were already 70 per cent solid with larger crystals that had been formed during the passage of the lava, and these were soon made fast by the rapid cooling of the remaining molten material in a fine-grained mass.

The Taal volcano (Philippine islands) eruption of 30 September 1965. Volcanic ash was hurled 500 m into the sky. (USGS)

It can be seen, therefore, that the type of rock, landforms produced, eruption intensity and general position of the volcanic outburst with respect to plate boundaries are all closely related. The more explosive volcanoes occur where the molten rock has to pass through greater thicknesses of crustal rocks. By the time it reaches the surface it may have cooled into a solid blockage of the vent. It requires pressure from beneath to shift this, and the gases build up to a major explosion which clears the obstruction after a period of dormancy. Oceanic volcanoes, like those of the Hawaiian islands, experience few interruptions to the outpouring of lava.

Volcanic areas are also characterised by **geysers** and hot springs. These are caused by the effect of hot volcanic materials on groundwater, superheating it until the expanding steam gushes out through any tiny crack (Fig. 6.5). Any minerals caught up in the rush of steam are deposited around the geyser mouth: if calcium carbonate is dominant the deposit is known as **travertine**; if silica it is **sinter**.

Monitoring and forecasting volcanic activity

Many millions of people live within range of devastating volcanic activity. The burial by ash of Pompeii, near Naples, Italy, in Roman times is well known, but there are many other examples. Taal, in the Philippine islands, is a particularly dangerous volcano (Fig. 6.6). It erupted in 1572 and 1754, but few people lived near it at that stage. By 1911 it had become the centre of a lakeside community, and a major eruption in that year killed nearly 500 people living on the island and 800 along the lake shores. The ash cloud from this eruption was visible 400 km away. Despite the survivors' tales of destruction, however, people moved back to the area until the next eruption in 1965. In addition to a vertical column of ash and steam, there were powerful horizontal blasts, uprooting trees and breaking them into small pieces. Ash accumulated to over a metre deep in places. Although 190 people died on this occasion, the figure would have been much higher had it not been for warnings from the local observatory as the lake temperatures rose.

The ability of man to forecast and modify natural events such as volcanic eruptions (or the weather) depends on how much he understands of the processes involved, and on the scale of the event. In the case of eruptions like that at Taal in 1965 there was an observatory at hand which was making

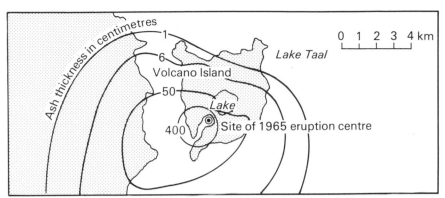

Fig. 6.6 Lake Taal and Taal volcano. The site of the centre of the 1965 eruption and the thickness of ash deposits laid down are shown. The amount of ash is quite small – unrelated to the ferocity of the eruption. (*After J. G. Moore*)

regular observations of a related feature (lake temperature), and the warning was heard and heeded by most of the people put at risk by a possible eruption. No one could do anything to stop or divert the eruption. Man's relationship with volcanoes has reached the stage where a number of methods are available for measuring the surface effects of gathering underground momentum, but the means of altering the eruption are extremely crude. Events which give warning of volcanic activity include the rising temperatures of nearby water bodies, increasing numbers and sizes of lesser earth tremors, and the swelling of volcanic slopes as measured by a tiltmeter. Some volcanoes, as in Hawaii, are monitored thoroughly, although even there some of the signs may be misleading. Earthquake activity may or may not be followed by lava rising to the surface.

Many of the world's volcanoes, however, are not monitored at all, and eruptions have been known to occur where the volcanoes had been thought to be extinct. Thus Mount Lamington in New Guinea, which erupted in 1951, was not even considered to be a volcano, but its outburst was so violent that the surrounding area was completely devastated over an area of 100 km^2 within minutes.

The only ways devised for mitigating the effects of volcanic activity on human occupation of the surrounding area are related to lava flow diversion. This may be accomplished by damming the flow, or by making a breach in the side of a flow to divert it in a harmless direction, or by bombing it! The 1973 eruption on Heimaey, just south of Iceland, was tackled with some success by a combination of bulldozed barriers and water jets to cool and divert the lava front.

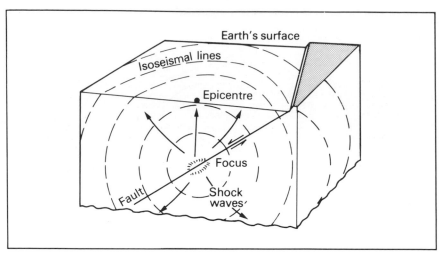

Fig. 6.7   The features of an earthquake.

Earthquakes

Earthquakes – what happens?

Earthquakes are movements of the surface ground caused by events within the earth's crust or upper mantle (Fig. 6.7). In the zone immediately above the earthquake source, or focus, the ground motion may be so great that people cannot stay on their feet and buildings collapse. Such strong movements may last from a few seconds up to four minutes, and their effects may be felt up to 130 km from the epicentre in swaying of the ground and buildings. Earthquakes are measured by scales of intensity and magnitude (Fig. 6.8). A large or moderate earthquake may give rise to similar effects at the epicentre, but the large one will last longer and affect a greater area.

A MODIFIED MERCALLI INTENSITY SCALE OF 1931* (1956 VERSION)†

I   Not felt. Marginal and long-period effects of large earthquakes.
II   Felt by persons at rest, on upper floors, or favorably placed.
III   Felt indoors. Hanging objects swing. Vibration like passing of light trucks. Duration estimated. May not be recognized as an earthquake.
IV   Hanging objects swing. Vibration like passing of heavy trucks, or sensation of a jolt like a heavy ball striking the walls. Standing motor cars rock. Windows, dishes, doors rattle. Glasses clink. Crockery clashes. In the upper range of IV wooden walls and frame creak.
V   Felt outdoors; direction estimated. Sleepers wakened. Liquids disturbed, some spilled. Small unstable objects displaced or upset. Doors swing, close, open. Shutters, pictures move. Pendulum clocks stop, start, change rate.
VI   Felt by all. Many frightened and run outdoors. Persons walk unsteadily. Windows, dishes, glassware broken. Knickknacks, books, etc., off shelves. Pictures off walls. Furniture moved or overturned. Weak plaster and masonry D‡ cracked. Small bells ring (church, school). Trees, bushes shaken visibly, or heard to rustle.
VII   Difficult to stand. Noticed by drivers of motor cars. Hanging objects quiver. Furniture broken. Damage to masonry D, including cracks. Weak chimneys broken at roof line. Fall of plaster, loose bricks, stones, tiles, cornices, also unbraced parapets and architectural ornaments. Some cracks in masonry C. Waves on ponds; water turbid with mud. Small slides and caving in along sand or gravel banks. Large bells ring. Concrete irrigation ditches damaged.
VIII   Steering of motor cars affected. Damage to masonry C; partial collapse. Some damage to masonry B; none to masonry A. Fall of stucco and some masonry walls. Twisting, fall of chimneys, factory stacks, monuments, towers, elevated tanks. Frame houses moved on foundations if not bolted down; loose panel walls thrown out. Decayed piling broken off. Branches broken from trees. Changes in flow or temperature of springs and wells. Cracks in wet ground and on steep slopes.
IX   General panic. Masonry D destroyed; masonry C heavily damaged, sometimes with complete collapse; masonry B seriously damaged. General damage to foundations. Frame structures, if not bolted, shifted off foundations. Frames racked. Serious damage to reservoirs. Underground pipes broken. Conspicuous cracks in ground. In alluviated areas sand and mud ejected, earthquake fountains, sand craters.

X   Most masonry and frame structures destroyed with their foundations. Some well-built wooden structures and bridges destroyed. Serious damage to dams, dikes, embankments. Large landslides. Water thrown on banks of canals, rivers, lakes, etc. Sand and mud shifted horizontally on beaches and flat land. Rails bent slightly.
XI   Rails bent greatly. Underground pipelines completely out of service.
XII   Damage nearly total. Large rock masses displaced. Lines of sight and level distorted. Objects thrown into the air.

*Original 1931 version in H. O. Wood, and F. Neumann, 1931, Modified Mercalli intensity scale of 1931, *Seismological Society of America Bulletin*, vol. 53, no. 5, pp. 979-987.

†1956 version prepared by Charles F. Richter, in 1958, *Elementary seismology*, W. H. Freeman & Co., San Francisco, pp. 137-138.

‡Masonry A, B, C, D. To avoid ambiguity of language, the quality of masonry, brick, or otherwise, is specified by the following lettering.

Masonry A. Good workmanship, mortar, and design; reinforced, especially laterally, and bound together by using steel, concrete, etc.; designed to resist lateral forces.
Masonry B. Good workmanship and mortar; reinforced, but not designed in detail to resist lateral forces.
Masonry C. Ordinary workmanship and mortar; no extreme weaknesses like failing to tie in at corners, but neither reinforced nor designed against horizontal forces.
Masonry D. Weak materials, such as adobe; poor mortar; low standards of workmanship; weak horizontally.

Fig. 6.8   Earthquake scales of intensity.

The events at depth giving rise to earthquakes vary in nature. At times they are related to the upward movement of volcanic lava and the stresses this imposes as it forces apart the crustal rocks. Most of these earthquakes have a shallow focus, and are local and of small size. Others are caused by a sudden fracture of rocks in response to an increasing build-up of pressure. Such fractures, or faults, send out shock waves to the surface and through the earth. The San Andreas fault is the world's best-known zone for earthquakes. It is part of a complex system of fractures in California (Fig. 6.9), which, in the last 200 years, have been associated with three very large earthquakes (e.g. San Francisco, 1906), twelve major earthquakes and over sixty moderate shocks (e.g. San Fernando, 1971). The largest earthquakes are related to the deepest-seated foci, and these occur along the plate subduction zones: the epicentres of such earthquakes are thus on the landward side of the ocean trench (Fig. 6.10).

Other effects of earthquake shocks at the surface include the setting off of landslides, the 'liquefaction' of fine water-saturated sands, fissuring of the

B   RICHTER MAGNITUDE

In 1935, Dr. Charles F. Richter of the California Institute of Technology devised a scale to indicate the "size" of an earthquake from the measured amplitude of an earthquake wave recorded on a *seismograph*. The Richter magnitude, or just *magnitude* (M), is related to the total energy of the earthquake measured. This results in an open-ended scale based on energy released at the source and should be the same for a given earthquake, wherever measured. Near its epicentre an earthquake of M = 2 is the smallest felt, earthquakes of M = 4.5 to 5 may cause local damage, magnitudes of 7 or more are associated with "major" earthquakes, and those of 7.75 and over, with "great" earthquakes.

On this scale, the earthquake at San Fernando, California, in 1971 with M = 6.4 was a "moderate" one, the Arvin-Tehachapi, California, earthquake of M = 7.7 was a "major" one, and the Alaskan earthquake of 1964 at M = 8.4 was a "great" earthquake. Because of the way the scale is set up and its relation to energy, each higher number means that the earthquake is about 30 to 35 times as great as the next number below. For example, a magnitude 6 earthquake would involve roughly 30 to 35 times as much energy as a magnitude 5 earthquake. Thus the Alaskan earthquake was about 1,000 times as great as the San Fernando earthquake. The greatest known earthquake had a magnitude of about 8.9 on the Richter scale (Lisbon, 1755;). Fortunately for the safety of all of us, minor earthquakes less than M = 2 are thousands of times more frequent than the great earthquakes.

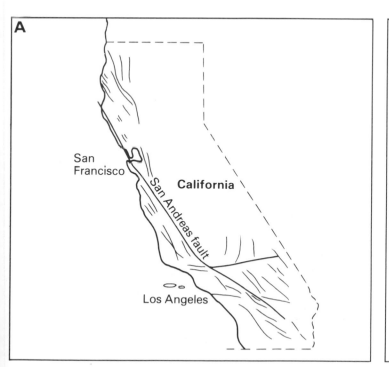

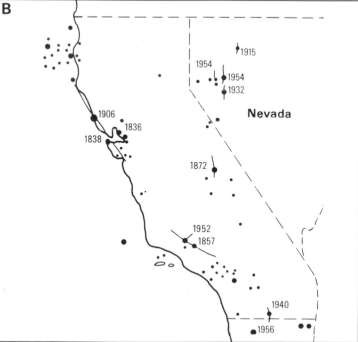

Fig. 6.9 The San Andreas fault and associated fractures in the western USA. A The network of faults in coastal California. B Epicentres of some large earthquakes in the region. Heavy lines show where the ground was broken, and the dot is proportional to the magnitude of the 'quake.

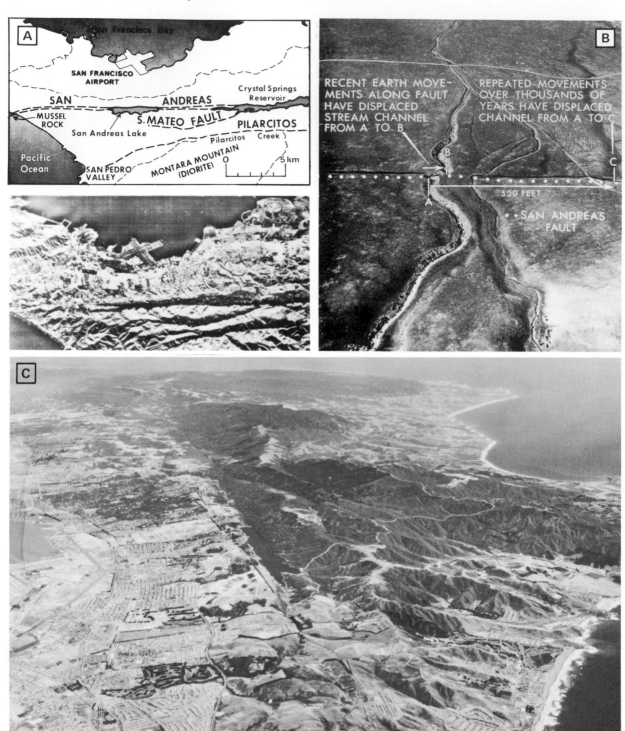

The San Andreas fault, California. **A** Radar image of the effect of the fault on the relief of the San Francisco peninsula. **B** The Carrizo plain area in southern California, showing displacements of stream channels. **C** The fault as it crosses San Francisco suburbs: locate this photo on **A**, and examine the relationships of the buildings to the line of the fault. (USGS)

ground, subsidence or uplift and seismic sea waves (tsunami). These often cause more damage than the vibrating ground surface at the epicentre. Thus the seismic sea waves of the 1964 Alaska earthquake caused great coastal damage and loss of life, whilst the 1970 Peruvian earthquake caused most damage by dislodging an avalanche on the slopes of Mount Huascaran.

Forecasting and controlling earthquakes

The sheer scale of energy and size involved in earthquakes is far greater than man could summon to his aid in attempting to prevent them. The energy of a 100-megaton fusion bomb is approximately equal to that involved in the largest earthquake, but less than the most violent individual volcanic outbursts. We are, however, at the stage where increasing knowledge enables man to predict when and where an earthquake will occur with greater confidence. A worldwide network of seismograph stations, measuring and recording the movements of the earth's crust with delicate instruments (Fig. 6.11), provides us with information about the zones most likely to be affected (Fig. 6.12). Within these zones information is collected about smaller-scale movements so that forecasts may be made (Fig. 6.13). It is salutary to remember that the Chinese were regarded as being ahead of others in this field until the unheralded and disastrous earthquakes near Peking in 1976.

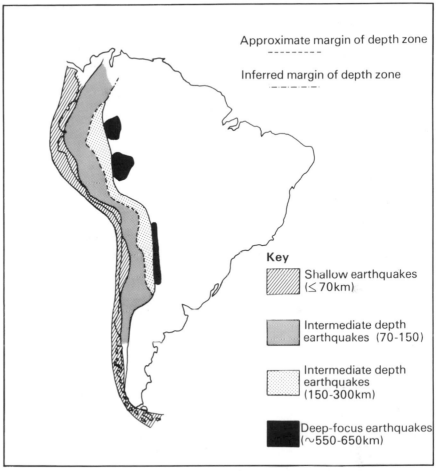

Approximate margin of depth zone

Inferred margin of depth zone

Key

Shallow earthquakes (≤ 70km)

Intermediate depth earthquakes (70–150)

Intermediate depth earthquakes (150–300km)

Deep-focus earthquakes (∼550–650km)

Fig. 6.10 Focal depth zones of earthquakes in the Andean region having magnitudes equal or greater than 6.0, 1906–72. (*USGS*)

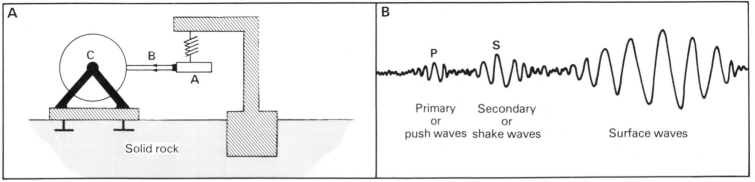

Solid rock

Primary or push waves

Secondary or shake waves

Surface waves

Fig. 6.11 (**A**) A seismograph. A motionless pendulum – X – is suspended from a metal frame anchored in rock. A beam of light – Y – is projected on a rotating drum – Z – covered with photographic paper. The drum is anchored to the rock and vibrates when an earthquake occurs, leaving a wave-like record on the paper. (**B**) A seismograph record. The recording moves from left to right. The P and S waves arrive first because they travel directly through the earth.

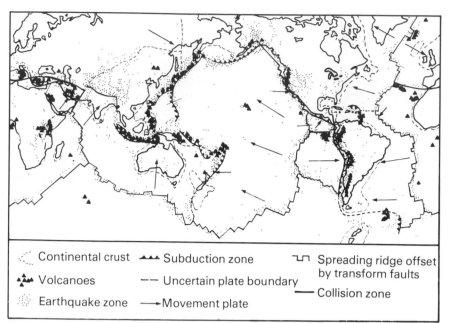

Continental crust   Subduction zone   Spreading ridge offset by transform faults

Volcanoes   — — Uncertain plate boundary   Collision zone

Earthquake zone   ——▶ Movement plate

Fig. 6.12   Map of the world showing the distribution of volcanoes and earthquakes in relation to major tectonic plates.

A collapsed reinforced concrete school building at Varto, Turkey, following the 1966 earthquake. (USGS)

A start has also been made on the possible means for modifying the worst effects of earthquake activity with the idea of lubricating the fault-line so that shorter, more frequent shifts take place with less resulting damage. It seems that major earthquakes occur when movement along a fault 'sticks' for a long period in an area of increasing stress. The eventual movement is so great that it causes extensive damage. It was found by accident that liquids pumped into old mines in Colorado gave rise to small local earthquakes, and further research is taking place into the use of this technique in areas where earthquakes threaten major population centres.

Other approaches towards lessening the effects of earthquakes emphasise the educational side: the general public does not seem to realise the hazard of building on a fault zone, or the precautions which may help if one has to locate buildings close to such an area.

Earthquakes, volcanoes and plate tectonics

The world distribution of earthquakes and that of volcanic activity are related closely to each other and to plate boundaries (Fig. 2.6). Earthquake zones include constructive margins, such as ocean ridges, transform faults (e.g. the San Andreas zone) and especially destructive margins. The last of these experiences nearly all the major earthquakes, since one plate is being forced beneath another, leading to high levels of frictional stress (Fig. 6.10).

We thus see that the plate margins are the major world zones of volcanic and earthquake activity. It is, in fact, types of activity like the eruption of volcanoes and earthquakes which reveal to us that the earth is an extremely dynamic planet with constant changes taking place at its surface. It was not until plate theory was put forward that the overall pattern could be discerned, and this is a further major contribution of the theory to our understanding.

Hazards or helpers?

So far the emphasis in our studies of volcanoes and earthquakes has been on the disasters they may cause, killing people and destroying possessions. They also have their positive side. Volcanic processes may be violent and destructive for local areas over a restricted period of time, but also provide many benefits. This is illustrated by the fact that people crowd back into volcanic areas after eruptions in order to make the most of the fertile soils or tourist opportunities. Volcanic islands are the main form of landfall throughout the Pacific ocean. Volcanic gases are added to the atmosphere continually, helping to maintain the balance of elements there. In Iceland volcanic heat is used in homes and glass-houses; in New Zealand and California geothermal power from volcanic sources is tapped to produce electricity.

Earthquakes and faults may have a less immediate positive contribution to make, but are experienced in regions where land is subsiding or being uplifted: they are responsible for creating, for instance, much of the detailed topography of the San Francisco Bay area, one of the world's major harbours. They are also associated with mineral resources, helping to trap oil and water, or forming lines along which mineralised fluids may penetrate and solidify in veins. When geologists search aerial or satellite views as a preliminary activity in the search for new sources of minerals, they look for lines which could be faults, and especially points where several converge, since it is in these places that major ore deposits have been found previously. In addition, the study of earthquakes has provided us with our major source of information concerning the earth's interior, and the techniques developed for analysing earthquake shock wave records have been used to locate hidden resources, such as the North Sea oil deposits, buried beneath the sea and several thousand metres of rock.

The study of both volcanoes and earthquakes has often been initiated and assisted by government programmes, or general grants, so that future disasters may be averted or warnings given in time. Such intensive studies assist our understanding of the earth in many ways. Other natural hazards, like violent storms (e.g. hurricanes or tornadoes) or flooding, can be seen in a similar light. Their danger to man's activities has stimulated attempts to understand the natural environment to a greater extent. We now know more about the relative frequencies of such events and the factors giving rise to them, and can plan to avoid the risks involved in living in areas prone to such hazards.

There is still a considerable way to go, however, before man can boast that he has mastered the natural environment. There have been too many occasions in the past where this has been assumed, and where larger disasters have been caused by too simple an approach to earlier problems. There is still plenty of scope for scientists entering this field.

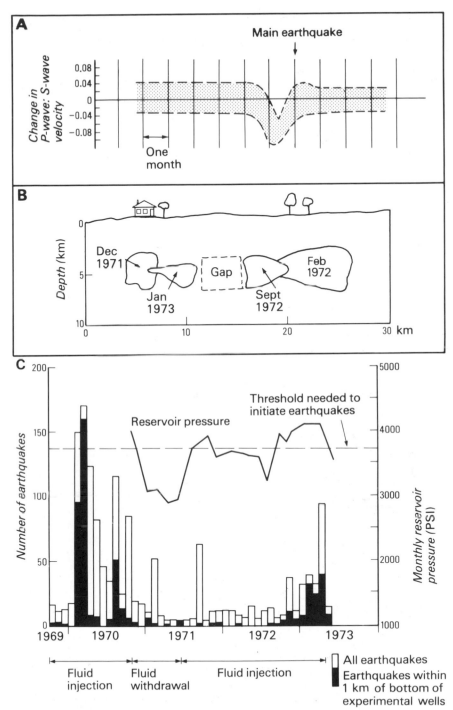

Fig. 6.13 **A** Predicting earthquakes: small foreshocks are plotted as a ratio of P:S wave velocities. This ratio often drops some weeks before a major earthquake, and then rises sharply. **B** Spotting the epicentre. Four moderate earthquakes along the San Andreas fault. The next may occur in the gap. **C** Controlling earthquakes. Earthquake frequency has been related to reservoir pressure at Rangely Oilfield, Colorado, USA. Pumping was controlled so that pressure was above or below the critical level.

The earth's interior and isostasy

Our knowledge of the earth's interior is of a different order to that of the surface rocks. No one has been more than a few kilometres down into the earth, and we must use a number of indirect methods of observation and infer the possible nature of the interior from the results gained.

Men have studied volcanoes spewing out molten lava at extremely high temperatures, and have noticed that the temperature rises as they descend very deep mines. Direct evidence of this type has led to the conclusion that the interior of the earth is much hotter than the surface: temperatures may be as much as 500°C 30 km beneath the surface, and 6000°C in the deep interior. We must remember, however, that pressure also increases towards the earth's centre because of the weight of the overlying rocks. This raises the melting points of the rocks, and one which melts at 800°C at the surface may not do so until several thousands of degrees in the deepest parts of the planet.

Most of our knowledge of the earth's interior has been obtained by indirect methods, and the study of earthquake records has been of particular value. Each of the different types of shock waves recorded (Fig. 6.11) has different characteristics, though the P and S waves are most important because they travel through the earth. The P waves travel fastest, and therefore arrive at the seismograph station first. Whereas the P waves are compressional, and the particles in their path vibrate back and forth, the S waves are transverse with side-to-side movements, and will not pass through liquids. Fig. 1 shows what four seismographs distributed around the world record when an earthquake takes place. Stations A and B receive all the shock waves first with the most powerful effects, and they have records similar to Fig. 6.11. Station C only receives surface waves, since the S waves have been obstructed and the P waves slowed down and refracted by a different material in the earth's centre. Station D receives P and surface waves, but with weakened signals. From such records we can conclude that there is a central mass inside the earth, known as the **core**, which will not allow S waves to pass through, and which slows down the P waves. It is very possible that this is liquid. Sir Isaac Newton's Law of Gravitation has helped us to calculate the mass of the earth. Since we also know its volume, we can work out the average density to be 5·527 gm/cm³. But we know that the surface rocks have a density of less than 3, and have to conclude that the rocks of the core must be much denser – about 12: they are probably a mixture of nickel and iron like many of the meteorites which have been found.

Seismic waves are deflected (refracted) and their velocities changed when there is a change of density in the rocks through which they are passing. Seismograph records are

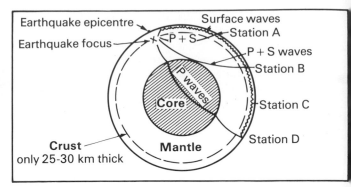

Fig. 1   The earth's interior. The use of earthquake shock waves to work out the nature of the earth's interior. A, B, C and D are four seismograph stations recording shocks from the earthquake at X. Note the type of wave each station receives, remembering the characteristics of the three groups.

usually more complicated than the one we have considered so far, with two or three sets of P and S waves arriving at the receiving station – indicating a layered structure above the earth's core. The most pronounced change in the velocity of the seismic waves occurs between the rocks of the thin earth's **crust** (with densities between 2·7 and 3·0 gm/cm³) and the underlying **mantle** (density 3·3 gm/cm³). The boundary between the two is known as the Mohorovičić Discontinuity (or just 'Moho'), after the Yugoslav seismologist who discovered it in 1909. A study of Moho tells us that the crust is thickest under the continents and very thin under the oceans. One of the more minor divisions shown up by earthquake records is one within the crust. There is a lower zone, similar in average composition to the rock basalt and often known as the **sima** (it is composed largely of silica and magnesia) or **basaltic crust**, and an upper zone of lighter rocks similar to granite known as the **sial** (silica

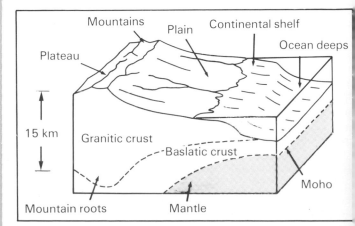

Fig. 2   The earth's crust. Note the relationship between the surface relief and the thickness of the crust.

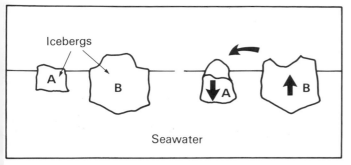

Fig. 3 Isostasy: (1) Icebergs. Two icebergs, A and B, have a similar proportion of their mass out of the water, even when some ice is transferred from one berg to the other – the state of equilibrium between ice and water is maintained.

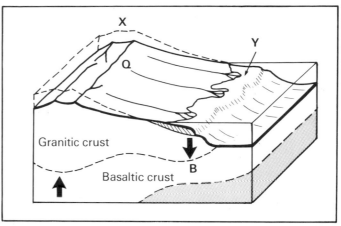

Fig. 4 Isostasy: (2) Erosion of mountains. Erosion and deposition lead to changes in the pattern of crustal thickness. The tops of the mountains at X are removed by erosion and the debris is deposited at Y. This leads to the removal of the load above A, which rises, and an increase of the load at B, which subsides. It is probable that some of the basaltic crust moves slowly from B to A.

and alumina) or **granite crust**. These divisions are illustrated in Fig. 2. The earth is thus composed of a series of shells getting lighter and lighter from the centre to the surface.

The crust is thickest beneath the continents, and especially beneath the mountain ranges. This discovery was first made by surveyors who mapped differences in gravitational pull over the earth's surface. It is to be expected that high mountain ranges like the Himalayas will attract a plumb-line because of their extra mass. Scientists were puzzled when they found that the deflection of the plumb-line towards the mountains was only one-third of the value they had calculated. They realised there were two possible explanations. Either the mountains were formed of much lighter rocks than the surrounding areas, or the lighter surface rocks extended to greater depths beneath the mountains. It was soon shown that the surface rocks do not vary in this way, and the latter theory has been adopted. There are deep 'roots' under each range of mountains.

We would notice a similar result if we examined two icebergs of different sizes. Icebergs float on water because they are less dense, but only one-eighth appears above the surface. The small and the large iceberg will both project above the surface in the same proportion (Fig. 3), and we can say that both are in a state of equilibrium. If we took some earth-moving equipment on to the larger iceberg and removed large quantities of ice to the smaller, we would not be in danger of drowning because as we removed part of the ice the rest would rise to maintain a one-eighth part above the water. Similarly the smaller iceberg would sink under the additional weight, and would not have more than one-eighth above the water.

The granite crust is less dense than the basaltic crust and underlying mantle, and may be said to be 'floating', like the icebergs in water, in a state of gravitational equilibrium that is known as **isostasy**. Where the continental rocks rise to greater heights they are compensated at depth by roots extending into the mantle; if there is a shift in the load there

will be movements to adjust the thickness of the crustal rocks, although, as they are taking place in solid rocks, they will take longer than those in the case of the iceberg floating in water. For instance, when a range of mountains is worn away a lot of debris will be produced which is taken to the sea by rivers and glaciers and deposited. The mountains have lost a large mass of material, which has been added to the sea floor, and so we must expect the mountainous area to rise slightly to maintain the isostatic balance, and the sea floor to subside under the extra weight, as shown in Fig. 4. A similar process takes place when large masses of ice accumulate on the land, as they did in the Ice Ages. The weight of ice causes the land to sink, and when the ice melts the land rises again. Fig. 5 illustrates what happened when Scandinavia was covered by a thick ice-sheet: the isostatic 'recoil' is still taking place and the floor of the Baltic Sea is rising to resume its position before the ice formed.

Although we have built up a general picture of what lies beneath our feet inside the earth, and of one of the very important effects of this arrangement, it would be an advantage to be able to confirm our indirect observations by a more direct method. The Americans attempted an interesting project some years ago in trying to drill through the crust to find out the actual composition of the mantle. As you have seen in Fig. 2 the crust is thinnest under the oceans and this is where they hoped to drill. The aim was to drill through the Mohorovičić Discontinuity and was known as the 'Mohole Project'. The engineering difficulties, which were immense, held it back, and the rising costs led to the cancellation of the programme in 1969. It is likely, however, that further attempts will be made in the future.

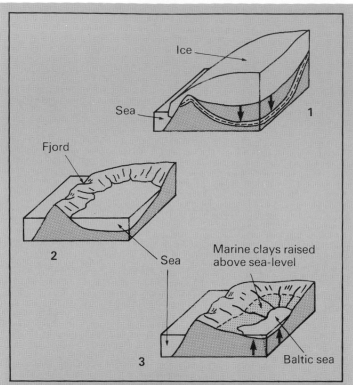

Fig. 5 Isostasy: (3) Ice on Scandinavia. These generalised diagrams summarise the recent history of Scandinavia. 1. The peninsula at the beginning of the Ice Age. The extra load of ice caused the crustal rocks to sag. 2. When the ice melted the sea rose and drowned the depressed area of land. 3. Today land in the Baltic Sea area is rising by a process of isostatic recoil: the load of ice has been removed, and the land is rising gradually to its former position.

Thus some streams have been regulated in parts of their courses and the area of flooding has been shifted to other areas: the river Rhine in its rift valley section is a good example of the effects of massive engineering works, and the effect has been to double flood levels at points downstream. In other cases flood protection schemes have been related to a particular flood level, but have given people a false sense of security, so that they have sited factories and homes on lowlying land, increasing the damage done when an extra-high flood overflows the defences. In many situations the inhabitants of high risk zones do not seem to understand the potential problems, and it is a most important part of local education to see that they do. These points further emphasise the necessity for the peoples of an increasingly crowded world to relate more closely to their physical environment.

Summary: geological processes at the earth's surface today

Throughout this section of the book (Chapters 4–6) we have emphasised the nature of the interactions of different processes with the earth's surface features. Since the range of processes – chemical disintegration, landslides, stream-flow, glacier movements, waves in the sea, volcanic eruption of new materials and the uplift and subsidence of the land – is proceeding continually, many changes are taking place. Some are noticeable over a few weeks or years; others within a lifetime; and still others take place over thousands or even millions of years. Thus hundreds of tonnes of silt may be excavated from a gully and dumped in the sea after a single storm. At the other end of the scale mountains are uplifted gradually with a series of earthquakes over millions of years, and are then worn down slowly over a similar period of time, with their materials reconstituted as layers of sediment on lowland plains or sea floors. In this sequence new minerals are formed by chemical changes, and environmental changes occur which cause natural changes or mutations in living organisms to be selected if they are advantageous to them in the new situation.

Whilst we can imagine such changes occurring in a study of the present, the major endeavours of the geologist are aimed at interpreting the sequence of 'presents' which make up the history of the planet earth. The formation of lava flows and of layers of sediment on the ocean floor at the present moment are vital aspects of the evidence available to him as he studies similar rocks formed millions of years ago, or as he studies rocks which were formed at too great a depth in the crust to be studied in the course of their genesis. Some rocks formed in the past are quite different from anything we can recognise today!

The detailed approach to geological history—and this includes an understanding of relationships between rocks which will enable us to discover new mineral resources – is based on a study of minerals, rocks, fossils and rock structures and sequences. At every step of the way the observations from groups of rocks formed in the past are compared with observations of the formation of similar materials under present conditions. The present thus becomes the key to the past.

Part III
Interpreting the past

7
Studying minerals

The study of geology has been likened to a detective story. The geologist finds his clues in the rocks which have an outcrop at the earth's surface, and from his studies of the evidence provided there he is able to suggest solutions to the puzzling mystery of what happened in the past, long before man recorded anything in written form.

You have probably been struck by the variety of rocks you see in the cliffs at the seaside, or in the museums you have visited. They are of different colours and have a variety of grain-sizes; some are extremely hard and compact, others soft and crumbly; some contain fossils and others do not; and there may be a variety of structures crisscrossing the rocks. All these features are significant when we attempt to work out the past history of the earth, and it is important for us to be trained to observe them efficiently, so that our records are intelligible to other geologists.

Rocks and minerals

The basic building blocks of earth materials are atoms, and the molecules they combine to form. It is not possible to see these, however, and the geologist would regard the mineral as his fundamental unit of study. Minerals are brought together in various ways to form rocks, and so it is important to see links between the study of minerals and the broader study of rocks.

Rocks are composed of minerals, or of fragments of older rocks which are also made up of minerals. Sometimes we cannot see the minerals with the naked eye, and they can be identified only under a microscope. It is important for us to be able to recognise these minerals, because they tell us what type of rock we are examining, and help us to begin to understand the rock's origin.

A geological definition of a mineral is 'an inorganic substance which occurs naturally and has a definite chemical composition and physical properties which vary within known limits'. Most minerals are compounds of several elements, and it is rare to find pure elements occurring naturally. The main elements combining to form minerals in the rocks of the earth's crust are oxygen (46·6 per cent of the total), silicon (27·7 per cent), aluminium, iron, calcium, sodium, potassium and magnesium: these eight make up 99·3 per cent by weight of all the rock-forming minerals (Fig. 7.1).

Most minerals have a definite internal arrangement of atoms, which is occasionally expressed for us to see in the shape of a **crystal**. Crystals only develop their distinctive shapes where they can grow freely, and as minerals usually have to compete for space as they are solidifying these are rarely found. Crystals of one mineral always form similar geometric shapes, and the angles between corresponding faces of the crystal are always the same (Fig. 7.2A). Although we cannot see the internal atomic structure of a mineral, we can learn a lot

A

Atomic number	Element	Crustal average (ppm)	Atomic number	Element	Crustal average (ppm)
1	H	1 400	45	Rh	0·005
3	Li	20	46	Pd	0·01
4	Be	2·8	47	Ag	0·07
5	B	10	48	Cd	0·2
6	C	200	49	In	0·1
7	N	20	50	Sn	2
8	O	466 000	51	Sb	0·2
9	F	625	52	Te	0·01
11	Na	28 300	53	I	0·5
12	Mg	20 900	55	Cs	3
13	Al	81 300	56	Ba	425
14	Si	277 200	57	La	30
15	P	1 050	58	Ce	60
16	S	260	59	Pr	8·2
17	Cl	130	60	Nd	28
19	K	25 900	62	Sm	6·0
20	Ca	36 300	63	Eu	1·2
21	Sc	22	64	Gd	5·4
22	Ti	4 400	65	Tb	0·9
23	V	135	66	Dy	3·0
24	Cr	100	67	Ho	1·2
25	Mn	950	68	Er	2·8
26	Fe	50 000	69	Tm	0·5
27	Co	25	70	Yb	3·4
28	Ni	75	71	Lu	0·5
29	Cu	55	72	Hf	3
30	Zn	70	73	Ta	2
31	Ga	15	74	W	1·5
32	Ge	1·5	75	Re	0·001
33	As	1·8	76	Os	0·005
34	Se	0·05	77	Ir	0·001
35	Br	2·5	78	Pt	0·01
37	Rb	90	79	Au	0·004
38	Sr	375	80	Hg	0·08
39	Y	33	81	Tl	0·5
40	Zr	165	82	Pb	13
41	Nb	20	83	Bi	0·2
42	Mo	1·5	90	Th	7·2
44	Ru	0·01	92	U	1·8

B

Element	Abundance, mg l⁻¹	Element	Abundance, mg l⁻¹
Li	0·17	Y	0·000 3
Be	0·000 000 6	Nb	0·000 01
B	4·6	Mo	0·01
C	28	Ag	0·000 04
N	0·5	Cd	0·000 11
F	1·3	In	<0·02
Na	10 500	Sn	0·000 8
Mg	1 350	Sb	0·000 5
Al	0·01	I	0·06
Si	3·0	Cs	0·000 5
P	0·07	Ba	0·03
S	885	La	0·000 002 9
Cl	19 000	Ce	0·000 001 3
K	380	Pr	0·000 000 64
Ca	400	Nd	0·000 002 3
Sc	0·000 04	Sm	0·000 000 42
Ti	0·001	Eu	0·000 000 11
V	0·002	Gd	0·000 000 60
Cr	0·000 05	Dy	0·000 000 73
Mn	0·002	Ho	0·000 000 22
Fe	0·01	Er	0·000 000 61
Co	0·000 1	Tm	0·000 000 13
Ni	0·002	Yb	0·000 000 52
Cu	0·003	Lu	0·000 000 12
Zn	0·01	W	0·000 1
Ga	0·000 03	Au	0·000 004
Ge	0·000 07	Hg	0·000 03
As	0·003	Tl	<0·000 01
Se	0·000 4	Pb	0·000 03
Br	65	Bi	0·000 02
Rb	0·12	Th	0·000 05
Sr	8·0	U	0·003

Note mg l⁻¹ is approximately the same as ppm.

Fig. 7.1   The abundance of elements in (**A**) the earth's continental crust, (**B**) seawater. What proportion is made up by the main metallic elements – tin (Sn), copper (Cu), zinc (Zn), lead (Pb), iron (Fe)?

Crystal structure. On the right is a large single crystal of potassium alum: notice the eight-sided form (octahedron), and the way in which smooth place faces cut in sharp edges. On the left is a pile of polystyrene balls: what do these suggest about the arrangement of molecules within the upper half of the crystal, and the formation of the crystal faces?

about it by looking at the crystal form, which is a direct reflection of the way in which the atoms are packed together (Fig. 7.2B). Thus the sodium and chlorine atoms in common salt form a cube-shaped pattern, and the mineral crystallises in cubes. Crystals of one mineral may be joined together in many ways, but when a regular pattern is visible they are said to be **twinned**. Twinning is very common in the feldspar minerals, and is an important diagnostic feature (Fig. 7.2C).

How do we identify minerals?

Most of the common minerals are so distinctive that we can recognise them just by looking at specimens and noting various physical properties. For identification of some of the rarer minerals, and confirmation of the identity of the common ones, it is necessary to examine them under a microscope or test them using a blowpipe and chemical reagents. It may even be necessary to study them by X-ray diffraction. Fortunately all the minerals we shall be interested in can be identified without using special equipment.

All the properties we will be using to identify minerals are related directly to the internal atomic structure of the minerals concerned. The forces between the atoms, and the ways in which the atoms are 'packed', determine the characteristics like cleavage and hardness, whilst the atomic weights of the constituent elements affect the density.

One of the best ways to approach the description and identification of minerals is to ask a series of questions.

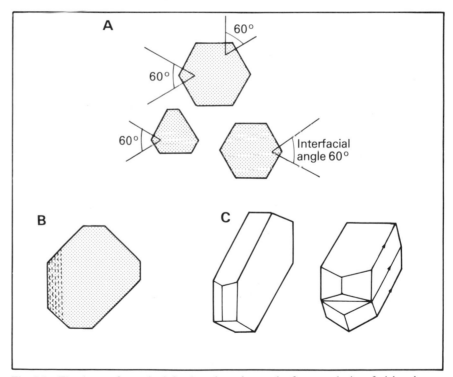

Fig. 7.2 The shapes of crystals. **A** Sections through crystals of quartz: the interfacial angles are always constant, although the crystal shape may change. **B** The relationship of the crystal faces to the internal atomic structure: the faces are parallel to lines of atoms. **C** Untwinned (left) and twinned crystals of orthoclase feldspar. In the twinned variety the lower half of the crystal has been turned 180 degrees.

A First Look At Rocks

At this stage we shall be concerned only with looking at some of the main characteristics exhibited by rocks: in Chapter 8 the various groups will be examined in greater detail.

When you handle a piece of rock with the aim of making a geological investigation it is important for you to make some definite observations of the features it shows. These should include some actual measurements and should use precise terms.

(1) **A rock is composed of various particles.** Can you distinguish them, or are they so small that the rock appears to be formed only of one type of material? A useful distinction can be made between rocks where the particles are easily visible with the naked eye (coarse-grained), those which have a rough feel to the touch and whose particles can only just be distinguished (medium-grained) and those which are smooth (fine-grained). Use rocks in your school collection to identify these types, and then measure the sizes of the grains in each type of rock. Where would you place the division between fine, medium and coarse?

(2) **If you cannot see the individual particles clearly** (i.e. if the rocks are medium- or fine-grained) there are other questions you can ask about the rock specimen.

(a) What is the **colour** of the whole rock? Certain rocks have distinctive colours: chalk (sedimentary rock) is white; basalt lava (igneous rock) and some fine shales (sedimentary rocks) or slates (metamorphic rocks) are black; limestone (sedimentary) may be grey or cream; rocks formed on the land containing iron oxide (sedimentary) are often

Four rocks. These photographs show how rocks are composed of minerals or rock fragments. The scale line represents 1 cm in each case.

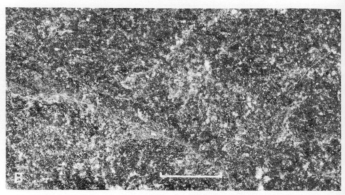

Rock B was also formed by cooling from a melt, but the crystals are much smaller than in rock A and include more dark minerals, like augite and hornblende. The lighter crystals are plagioclase feldspar. The rock is basalt.

Rock C is composed of rounded flint pebbles, cemented by a matrix of fine sand and silica. It is known as a conglomerate.

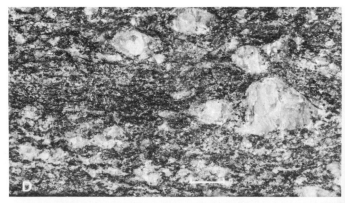

Rock D is formed of crystals of similar minerals to those occurring in rock A, but they have been squashed and drawn out in lines: this is typical of rocks formed under conditions of heat and pressure due to deep burial in the earth's crust.

Describe these rocks in terms of the grain-sizes, minerals contained and other visible characteristics. Name them: igneous, metamorphic or sedimentary.

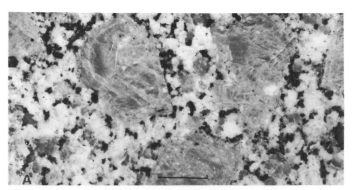

Rock A is formed of interlocking crystals: the large grey crystals are pink potassium-rich feldspar (orthoclase); the smaller grey crystals are glassy quartz; the white crystals are sodium/calcium-rich feldspar (plagioclase); and the black crystals are mica. This rock is a granite, formed by cooling from molten rock material.

red. By itself, however, colour is not such a useful diagnostic feature of rocks as it is of minerals. It should be used as a part of the total evidence.

(b) Does the rock contain animal- or plant-like markings (i.e. **fossils**)? These are found only in the major group known as the sedimentary rocks, which are formed by the accumulation of layers of mud, sand, shell or plant debris. (Igneous rocks originate from the earth's interior and solidify as they cool after being molten and fluid; metamorphic rocks have been subjected to great heat and pressure so that any fossils they once contained as sediments may have been distorted or destroyed.)

(c) What **small-scale structures** can be seen? If there are partings between the rock layers it is most likely a sediment (though igneous rocks are sometimes layered and metamorphic slates show fine cleavage structures).

(d) Is the rock a **carbonate**? Rocks containing a proportion of calcium carbonate are normally sedimentary rocks, and can be identified because they have a fizzy reaction with cold, dilute hydrochloric acid. If calcium carbonate is the main component of the rock this is known as a limestone, but it may form only the cementing medium for other materials such as sand grains – when the rock would be known as a calcareous sandstone.

(3) **If you can see the individual particles** of which the rock is formed, they may be of two varieties. Some may be rounded or angular rock-fragments (i.e. pebbles, sandgrains) set in a finer-grained cement: this will be a sedimentary rock. Or they may be a series of interlocking minerals, which formed from a mobile mass of molten rock material (i.e. igneous), or by recrystallisation under conditions of intense pressure and heat (i.e. metamorphic – these will often have the crystals drawn out in marked alignments).

These questions will have shown you that rocks have certain properties which can be observed, measured and recorded: each type of igneous, sedimentary or metamorphic rock always has similar features. At the same time you will have noticed that it is not so simple a matter to divide a series of unconnected pieces of rock into even the major groups. Before you are able to identify and interpret the significance of the varieties of rock with confidence, you will need to understand more about a number of characteristics and processes.

(a) You should be able to identify the minerals which make up the rocks (this chapter).

(b) You should understand something of the processes which lead to the formation of the rocks (Chapters 4–6).

(c) It is also important to examine the associations of the different rock-types in their natural setting, and the distinctive patterns produced by their outcrops (Chapter 8).

(1) What is the mineral's **colour**? Some minerals like pyrite (light, bronzy-yellow), chalcopyrite (brass-yellow) and malachite (green) are distinguished by their colour. We have to be careful in using colour, however, because a number of the metallic minerals may have developed a tarnish on exposure to the atmosphere: thus chalcopyrite often has a blue-black surface. In addition a number of minerals may occur with a variety of colours, due to the presence of tiny quantities of impurities or to the reaction of small changes in the internal crystal structure to visible light. Quartz may occur as colourless (rock crystal), white (milky quartz), dark brown (smoky quartz), mauve (amethyst), pink (rose quartz) or dark yellow (citrine), whilst fluorite may be mauve, colourless, blue, green or yellow. With experience, however, mineral colour can become one of the main diagnostic properties.

(2) What is the **streak** like? The streak is the colour of a line made by scratching the mineral on a piece of unglazed porcelain. It is also sometimes obtained by scraping a fine powder from the mineral surface, or by crushing the mineral, but these methods may destroy other characteristics of the mineral. Streak is more constant than colour: haematite may be red or black, but the streak is always blood-red; fluorite may have many different colours, but its streak is always white.

(3) What is its **lustre**? This is the appearance of the mineral in ordinary light. It can be shiny and metallic, shiny and brilliant like a diamond (adamantine), or shiny and glassy (vitreous). It can have a dull surface, and it can be opaque or transparent.

(4) Does the mineral have a distinctive **cleavage** or **fracture**? Mineral cleavage (to be distinguished from slaty cleavage, see p. 000) is a line along which a crystal may break in relation to the internal atomic structure of the mineral: it is clean and flat and is parallel to a possible crystal face (Fig. 7.2B). Mica is one of the best examples, since it cleaves into thin sheets parallel to the base of the crystals; the well-developed nature of cleavage in mica is designated as 'perfect cleavage' – in comparison with other

minerals where it is just 'good' or 'poor'. Minerals without cleavage may break along irregular surfaces unrelated to the internal structure, and this is known as fracture. Quartz breaks with a conchoidal (shell-like) pattern, whilst asbestos has a fibrous, thread-like fracture.

(5) How **hard** is the mineral? Hardness is the resistance of a mineral to scratching or abrasion. It is greater if the atoms are smaller or more closely packed. Diamond and graphite are both composed of carbon atoms, but those in diamond are more densely packed making it the hardest known mineral. We use a scale devised by F. Mohs in 1812 to compare mineral hardness: this is the type of scale which merely provides an order, and diamond, for instance, is many times harder than corundum.

1	Talc (very soft)	6	Orthoclase feldspar
2	Gypsum	7	Quartz
3	Calcite	8	Topaz
4	Fluorite	9	Corundum
5	Apatite	10	Diamond (very hard)

None of us is able to carry a set of these minerals around with us, but a strong finger-nail, copper coin and pocket-knife are often useful. Talc and gypsum can both be scratched by a finger-nail (H = 2−2½), and calcite is roughly equal in hardness to a copper coin. One of the most important distinctions to make is that between quartz and calcite, which both occur as white veins in rocks: a pocket-knife with a hardness of about 5½, between apatite and orthoclase, will scratch calcite easily but it will not affect quartz.

Mineral	Composition	Colour	Hardness	Density
I. ROCK-FORMING MINERALS				
Olivine	$(Mg,Fe)_2SiO_4$	Colourless	6-7	3·2-4·3
Augite (Pyroxene)	Silicate of Ca,Mg,Fe,Al	Black	5-6	3·2-3·5
Hornblende (Amphibole)	Silicate of Ca,Mg,Fe,Al,Na	Black	5-6	3-3·47
Biotite Mica	Silicate of Mg,Al,K,H,Fe	Brown	2·5-3	2·7-3·1
Muscovite Mica	Silicate of Al,K,H	White	2-2·5	2·76-3
Orthoclase Feldspar	$KAlSi_3O_8$	White, pink	6	2·57
Plagioclase Feldspar	$(Ca,Na)AlSi_3O_8$	White, grey	6	2·6-2·76
Quartz	SiO_2	White, colourless	7	2·65
Calcite	$CaCO_3$	White	3	2·71
Garnet	see Fig. 7.7			
II. METALLIC MINERAL ORES				
Magnetite	Fe_3O_4	Black	5·5-6·5	5
Haematite	Fe_2O_3	Red/black 'skin'	5·5-6·5	4·9-5·3
Limonite	$2Fe_2O_3.3H_2O$	Bronze/yellow	5-5·5	3·6-4
Pyrite	FeS_2	Pale bronze-yellow	6·5	4·7-5·1
Chalcopyrite	$CuFeS_2$	Brass yellow	3·5-4	4·1-4·3
Malachite	$Cu_2CO_3(OH)_2$	Bright green	3·5-4	4
Azurite	$Cu_3(CO_3)_2(OH)_2$	Azure blue	3·5-4	3·7-3·8
Galena	PbS	Lead grey	2·5	7·5
Sphalerite	ZnS	Black/brown	3·5-4	4
Cassiterite	SnO_2	Black + pink powder	6-7	7
III. OTHER MINERALS				
Fluorite	CaF_2	Usually purple	4	3-3·25
Barite	$BaSO_4$	White/pink/blue	3-3·5	4·5 ('Heavy sp
Graphite	C	Black/dark grey	1	2-2·3
Halite	NaCl	Colourless (pure)	2·5	2·2
Gypsum	$CaSO_4.2H_2O$	Colourless (pure)	2	2·3

(6) Is the mineral **heavier** than similar-sized specimens of other minerals? This will reflect the densities of the minerals compared and thus the atomic weights of constituent elements: lumps of many metallic ores such as those of lead or iron are quickly recognised by their weight when compared with pieces of ordinary rock or non-metallic minerals.

(7) What **habit** does the mineral have? This is a comment on the mineral's shape. Does it show crystal faces, or does it have a rounded, nodular shape, or is it without any particular shape (massive)?

(8) Has the mineral any other distinguishing properties? Some may be **magnetic** (iron oxides like magnetite), and some may be **easily dissolved** in weak acid. Calcite can also be distinguished from quartz because it fizzes in weak hydrochloric acid, whereas quartz does not react.

The common rock-forming minerals

We have already noted that oxygen and silicon are the two most important elements in the rocks of the earth's crust. They combine to form silica (SiO_2), the commonest form of which is quartz. The great majority of the minerals found in the crustal rocks are silicates, which may be thought of as combinations of silica with the oxides of other elements. This is particularly true in the igneous and metamorphic groups of rocks, where they are absolutely dominant. The **igneous rocks**, which include all those which were once molten and mobile, have seven 'families' of minerals which make up 99 per cent of all the rocks in the group. They are the olivines, pyroxenes, amphiboles, micas, feldspars, feldspathoids and quartz. The **metamorphic rocks**,

ak	Other Properties	Commonest Modes of Occurrence	Importance, Uses
	Green alteration	Basic igneous rocks	
	Prismatic crystals	Basic igneous rocks	
	Prismatic crystals	Igneous and metamorphic rocks	Asbestos is one form
	Thin laminae	Igneous and metamorphic rocks	Electrical industry,
	Thin laminae	Many kinds of rocks	lubricants
		Many kinds of rocks	Pottery; glazes;
	Multiple twinning	Basic igneous rocks	mild abrasive
	Vitreous lustre	Many kinds of rocks; gangue	Building sands; pottery;
	Reacts with cold, dilute acid	Limestones; gangue	furnace-lining; electronics
k			
ry red	Strongly magnetic, metallic lustre	Igneous, metamorphic rocks	Iron ore
ow-brown		Veins, metamorphic rocks	Chief iron ore
nish-black	Dull lustre	Weathered iron ores	Iron ore
nish-black	Metallic lustre	Veins, nodules	Most common sulphide
		Veins	Copper ore
		Associated minerals formed by	
		underground waters	Copper ores (e.g. Katanga)
d grey	Cubic crystals, metallic lustre	Veins	Chief lead ore
te	Adamantine lustre	Veins	Chief zinc ore
		Veins, placer gravels	Chief tin ore
	Cubic crystals	Gangue; hydrothermal veins	Enamelling; steel flux
		Gangue; hydrothermal veins	White paint; paper
k	Metallic lustre	Veins; metamorphic rocks	Pencil 'leads'; electronics
	Taste	Evaporite sediments	Chemical industry
	Pearly lustre		Cement; plaster; paper

Fig. 7.3 The characteristics of some important minerals.

which have been altered in various ways by heat and/or pressure from their original nature, have a wider range of minerals, but they are still mostly silicates. The other major group, the **sedimentary rocks**, are formed by the accumulation of layers of rock debris such as sand and mud, or by the piling up of shells, decaying vegetation or chemical deposits. The minerals contained in this group include some 'inherited' from the other groups, like the more resistant quartz, and some newly formed by chemical reactions including the clay minerals (also complex silicates), the carbonates and evaporated salts.

All the common **rock-forming minerals** can be classified in these groups:

The silicate minerals
The Olivine group
The Pyroxene group, e.g. augite
The Amphibole group, e.g. hornblende
The Micas and Clay minerals
The Feldspar group
The Feldspathoid group
The Quartz group
The Garnet group

The non-silicate minerals
The Carbonates
The Halides, e.g. rock salt
The Sulphates, e.g. gypsum
The Oxides, e.g. magnetite
The Sulphides, e.g. pyrite

A major division we recognise amongst the silicate minerals is between the dark-coloured group containing a high proportion of iron and magnesium – known as the **ferromagnesian** (or **mafic**) minerals – and the lighter-coloured feldspars, feldspathoids and quartz – known as **felsic**. Fig. 7.3 summarises the diagnostic properties of the main rock-forming and ore minerals.

The silicate minerals

The Olivine group. These have the simplest chemical formula and the smallest proportion of silica, and this is why they are mostly found in dark-coloured igneous rocks in which there is less than 55 per cent silica. Most olivines are a mixture of iron and magnesium with silica: $(Mg,Fe)_2SiO_4$. The atomic structure is tightly bonded, and olivine forms compact crystals having a glassy colourless appearance. Olivine crystallises at very high temperatures, and therefore often has a good crystalline shape, but it is easily altered to the green mineral serpentine and black iron oxide. When examined under the microscope (PPL) olivine is colourless, but has a high relief with hard, black outlines and a pitted surface; it is crossed by arc-like fractures, and the greenish alteration products can often be seen eating into the margins of the mineral. When the polars are crossed (XP) olivine has bright polarisation colours varying from reds to blues and greens.

The Pyroxene group. Augite is the most common member of this group, which has a higher proportion of silica than the olivines. The pyroxenes are a large group, also found in the dark igneous rocks, and augite itself may vary in its composition. Its formula is $(Ca,Mg,Fe,Al)_2(Al,Si)_2O_6$, with atoms of aluminium replacing some of the silicon. Augite forms 4- or 8-sided crystals and the cleavages are a feature of the mineral when seen under the microscope. it is usually colourless or pale brown (PPL) with bright polarisation colours (XP).

The Amphibole group. Hornblende is the best-known representative of this group which has more silica and is an even larger group than the pyroxenes. One feature of the amphiboles is their long, fibrous crystals and many of the asbestos minerals occur amongst them. The composition of these minerals is very complex, but similar to the pyroxenes with the addition of hydroxyl (OH): they can be termed hydrous calcium-magnesium-iron silicates. Hornblende forms long, flat crystals which are dark in colour like augite, but which have a glittering, silky sheen in contrast to the dull surface of augite. In thin section it is usually pleochroic from green to yellow, with marked cleavage lines (PPL) and long crystals, but the polarisation colours (XP) are often obscured by the body colour of the mineral.

The Micas and Clay minerals. The minerals in this group are hydrous potassium–aluminium silicates, and the most distinctive feature of the micas is their flaky nature due to the cleavage planes along weak bonds between the strong sheets of silicate molecules. There are two main varieties: **muscovite**, which is lighter in colour, and **biotite**, which is darker because it contains iron. Both have the typical strong cleavage which gives ragged outlines to the crystals examined under the microscope.

The polarising microscope

The polarising microscope is a very important instrument which geologists use to help in the task of identifying minerals and examining rocks. It is often called the petrological microscope because of its use in the study of rocks (i.e. petrology). In order to make a section of rock thin enough for light to pass through a thin slice has to be cut with a diamond saw; one face of it is polished and fixed with Canada Balsam to a glass slide; the other side of the rock slice is then ground down and polished until it is 0·03 mm thick.

A simple type of polarising microscope is drawn in Fig. 1 to show the main features. It is just like the ordinary microscope you may have used in biology, but has two pieces of **polaroid**, which have the property of polarising the light passing through them. Ordinary light vibrates in all directions in a plane at right angles to the direction of travel, but polarised light only vibrates in one direction. One piece, the polariser, is below the stage; the other, the analyser, is above it.

When we use the polarising microscope we are able to examine a mineral in three different ways, during each of which we can record certain diagnostic features of the mineral as the light passing through it is modified.

First, we can take the Polariser and the Analyser out of the beam of light passing up the microscope, and examine the rock section as ordinary light passes through it into our eye at the top. We can note (a) that most minerals are transparent and colourless, but that a few are opaque and coloured; (b) that some minerals stand out from the others and have a high relief with strongly marked outlines and even a pitted surface; (c) that the first minerals to form have a good crystalline shape, and it may be possible to refer them to a definite crystal system; (d) the cleavage or fracture present; and (e) if fossils are present and attempt to identify them.

The polariser can then be inserted into the beam of light, and the mineral examined as plain polarised light (**PPL**) passes through it. The main property to be noticed is whether the coloured minerals change colour as the stage rotates. Some do so markedly, and are known as **pleochroic** minerals. Pleochroic minerals show different colours according to the orientation of their crystals with respect to the direction of vibration of the polarised light. The best example is biotite, which changes from brown to yellow as the rock section is rotated on the microscope stage.

The first two methods of examination are usually combined when one is using a polarising microscope, because the facts which we have suggested should be noted under ordinary light can also be seen under plain polarised light. When both the polariser and analyser are inserted, the pieces of polaroid are crossed (**XP**). We note that (a) a few minerals, mostly of the cubic system, are black however the stage is rotated; (b) most of the minerals have distinctive colours varying from bright to pastel shades and from greys to reds, greens and blues; and (c) some minerals show evidence of twinning, where half the mineral is extinguished when the other half is lit up and vice versa.

When you are examining a thin section of rock in this way it is best to recognise each mineral contained in the rock, noting down all the distinctive features which enable you to identify it. Then compare it with the other minerals in the slide. Is it bigger than the rest? Has it a better shape? What proportion of the slide does it occupy? When you have treated the other minerals in a similar way, make simple drawings of representative parts of the slide, and finally suggest a name for the rock from which it was cut.

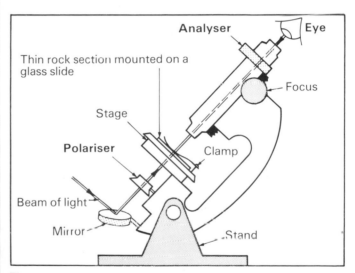

Fig. 1   A polarising microscope.

Investigation: minerals under the microscope. Make coloured drawings of the rock-forming minerals listed in Investigation 1 p. 000 as you see them in thin rock sections under the microscope (under plain polarised light (PPL), and under crossed-polars (XP)). This will give you at least two drawings for each mineral, and those which are pleochroic under PPL may need a third: these can be arranged neatly by drawing a cross on your page and inserting one mineral drawing in each of three sections. Compare the characteristics you notice under the microscope with those mentioned in connection with the mineral descriptions earlier in this chapter, and attempt to identify the minerals again in other thin sections. Note the rock-types in which they occur.

Muscovite is colourless (PPL) but bright pink and green (XP), whereas biotite is dark brown in colour (PPL) and very pleochroic (to pale yellow). Biotite does not change much under XP, as the strong body colour swamps the pinks and greens.

The clay minerals are related to the micas, and are formed when feldspars and ferromagnesian minerals break down. They are hydrous aluminium

silicates, which are easily transported by rivers to form layers of mud. The individual grains are so small that they can be distinguished only under a very powerful microscope. **Kaolinite** is one of the most important, and is the main constituent of China Clay.

The Feldspar group. These anhydrous potassium, sodium and calcium alumino-silicates are the most important of all the minerals occurring in the igneous rocks and may compose as much as 75 per cent of a granite. The two main varieties are **orthoclase** (potassium-rich) and **plagioclase**, which is a whole series of minerals with varying proportions of sodium and calcium. The sodium-rich varieties commonly occur with orthoclase in the light-coloured igneous rocks, whilst the calcium-rich plagioclases are found in the darker types. Orthoclase is white or pink in colour and has a strong cleavage which often exposes smooth, glistening surfaces. All feldspars are colourless under the microscope (PPL), and only have faint cleavage lines, but under XP have grey colours and can be distinguished by the patterns formed by twinning in the minerals: orthoclase crystals are sometimes found divided into two parts (simple twinning), but plagioclases almost invariably have a series of narrow parallel strips (multiple twinning).

The Feldspathoid group. These form a less common group of minerals which are found in rocks with a small percentage of silica and high proportion of alkaline elements like sodium and potassium. An example is leucite, which is an important constituent of the lavas erupted on Tristan da Cunha.

The Quartz group. Pure silica is found in important quantities only in the light-coloured igneous rocks, e.g. granite. The original molten material contained so much silica that some was left over when the other substances had combined with as much as they could, the residue then crystallising on its own. When seen in granite it forms colourless, rather glassy-looking (vitreous) crystals or white veins. Quartz is one of the hardest minerals, harder than steel, and very resistant to chemical action: it is often preserved as rounded grains in sedimentary rocks. When examined under the microscope quartz is recognised by the properties it does not have! It has no colour, cleavage or fracture, and as it is usually the last mineral to form it fits in between the others without any recognisable crystal shape (PPL). The polarisation colours are greys, whites

Minerals and crystals. Minerals occur as crystals having faces related to the internal atomic structure only when free to grow in that way: normally other minerals are growing at the same time and prevent this happening.
A Rock salt: a cubic crystal, cleaving parallel to the face.
B Fluorspar; two intergrown cubic crystals. Where does cleavage occur?
C Native copper: it is unusual for an element to occur in the pure form. Does the shape of this specimen indicate how it may have formed?
D Pyrite: intergrown crystals with pentagonal faces. These often have lined (striated) and pitted surfaces due to the imperfection of the natural processes of crystallisation and later chemical effects. (JS)

or yellows, unaffected by twinning patterns, although the wavy extinction pattern which is sometimes seen is evidence that the mineral has been subjected to stress.

The Garnet group. Most of the minerals in metamorphic rocks are silicates, but the more unusual members of groups like the pyroxenes and amphiboles tend to occur. The garnets are one of the most important distinct groups, forming dark red or green cubic crystals. Under the microscope they show up as large pink or grey crystals, but are always black under XP.

The non-silicate minerals

The Carbonates. Calcite and dolomite are the main non-silicates, and like the clay minerals are mostly found in sedimentary rocks and their metamorphosed equivalents like limestones and marbles. Calcite ($CaCO_3$) and dolomite ($MgCO_3.CaCO_3$) can be told apart because calcite effervesces with cold dilute acid, whereas dolomite needs warm acid.

The Halides and Sulphates. Rock salt ($NaCl$) and anhydrite ($CaSO_4$) are the most important minerals formed in evaporating lakes, and occur as soft crystalline deposits coloured by impurities. Rock salt is easily identified by its taste.

The Oxides and Sulphides. These are mainly iron minerals, like magnetite (Fe_3O_4) and iron pyrites (FeS_2). **Magnetite** occurs in small quantities in many igneous rocks, forming as black, diamond-shaped crystals. **Pyrite** is often found as crystals or nodules in sediments, where its silvery or bronzy-yellow colour is distinctive.

Rock-forming minerals. Such perfect crystals seldom occur in rocks. **A** Augite. **B** Biotite mica. **C** Potassium-rich feldspar. **D** Quartz. **E** Garnet. **F** Calcite, showing rhomb-like cleavage. **G** Calcite—'Dog-tooth' spar. Scale lines represent 1 cm. (JS)

Type	Manner of formation	Representative deposits
ORIGIN DUE TO INTERNAL PROCESSES		
Magmatic segregation	By settling of early formed minerals to the floor of a magma chamber during consolidation.	Magnetite, chromite and platinum-rich layers in the Bushveldt intrusion, South Africa.
	By settling of dense immiscible sulphide or oxide melts which formed in parts of the magma, and which either crystallized between earlier silicate minerals, or were injected along faults and fissures of the wall rocks.	Copper-nickel deposits of Norway and those of the Sudbury district, Canada. Injected bodies of magnetite in Sweden.
Contact-metasomatic	By replacement of the wall rocks of an intrusion by minerals whose components were derived from the magma.	Magnetite deposits of Iron Springs, Utah; some copper deposits of Morenci, Arizona.
Pegmatite (deposits from watery silicate fluids)	By filling fissures in the wall rocks and consolidated outer part of an intrusion.	Mica, and rare element deposits at Petaca, New Mexico.
Hydrothermal (deposits from hot, watery solutions)	By filling fissures in and replacing both wall rocks and the consolidated outer part of an intrusion.	Lead-copper-zinc deposits of Cornwall, North Pennines and Eire (Tynagh and Navan). Porphyry copper deposits.
ORIGIN DUE TO SURFACE PROCESSES		
Sedimentary	By evaporation of saline waters, leading to successive precipitation of valuable salts.	Salt and potash deposits of Northumberland.
	By precipitation of particular elements in suitable sedimentary environments.	Iron deposits of Northamptonshire; Kupferschiefer, Germany.
	By deposition of rocks in which the detrital grains of valuable minerals are concentrated because of superior hardness or density.	Placer gold deposits of Australia, California, Siberia, Nome (Alaska); titanium of Travancore (India) and Australia; diamond placers of South-West Africa.
Residual	By weathering, which causes leaching of soluble elements, whose removal leaves concentrations of insoluble elements in the residual material.	Bauxite (aluminium) ores of Arkansas (USA), France, Hungary, Jamaica and Guyana.
Secondary enrichment	Precipitation of soluble elements in groundwater at depth within a mineral deposit to give concentrations of valuable elements.	Copper deposit of Miami, Arizona.

Fig. 7.4 The origins of mineral deposits. Each of these processes carries out the concentration of the particular mineral into deposits which become worth mining.

Metallic ore minerals and gemstones

The earth's continental crust and the ocean waters are the normal sources from which we obtain minerals for industrial use, and they contain a great variety of elements with a wide range of concentrations (Fig. 7.1). Some of the most valuable, and even some of the common metals like copper (55 ppm) and tin (2 ppm) occur on average in tiny proportions. It would be difficult to justify the money expended on extracting these metals if the average concentration was the rule. Fortunately geological processes, both internal and external, have acted to concentrate these materials into deposits which can be exploited (Fig. 7.4). Thus the tin veins of Cornwall contain 30 000 ppm of tin and major copper deposits up to 5000 ppm. The processes are explained and amplified in Chapter 8, and it is sufficient to note here that the process of concentration has been necessary.

Most metallic ores occur in veins which have formed after the main mass of igneous rock has solidified: hot gases and fluids permeate the rocks of the surrounding country, and the minerals they contain are deposited after chemical reaction with these rocks (**metasomatism**), or as the **hydrothermal** fluids cool so that certain minerals are precipitated out of solution. Fig. 7.5 shows the relationships of this type of mineral deposit to the Cornish granites. The veins so formed are composed of the metallic ores (e.g. cassiterite containing tin, galena with lead, or zincblende) and other minerals, often without economic value at the time of mining, are known as the **gangue** (e.g. fluorspar, barite, calcite, quartz). Some of these gangue minerals, which used to be discarded by the miners, have recently become of value, and old mine tips are being re-worked.

Whilst dealing with metallic ores we must also mention that many are formed by non-igneous processes. Some are formed after weathering has enriched an original deposit (Fig. 4.6), or when deposition occurs in limestone cavities or river or coastal **placers**. Placers are sediments containing sufficient gold, tin, diamonds, or other high-density minerals to make them worth mining. Ironstones are often of sedimentary origin, formed by offshore accumulation. Metamorphism may also lead to the enrichment of the metallic content of a sediment, thus making it worth mining. A combination of processes will also lead to a sufficient concentration, and the origin of some of the world's largest ore bodies is extremely complex. The iron ores of the Lake Superior region in North America were originally banded Precambrian sediments of a type which is common all around the world, but the iron content of 15–40 per cent would not have been enough to warrant extraction: surface weathering or metamorphism has raised this percentage to 55–60 per cent iron. The Labrador deposits, those of Cerro Bolivar in Venezuela, Western Australia, West Africa, Minas Gerais in Brazil, and Krivoi Rog in the USSR – amongst the world's largest – are similar in origin.

Many problems are associated with the mining of ores and other economic minerals. Opencast pits and quarries take away valuable farming land; underground removal of rock can lead to subsidence and damage to housing, or the formation of areas of flooded land at the surface. More scientific planning has sought to reduce these dangers in recent years. Fig. 8.16 shows how one such scheme is applied.

Gemstones are crystals which are rare and resistant both to corrosion and abrasion. They also have high refractive indices and can be cut so that they sparkle. Many of the colours in them are due to rare impurities occurring in normally colourless crystals.

Diamonds are the most valuable gemstones, and for many years South Africa has been the main producer, although the USSR is now the scene of the majority of new discoveries. The diamonds often occur in river gravels, into which they have been washed from unusual volcanic pipes filled with blue-ground, or kimberlite, an ultrabasic igneous rock. The famous Kimberley pipe (South Africa) has been mined to nearly 1300 m and is illustrated in Fig. 7.6. The rock has to be crushed to extract the diamonds, and a huge volume of rock has to be examined for each stone of value. The beauty of a diamond is largely due to its high refractive index, which is used to advantage in the 'brilliant-cut' treatment. Since diamond is also the hardest substance occurring naturally, the poorer quality stones are used for cutting hard metals, and for studding the drill-bits used in rock-borings.

The other main gemstones, and their characteristics, are listed in Fig. 7.7.

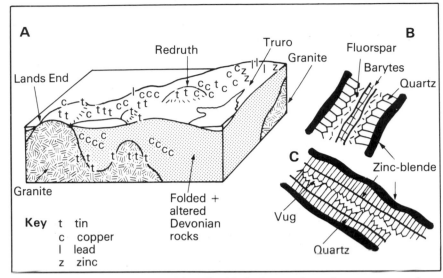

Fig. 7.5 Minerals and igneous activity. The block diagram shows the arrangement of ore-bearing veins around the Cornish granites. Can you detect any zoning? (Which ores occur nearest to the granite?) The upper vein structure (**B**) is known as crustification, the lower (**C**) as comb-structure. Which are the gangue minerals? N.B. Tin also occurs as a vein mineral.

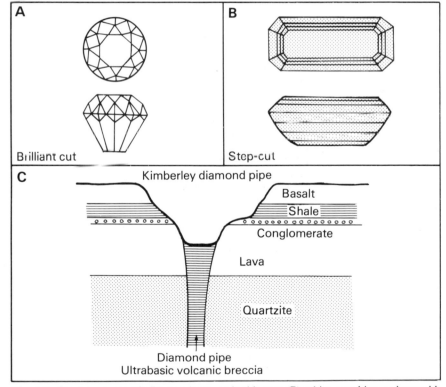

Fig. 7.6 Precious stones. (**A**) Diamonds are cut in this way; (**B**) rubies, sapphires and emeralds are cut like this. The Kimberley Diamond Pipe in South Africa (**C**) has been the world's richest source of diamonds.

Mineral	Composition	Hardness	Density	Colour	Mode of Occurrence	Major Producing Areas
Precious Stones						
Diamond	Carbon, C	10	3·52	Colourless	Volcanic pipes; river gravels	South Africa; Congo; USSR
Ruby Sapphire	Corundum, Al_2O_3	9	4	Red (Chromium impurity) Blue (Titanium)	Metamorphic rocks; gravels	Burma; Siam; Ceylon; Kashmir
Emerald Aquamarine	Beryl, $Be_3Al_2Si_6O_{18}$	7·5	2·7	Green Blue-green	Metamorphosed limestones. Pegmatite veins near granite	Colombia; Ecuador; Peru; South America; Asia
Semi-Precious Stones						
Topaz	$(AlF)_2SiO_4$	8	3·5	Usually yellow, brown	Pegmatite veins with tinstone	Brazil; USSR
Tourmaline	Boro-silicate of aluminium	7·5	3·2	Yellow/green/blue/pink/red	Pegmatite veins	Brazil; USSR; USA
Garnet	Complex silicate	7	4·2	Green/pink/red/yellow	Metamorphic rocks	Widespread
Peridot	Olivine, $(Mg,Fe)SiO_4$	7	4·2	Green	Dark-coloured igneous rocks	St. John's Island (Red Sea)
Zircon	$ZrSiO_4$	7·5	4·7	Colourless (cf. Diamond)/green/red/yellow/blue	Light-coloured igneous rocks, river gravels (gems)	Ceylon
Spinel	$MgAl_2O_4$	8	3·7	Red/sometimes blue	With corundum in metamorphic rocks	Burma, Ceylon
Quartz	SiO_2	7	2·6	Colourless: Rock Crystal Purple (Manganese): Amethyst Brown (Iron): Cairngorm Pink (Titanium): Rose Quartz	Widely distributed in the rocks; gems from pegmatite veins	Brazil; Uruguay; USSR; India; Ceylon
Turquoise	Phosphate of aluminium, copper	6		Green-blue	Amorphous cavity fillings	Persia (in lavas); Mexico; USA
Lapis-Lazuli	Silicate of sodium, aluminium	6·5	3	Ultramarine	Metamorphosed limestone	Afghanistan
Jade	Nephrite; silicate of magnesium, calcite	6·5	3·3	Green	Nephrite needles in schists	China; Turkestan
	Jadeite: silicate of sodium, aluminium			Green	Jadeite in altered feldspars	Burma

Fig. 7.7 Gemstones and their characteristics.

The world distribution of minerals

Many of the processes involved in the formation of mineral deposits – especially those associated with earth interior processes – are related to plate margins, and it is common to associate mineral deposits with such situations. It is not so easy to relate surface processes and the deposits formed by them to plate margins, but it is clear that, for instance, the Cheshire salt deposits were laid down when the British Isles were farther south in the belt of subtropical aridity approximately 200 million years ago. Ancient metal deposits have to be studied in connection with reconstructed continental positions of the time in relation to ancient plate margins. Thus the tin and copper deposits of south-west England were formed along a destructive margin in late Carboniferous times (280 million years ago), but there is no sign of that plate margin today.

Ores, resources and mining

A resource is essentially something which man finds useful in providing him with food, shelter, clothing or the less needful items of the more affluent life in developed countries. Water, soil and metal deposits are all resources. All are limited to the extent that there is a finite quantity available on the earth, but some are recycled and are therefore **renewable**, whereas others are **non-renewable** (once we have used them, they cannot be used again). Thus once we have consumed the world's oil stocks there will be no more left. Water is renewable since it is recycled by the hydrological cycle. There is a gradation between those resources that are non-renewable and those that are renewable. It is possible that coal is being formed in deltaic swamps at the present moment, but the speed at which the process is operating is too slow to replace the coal man is using. Water is renewable, but it is possible to use too much and so reach the limits of what is available. This would interfere with the hydrological cycle so that more water is diverted underground, or evaporation is reduced by a thin film of oil over the ocean surface.

In talking about resources we must also be aware of the **economic factors**. There may be a source of a particular material, but it will not become a usable resource until the costs of extracting it and bringing it to the point of usage will compete with other sources of the same material. The Athabasca Tar Sands of Alberta, Canada, and the oil shale deposits of western USA contain as much oil as has been extracted from the rocks of North America so far. Extraction from the tar sands and oil shale, however, is extremely expensive and so the deposits will not be developed to a great extent until the higher prices can be justified when present sources of oil are depleted. Some minerals have well-known, or

proven, reserves, and may have further potential reserves in similar situations which have not yet been examined. These will be used only as the cost becomes reasonable. We are now seeing, for instance, that the search for new deposits of oil and natural gas is being extended to offshore areas in the North Sea and around the coasts of North America. Costs are high, and are only feasible whilst there is a world shortage of oil and gas.

Man's use of earth resources has increased rapidly in the twentieth century as the world population has grown and the wealth of developed countries has risen, leading to raised levels of demand. We are now getting to the point where questions are being asked about the possibility of limiting the growth of wealth in developed countries so that those in developing countries may catch up a little before the resources are exhausted or the costs of obtaining them become too high. There is also a need to develop a more adequate technology for re-

Gold mining in Alaska, c. 1900. Gold-bearing placer gravels were shovelled into the upper end of the sluice box, and a set of riffles (groups of stones) set across the bottom of the box trapped the gold. (USGS)

cycling resources, so that the crisis of 'exhaustion day' may be postponed.

Metal ores are those metal resources which it is economic to extract from the earth: changing world prices may alter the status of a particular deposit. In the mid-nineteenth century the copper deposits of Cornwall and west Devon were regarded as the world's best ores, but cheaper sources elsewhere and lowered costs of transport brought copper mining to an end in the area before the end of the century. Tin was mined in Cornwall at this period, and nearly met the same fate, but two mines kept going at South Crofty near Camborne and at Geevor near Lands End (the latter with government assistance). Rising tin prices and exhaustion of the cheaper world sources in the late 1960s and 1970s have resulted in an expansion of mining in the area again. The costs of pumping water and the fact that the grade of ore was lower than expected led to closure of two of the recently re-opened mines at Mount Wellington and Wheal Jane in 1978.

The availability of mineral resources and their exploitation are thus to be distinguished: economic considerations are all-important to the mining company which has to invest heavily in sinking shafts and in equipment before it gets any returns. Thus Consolidated Goldfields Ltd invested £6½ million at the Wheal Jane mine in Cornwall.

Mining also means cutting into the earth's surface, or dumping quantities of waste. This has often led to unsightly landscapes, and recently measures have been taken by many countries to ensure that mining companies should return the landscape to some form of productive use (Fig. 8.14). This adds costs to the price of extracting the minerals, and may even make some localities too expensive. We have to make a compromise decision at some stage if we are going to have the use of the raw materials and also retain some of the beauty of the natural landscape as well as continued productive use of the land. Further examples of the use of earth resources will be found in Chapter 8.

Investigation 1: some common rock-forming minerals. Refer to the questions on page ooo and describe specimens of as many of the following common rock-forming minerals as possible: quartz, calcite, augite, hornblende, mica, feldspar, garnet, olivine. Hand specimens of the individual minerals should be compared with their occurrence in the rocks (e.g. quartz, feldspar and mica in a granite).

Construct a chart to summarise your description: you will not be able to complete every section for all the minerals. Write what you observe in blue ink, and what you have added from the chart, Fig. 7.3, in red ink or pencil.

Investigation 2: metallic ores and other minerals of economic value. Carry out the insructions as for Investigation 1 in the case of some of the following minerals (your list may be determined by the school collection or examination syllabus): galena; haematite; pyrite; chalcopyrite; cassiterite; halite (rock salt), fluorite; zincblende; limonite; magnetite; malachite; azurite; barite; graphite; gypsum.

COMPOSITION	FORM	STREAK	COLOUR	HARDNESS	OTHER CHARACTERISTICS	NAME OF MINERAL

8
Studying rocks

Rocks in the field

Rocks are the building blocks from which continents, mountains and the ground beneath our feet are made. They are made up of characteristic minerals, and are themselves often grouped in distinctive associations, so that a consistent study of rocks leads to a reasoned understanding of the ways in which rocks were formed in the past. We need to be able to describe rock specimens accurately and in detail before we can get the most out of this process, and we must always relate the type of rock to its immediate surroundings and associations with other rocks. Some basic questions were introduced in Box 7.1, but this was a 'first look' at rocks. As you become more knowledgeable you will be able to relate what you see to the precise geological terms. It is most important that you do not jump to conclusions, but describe rock specimens as fully as possible before giving them names or deciding on the way in which they were formed. It is also important to know how rocks are formed so that the detailed description can be related to a particular origin: each rock has its own history, and there is little point in merely describing them.

Environments of rock formation

The geological cycle (Fig. 3.6) summarises the major geological processes and emphasises the cyclical nature of many of them. It also shows that new rocks may form under a great variety of conditions: they may originate either when (1) gravel, sand or silt worn from mountains is deposited in layers on the sea floor (sedimentary); (2) when rocks at depth in the crust (or below its base) are subjected to localised heating and then melt, finding their way towards the areas of lower density near the earth's

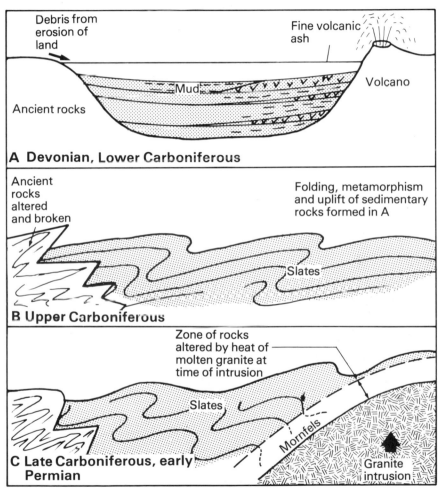

Fig. 8.1 Relationships between the major rock types. **A** The rocks in south and central Devon were firstly muds deposited in a sea which covered the area over 350 million years ago. (The clay, silt and sand of these deposits came from even older rocks, but there is no trace of these at the surface, except, perhaps, in the ancient granites and metamorphic rocks of Britanny.) These muds were compacted by burial and hardened to form shales. **B** Later, about 300 million years ago, they were folded and subjected to high pressure, and this altered the shales to slates. **C** When the granite pushed up into these folded slates about 280 million years ago, the heat destroyed most of the slaty cleavage in a zone up to 500 m wide, and produced a new, flinty type of rock known as hornfels.

surface (igneous); (3) when intensive crushing or heating gives rise to such changes that the original characteristics of the rock are lost and a new rock with new minerals is formed (metamorphic). Rocks from such different environments of origin behave in contrasting ways when subjected to the conditions of the earth surface environment as they are exposed to the atmosphere. Sandstones and shales, formed after earlier phases of prolonged weathering, erosion and transport at the surface, will be subject to only slight further chemical activity. On the other hand a volcanic rock like basalt is composed of minerals which originated within the earth at conditions of high temperature and pressure, and these soon succumb. Whilst a mud or shale may be weak and easily eroded, alteration to slate under high pressure stresses will make it a much more resistant rock.

The different environments outlined above are chiefly responsible for the three major rock types: igneous, sedimentary and metamorphic. Fig. 8.1 is a case study showing relationships between the three types in one area.

(1) **Igneous rocks** are those which have formed from molten, mobile rock material (or magma),
and this often migrates from below or within the crust towards the surface. Basalt and granite are the main varieties of igneous rocks.

(2) **Sedimentary rocks** are those formed in layers on the sea floor or land surface. Sandstone, shale and limestone are the main types.

(3) **Metamorphic rocks** are formed when any other type of rock is altered by intensive heat, or pressure, or a combination of both. Typical examples of metamorphic rocks are slate, marble, schist and gneiss.

Transformations take place as materials are transferred from one rock-forming environment to another. As the rocks formed in the igneous environment reach the surface, weathering attacks them and the mineral composition is changed, whilst the structures are broken down by erosion. Igneous rocks formed of minerals like olivine, augite, feldspar and quartz end up as soluble ions, clay and sand which may be the basis of new sedimentary rocks (limestone, sandstone, shale). If these are dragged down to the igneous environment at a destructive plate margin, the heat rapidly transforms them to molten and mobile magma which recrystallises as new igneous rocks.

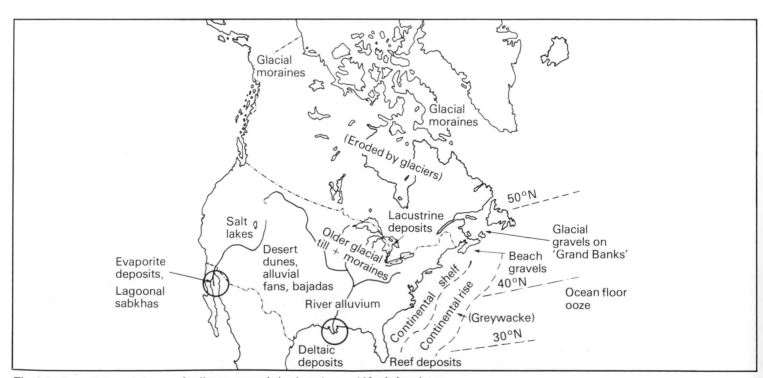

Fig. 8.2   Some of the varied types of sediment accumulating in and around North America at present, or in the recent past.

Sedimentary rocks: product of surface processes

Formation and characteristics of sedimentary rocks

The sedimentary rocks cover 75 per cent of the continental surfaces, and are thus the most accessible group of rocks for study. They are the only group of rocks to contain fossils, which are the main basis of the decisions we make as to whether a rock is older, younger or the same age as one in another region. Most sediments are formed as layers of debris on the sea floor, but some accumulate on the land, and many types may be forming at any one moment in time: desert sands, glacial moraines, river alluvium, lacustrine and lagoonal salts and muds, deltaic and estuarine muds, beach gravels and sands, continental shelf deposits and ocean-floor oozes (Fig. 8.2). Each sedimentary environment has distinctive sediments associated with it; different types of rock formed at the same time, but in a variety of environments, are often known as different **facies** (e.g. reef facies, deltaic facies).

So far we have traced the geological story from the destruction of rocks by weathering and erosion, through the stages of transport and deposition, and have seen how sediments form at the present day (Chapters 4 and 5). Different climatic conditions (with related variations in conditions of weathering and type of erosion), distances of transport and conditions of deposition result in different associations of minerals with varied sizes and shapes. A **'mature' sediment** is one which has been well sorted out by these processes – it has a consistent grain size, the grains are well-rounded (i.e. textural maturity), and few materials easily dissolved by chemical activity are included (chemical maturity). This is, then, a reflection of the extent to which surface processes have modified the original materials (Figs. 8.3 and 8.4).

Another important step, known as **lithification** or **diagenesis**, is necessary before a soft sediment becomes a hard, resistant rock. A series of changes affect the sediment at conditions of low temperature and moderate pressure, and cause alterations over a long period of time.

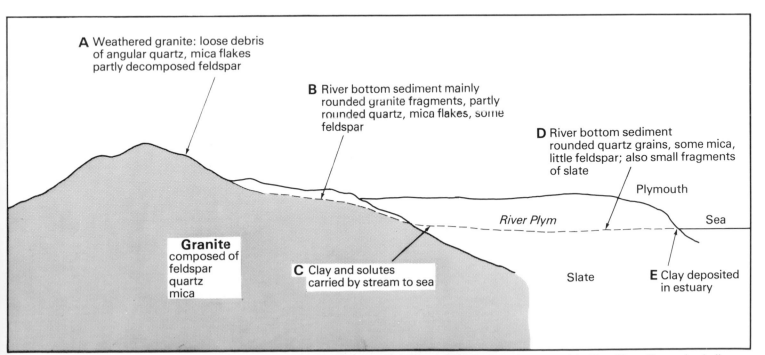

A Weathered granite: loose debris of angular quartz, mica flakes partly decomposed feldspar

B River bottom sediment mainly rounded granite fragments, partly rounded quartz, mica flakes, some feldspar

D River bottom sediment rounded quartz grains, some mica, little feldspar; also small fragments of slate

Plymouth

River Plym

Sea

Granite composed of feldspar quartz mica

C Clay and solutes carried by stream to sea

Slate

E Clay deposited in estuary

Fig. 8.3 Different types of sediment along the course of the River Plym. Draw a simple diagram labelled 'mature' at one end and 'immature' at the other, and place the sediments of **A**, **B**, **D**, **E** in their relative positions. (See main text for definitions.)

(1) **Burial.** Sediments are covered by other layers of sediment, and the weight on top increases.

(2) **Compaction.** The individual grains are packed firmly together until a very compact deposit results having a much lower porosity. The change from mud to mudstone may result in a decrease in thickness to one-tenth of the original, whereas sand may hardly change its thickness in becoming a sandstone (Fig.8.5).

(3) **Dewatering.** Sediments often contain a high proportion of water, some muds holding as much as two-thirds of their volume, including a good percentage of dissolved matter. As other sediments are formed on top the underlying layers are compressed, and this water is squeezed out.

(4) **Cementation.** The material between the grains is known as the **matrix** of the deposit. The chemicals included in the water of the original sediment are often left behind, and may help to weld or cement the grains or shells together. In some cases the cement is introduced later, and is deposited in the pore-spaces. The chief types of cement in sedimentary rocks are calcite, silica and compounds of iron.

(5) **Replacement.** Another process which takes place at this stage is replacement of the original minerals in the rock, especially if that rock was highly soluble. Seawater percolating through a porous limestone will dolomitise it by replacing the calcite with magnesium-rich dolomite.

The processes of burial, compaction, dewatering, cementation and replacement gradually cause the rock to become hard. They may continue until the time when the rock in its turn is attacked by the atmosphere and broken up. We often find that the oldest rocks are extremely hard, whilst the younger ones are soft and poorly cemented.

Sedimentary rocks: evidence of the past. The ways in which sediments are deposited and then altered to sedimentary rocks reflect the processes and the environment in which they were formed. An examination of sedimentary rocks will give us much information concerning these factors. There are certain features for which we should be looking.

(1) The most distinctive feature of sedimentary rocks is the fact that they are normally arranged in layers, known as beds or **strata**. Almost all of these layers were formed in the horizontal position, although sometimes scree may be formed at an angle of 30 degrees to the horizontal. Each bed of rock is bounded at the top and bottom by **bedding planes** (Fig. 8.6), which mark breaks in the deposition and slight changes in the general conditions.

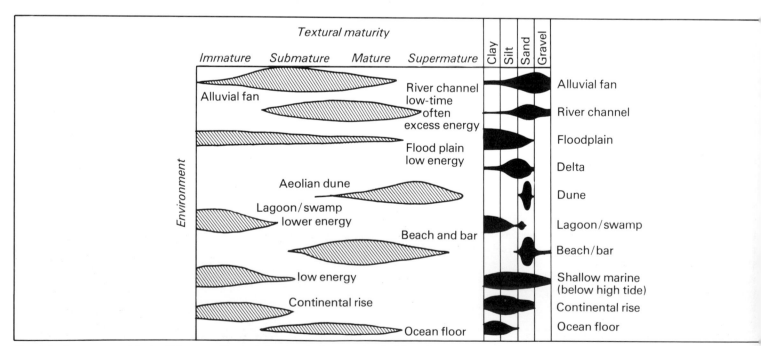

Fig. 8.4 Attempt to explain the differences of 'maturity' and grain sizes associated with particular deposits shown here in the light of your knowledge of their accumulation (Chapters 4 and 5).

The top surface of each bed is often hardened and may be marked by mudcracks, ripple-marks, fossil shells in their living positions and even raindrop pits. It may also be turned over by animals burrowing through the surface sediment in search of food particles. Very fine bedding divisions are known as laminations.

The layers of sedimentary rocks are not always regular in appearance (Fig. 8.7). Some may be strikingly lens-shaped; this usually indicates a highly variable environment of deposition, such as that at the mouth of a river, or on a beach (Fig. 5.12). **Graded bedding** is formed when a whole mass of debris is deposited rapidly, and the heavier fragments sink to the bottom whilst finer particles remain suspended in the water for some time before settling: it is often associated with turbidity current conditions (Box 8.1). **Cross-bedding**, or false-bedding, is found within a bed of rock and indicates the effect of wind or water currents on the movement and deposition of debris. A mass of grains is carried along as a surface carpet, tumbling over each other and rolling or avalanching down the front of an advancing mass of sediment. Greater thicknesses reflect the depth of water in which they were formed, but the thickest cross-bedded units are thought to have been formed by the wind. This internal structure is called false-bedding because it sometimes occurs on such a large scale that the dips of the internal laminae may obscure the true dip of the bed as a whole. **Slump bedding** occurs when unconsolidated sea-floor deposits are disturbed and crumpled, often due to collapse down a steep slope. All of these structures help us to determine the correct succession in a group of rocks which has been turned on end, or upside down (Fig. 10.1).

(2) Within each bed of rock the **sizes, shapes and arrangements of the individual grains** also tell us important facts. The grain size is one factor often used to divide different groups of rocks, although some sediments contain grains of widely varying sizes (e.g. boulder clay). We say that such sediments are poorly sorted: they were deposited too rapidly to allow sorting to take place. In a well-sorted deposit (cf Fig. 8.8) all the larger fragments would have been dropped or broken into smaller fragments, and the finer particles winnowed away, leaving all the debris of a similar size. Such sorting is a feature of sediment maturity.

Angular grains indicate that they have not been transported very far, and rounded fragments are evidence of considerable travel or long periods of

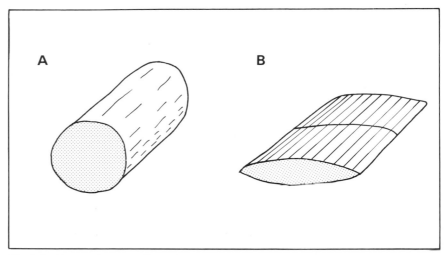

Fig. 8.5 Compaction. The effects of sedimentary compaction on fossil tree stems preserved in sandstone (**A**), and in shale (**B**).

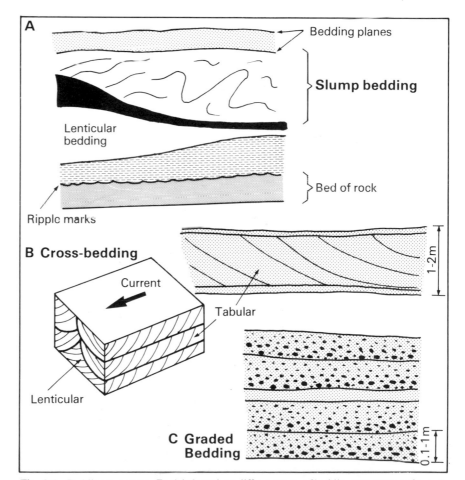

Fig. 8.6 Bedding structures. Explain how these different types of bedding structure are formed.

abrasion on a beach (though roundness also depends on the size and composition of the fragments involved); wind deposits are characterised by rounded, polished sand grains and faceted pebbles. On the other hand oolitic limestones with their characteristic egg-like structure are commonly found in warm, shallow seas. Thus many tiny details have to be investigated to enable us to make the most of our rock studies.

(3) The **composition** of the sediment is also related closely to the conditions in the source area from which it is derived (i.e. parent rock, weathering processes, speed of erosion), and the conditions at the site of sedimentation. A river may be eroding an ancient desert sandstone, and the sediment it deposits in the sea will have a strong resemblance to one formed by the wind because it is still formed of the same wind-polished, 'millet-seed' grains which have been altered little by river transport. Minerals in a rock can often be traced back to a distinctive source region: thus the Millstone Grit rocks of the southern Pennines contain unusual minerals which suggest that they have been derived from rocks similar to those at the surface today in northern Scotland and Norway.

Rapid erosion and deposition will produce a sediment with a wide variety of rock and mineral fragments in terms of composition and size, whilst slower processes will sort out the detritus, leaving behind the larger and heavier materials. Strong currents in the area of deposition will keep all but the coarsest material moving. It is helpful to think of high energy environments with strong currents, in which only material of sand grade and above is deposited, and of low energy environments where the stillness enables mud and dust to fall out of suspension (Fig. 5.17).

Classifying and identifying sedimentary rocks.

Classifications are designed to mean something to those who use them. Several methods of classifying sedimentary rocks are available, but the best one for our purposes is that which can be easily related to our field- and hand-specimen observations.

Fig. 8.9 outlines the classification we shall use here, and yet another approach is suggested at the end of this chapter. As with all rock classifications we are picking out types, between which there are many intermediate varieties: there are very few 'pure' sediments like chalk (99 per cent calcium carbonate). The major component determines the name (e.g. calcareous sandstone, muddy limestone).

We make a basic division between the sediments which are largely composed of material that has been transported before deposition – the fragmental sediments – and those which are precipitated directly on the sea floor. The fragmental components come from erosion, weathering and transport on the land (**clastic**), or from the accumulation of largely carbonate material from shell-bearing and skeleton-forming creatures (**bioclastic**) (Fig. 8.10). If erosion and transport are rapid on the land the clastic material poured into the sea dominates all the other sediments; if the land is low-lying and little clastic material reaches the sea, the slower bioclastic accumulation results in lime-rich rocks.

The sediments formed without horizontal transport are largely of biogenic or chemical origin. Organic action produces a variety of reef limestones, deposits of algal and foraminiferal oozes, most coals and phosphate rocks. Chemical rocks are formed by direct precipitation of iron or salts, or by later (secondary) chemical reactions in which the original materials are replaced. Thus dolomite replaces calcite in limestones and siliceous rocks like flint and chert form as nodular layers after water percolation, whilst aluminous rocks result from chemical weathering.

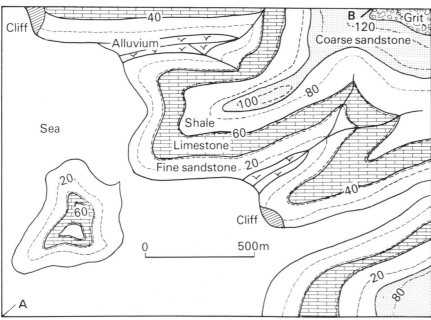

Fig. 8.7 Horizontal rocks. Calculate the thicknesses of the limestone, shale and coarse sandstone. Why can you not work out the thicknesses of the fine sandstone and grit? (All measurements in metres.)

The following treatment of the main rock types recognises the overlapping nature of such a classification. After a consideration of the main clastic groups, the limestones are taken together, rather than treating each type separately as it occurs in the classification. The bioclastic and non-transported sediments are therefore treated according to their major chemical constituent. Three rock groups – shales, sandstones and limestones – make up 98–9 per cent of the crustal sedimentary rocks by volume.

Rudites: conglomerates and breccias. The fragmental group of rocks is divided according to the size of the grains they contain. When most of the grains are over 2 mm across the rocks are known as conglomerates if the fragments are rounded, and breccias if they are angular. American geologists use the descriptive term, sharpstones, when they refer to breccias. Within the whole group are rocks formed of large boulders, pebbles and small granules bound together by a cement which may be either a mineral like silica, or a matrix of fine mud (Fig. 8.11).

Conglomerates are produced when the fragments have been rounded after transport, or after some time on a pebbly beach. We have all seen the results of coastal wave-action, which throws pebbles against each other and soon rounds off the corners: the same thing goes on in rivers though it is not as a rule so obvious. Some conglomerates are formed of one type of resistant pebbles like those of flint or quartzite, which have been well-rounded. They form thin layers in places, and record the slow encroachment of the sea over the land. An example is the Hertfordshire 'Puddingstone' of Tertiary age, which contains small, round flint pebbles cemented by silica. Another type of conglomerate may be formed from a mixture of different rock fragments of all sorts of shapes and sizes, including pebbles of limestone and other rocks which are usually soon lost in transport. These conglomerates form thicker layers, and were the product of rapid mountain erosion, such as occurred in Britain during the early Old Red Sandstone period. These types illustrate the idea of maturity again: the 'Puddingstone' is highly mature in chemistry, and in texture, whilst the ORS conglomerate is not. Maturity is thus a function of the minerals present in the rock and their liability to chemical decomposition, plus the texture of the deposit.

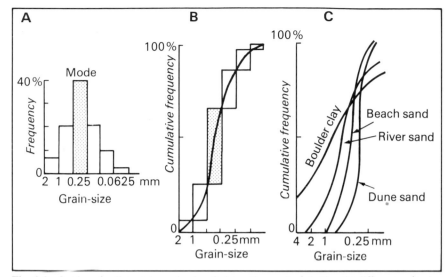

Fig. 8.8 Sorting of sediments by grain-size. **A** shows the result of plotting the different grain-size grades of a sediment after sieving it – a simple histogram. **B** The information shown on **A** is transferred to a cumulative frequency graph, and a curve drawn. **C** The characteristic cumulative frequency graph curves for sediments formed under four different conditions. Can you explain how these differences occur?

Origin of most important components		Rock types	Common names
Transported materials	Clastic	Rudites (Psephites)	Conglomerates, breccias
		Arenites (Psammites)	Sandstones, greywackes, arkoses
		Lutites (Pelites)	Mudstones, shales
	Bioclastic (mainly carbonates)	Calcirudites Calcarenites (+ Oolites) Calcilutites	Bioclastic limestones
		Some coals	Cannel coal
Untransported materials	Biogenic	Reef limestone Foraminiferal, Coccolith limestones Oozes — Calcareous — Siliceous	Biogenic limestones
		Most coals Phosphates	
	Chemical precipitation	Lithographic limestones Dolomite, Magnesian limestone	Chemical limestones
		Ironstones	
		Evaporite salts	Rock salt, potash
		Aluminous deposits	Bauxite
		Some siliceous deposits	Chert, flint

Fig. 8.9 A classification of sedimentary rocks based on their origin and composition.

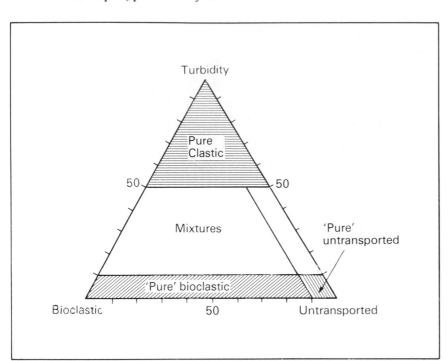

Fig. 8.10 A ternary diagram. This diagram can be used to classify sedimentary rocks, which are all essentially mixtures.

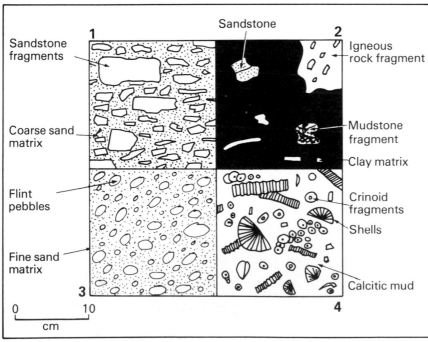

Fig. 8.11 Rudites. Use the information provided on these diagrams representing drawings of hand specimens of the rocks to suggest names for the rocks. How was each formed? (The information available includes size, composition, roundness of fragments.)

Boulder clay has been mentioned as a glacial deposit. It is usually a mixture of pebbles in a fine-grained matrix, though the actual composition depends on the source of the rock fragments contained: the chalky boulder clays of East Anglia are very different from the deposits formed from clay and sandstone rocks in the Midlands; the boulder clay of the Yorkshire coast has a few large boulders set in a chocolate-coloured matrix of clay, whilst the Old Red Sandstone rocks of Pembrokeshire gave rise to a reddish deposit with many more rock fragments. When a boulder clay is compacted it will harden to a **tillite** and the finer matrix may recrystallise as a cement.

Breccias are formed of angular fragments, and are usually found close to the scene of the weathering which led to their formation. Most breccias are composed of scree accumulating at the foot of a mountain slope, like the Brockrams of New Red Sandstone age in the Eden Valley of Cumberland. This deposit contains a proportion of rounded fragments, and is sometimes called a breccio-conglomerate. When a fault moves, the rocks on either side are often crushed or broken, but the fragments may be recemented without travelling at all.

Arenites: sandstones, arkoses and greywackes. The members of this group, in which the main grain sizes are between 2 mm and 0·0625 mm, are often known as the **arenaceous** rocks. Most of them contain minerals and tiny rock fragments which can just be seen with the naked eye, and another characteristic feature of these rocks is that they have a rough feel when the surface is rubbed. The sandstones, and rocks of related grain size, are the largest group of sediments, but as in the case of the conglomerates the division of the fragmental rocks according to grain size masks several vital distinctions (Fig. 8.12). Rocks within the group may be formed under very different conditions, though they all require active currents to prevent finer materials from settling: they are common in high energy environments. We recognise four main varieties:

(1) **Orthoquartzite** is a sandstone composed almost entirely of quartz. Rounded quartz grains are cemented together by silica, which often forms crystal-shaped outgrowths around the transported grains. These grains are usually very well-sorted, and cross-bedding and ripple-marks are common. Whole fossils are rarely found in these rocks since they are either broken up during the formation of the sediment, or are dissolved away during diagenesis. They are called orthoquartzites to distinguish

them from similar quartz-dominated rocks which have been metamorphosed (metaquartzites). Sedimentary quartzites occur as widespread deposits up to a hundred metres thick, and are often associated with layers of limestone and thin underlying conglomerates. The conditions in which they were formed are illustrated in Fig. 8.13A. Present-day beaches provide similar environments. In such conditions only the most resistant grains (i.e. quartz) survive, and the resultant rock is often extremely durable. One British example is the Stiperstones quartzite of Ordovician age, which forms a distinctive ridge in Shropshire. We would thus interpret the origin of this rock as being an ancient beach or shelf deposit.

Many **other sandstones** are also formed largely of quartz grains, but may have a different cementing material such as calcite (Fig. 8.12(2)), or an iron compound. Desert sandstones, for instance, have rounded millet-seed quartz grains coated with red or mauve iron hydrate which helps to cement them together, many of the rocks in the New Red Sandstone group being of this type. The true greensands of southern England are formed of quartz together with a green iron silicate, glauconite, and when the latter is removed by weathering the soils produced are very poor because they consist almost entirely of ash-like quartz sand. The mineral glauconite is interesting because it is formed only in marine conditions, and combined with the frequent occurrence of cross-bedding in the greensands tells us that these rocks were formed in shallow seas, very much like those in which the orthoquartzites were formed, but with a greater proportion of iron salts entering the solution. Most sandstones contain a small proportion of distinctive **heavy minerals**. These are high density metallic and non-metallic minerals, and their occurrence helps us to fix the origin of the sediment debris: deposits in Devon, for instance, containing tiny fragments of tourmaline, would have come from the erosion of the Dartmoor granite. When the heavy metallic minerals increase in the proportion of the rock they occupy so that they are worth mining the whole deposit is known as a **placer**. Examples include the tin-bearing gravels of Malaya and the diamond-bearing beach deposits of south-west Africa: they are both recent deposits washed out of the original veins. The gold-bearing gravels of the Witwatersrand region in South Africa are placers in Precambrian clastic stream channels.

Another group of quartz-rich arenaceous rocks is found immediately beneath many coal seams. These

ganisters are almost pure quartz, but contain a few black rootlets clearly visible in the white rock.

(2) **Arkoses** are sandstones containing a high proportion of feldspar minerals, and very little matrix. With a smaller proportion of feldspar rocks may be known as feldspathic sandstones, and many of these occur in the Millstone Grit of the Yorkshire Pennines, and in the Middle Jurassic rocks of the North York Moors: in each of these cases the rocks were formed in deltaic and fluvial conditions. True arkoses are nearly always associated with products of the erosion of granite and metamorphic rocks of similar composition (e.g. gneiss) forming high mountains (Fig. 8.13B). Normally feldspar is soon decomposed by weathering in humid regions, and sediments containing a high proportion of this mineral (over 25 per cent) must have been formed very rapidly, or in deserts. Arkoses usually contain a mixture of angular and moderately rounded coarse and fine grains, and occur in thick wedge-shaped formations or in thin layers, depending on the rate of subsidence of the area in which they were laid down and are usually cross-bedded. They are thus immature chemically, and poorly sorted. They tend to be pinkish in colour because of the orthoclase feldspar, which is the variety most resistant to decomposition, and is therefore usually the dominant feldspar. The erosion of ancient mountains in north-west Britain led to the formation of Precambrian arkoses, which now form majestic mountains themselves rising above the levelled basement rocks

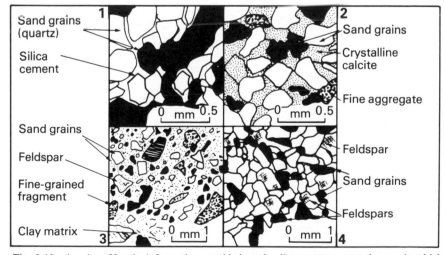

Fig. 8.12 Arenites. Use the information provided on the diagram to suggest the way in which each rock was formed, and to identify their names. The drawings were taken from magnified thin sections of rocks.

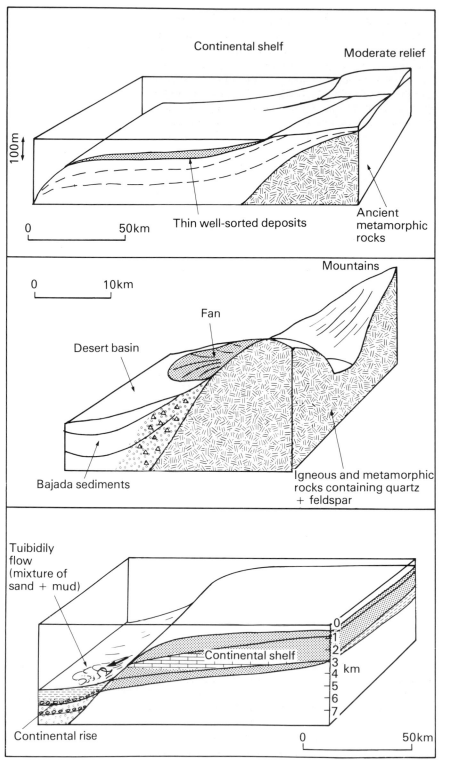

Fig. 8.13 The formation of sandstones. **A** Orthoquartzite; **B** Arkose; **C** Greywacke. Compare the differences of source conditions and the sedimentary environments.

along the west coast of northern Scotland. Thick arkosic deposits, known as **molasse**, were formed to the north and south of the Alps towards the end of the earth movements which resulted in their elevation.

(3) **The greywackes** form another contrast, since they have a high content of fine matrix material made up of clay minerals and tiny flakes of mica or chlorite. This surrounds the sand-size quartzes, rock fragments and feldspars which are characterised by poor sorting and angular shapes. Greywackes are formed where a sediment composed of a mixture of grain-sizes flows in a turbidity current to a point where the flow is stopped and the sediment settles out of suspension. The coarser debris is deposited first, followed by the finest mud in quiet conditions, and this produces graded bedding and uneven mixtures of fine and coarse materials. Greywacke-and-shale alternations are common in central and north Wales, the Southern Uplands of Scotland and central Devon. Seen in field exposures, greywackes are greyish, mottled rocks, weathering to rusty brown, and are characterised by sole marks on the base of each bed: these are formed by the weight of overlying coarser sediment pressing down unevenly on underlying soft mud, and by the movement of small pebbles across the muddy surface during deposition. Although the individual layers of greywacke may be up to 0·5 m thick, the total thickness of greywacke, shale and other associated rocks like volcanic lava and ash, may be of the order of 10 000 m. Present-day sediments of this type are forming in continental rise areas.

(4) **Sub-greywackes** are perhaps the most common of all the arenaceous group, but are not so distinctive as the other types, since they are often a general mixture of minerals. They are mainly composed of quartz grains, plus a proportion of rock fragments and feldspars, but have a smaller amount of clay matrix and more mineral cement than true greywackes. Better sorting, and a greater degree of rounding, characterise the grains, but the colour is similar to that of a greywacke. They are common in this country in the Coal Measures, and many of the sandstones formed between the coal seams are of this type. Current-bedding and ripple-marks are common on the upper surfaces of the rock layers, and where there is a higher quantity of mica present the rocks split easily and are known as **flagstones**.

Lutites: silts, shales and mudstones. The finest detritus and colloidal material includes the silts (between 0·0625 and 0·004 mm) and muds (less than 0·004 mm), and once again the constituents have varied origins. Fine 'rock-flour' comes from glacial scraping, or from abrasion by running water or wind. The new clay minerals have been produced by the chemical weathering of feldspars and ferromagnesian minerals, and are precipitated with other salts where fresh water enters the sea. The silts have a higher proportion of quartz grains than the muds. The minerals in muds are so tiny that they can be distinguished only under very high-powered microscopes. Such fine-grained sediments can settle out only in low energy environments, which exist in currentless lagoons, lakes, or in the sea beneath wave-base. The individual grains cannot be seen when we examine these rocks in hand-specimen, and they have a very smooth surface. The finest-grained rocks are often known as the **argillaceous** group.

As the original mud deposit has the water squeezed out, it changes to a sticky, compact clay, and, as the process continues, to a hard mudstone or laminated shale. Such lamination often makes a shale highly fissile (it will split easily along the laminations). These fine-grained sediments are also commonly subject to slumping. Some special types will illustrate the variety of argillaceous rocks.

(1) **Black shales** are very common in the lower parts of the geosyncline basins that were later filled in by greywackes, and they are plentiful in Wales and the Southern Uplands. They have a high proportion of organic debris, and often contain nodules of pyrite (FeS_2) formed as the sulphur from dead organisms combined with iron salts in the deep, oxygen-free waters. Conditions for the formation of these rocks must have been similar to those prevailing in the Black Sea at the present day (cf Fig. 5.24). There is little circulation of the deep waters (very low energy environment), and so the finest matter settles out of suspension.

(2) **Fireclays** are the fine-grained equivalents of ganisters. Both of these seat-earths occur beneath coal seams, and are rich in alumina (38–44 per cent), making them of use in the building of steel furnace linings. The most important horizon of this type in Britain is the Pot Clay of the Sheffield area: 80 per cent of casting pit hollow ware produced in Britain comes from this fireclay deposit in the Loxley valley. Fireclay is usually light grey in colour without any bedding structures, and is often crossed by carbonised rootlets.

(3) **Oil shales** are black or brown in colour, with an oily smell, and give a curved fracture when broken. They are found in central Scotland, where the hydrocarbons they contain were extracted for some years as a source of oil. There are also vast deposits in western USA, but the costs of extracting the oil are too great at present.

(4) **Ball clay** has a high proportion of the clay mineral kaolinite, formed in the later stages of granite cooling (p. 000). Ball clay deposits, such as those at Bovey Tracey and Petrockstow in Devon, are old lake sediments, where the clay has been deposited after transport by streams from the nearby Dartmoor granite. It is not so pure as the china clay itself, contains quantities of other clay minerals and iron compounds, and is interbedded with lignite. It is mined and used for coarser earthware products.

(5) Other types of argillaceous deposit include **marl**, which has a high proportion of calcium carbonate; **Fuller's Earth**, which is very absorbent and is used in the refining of oils and fats; and the **Alum shales** found in the Jurassic rocks of the Yorkshire coast.

The carbonate rocks: limestones. In most sediments there is a certain proportion of debris which is organic or chemical in origin, but it is only when these elements become dominant that they give rise to a distinctive rock. Such conditions are only found where the supply of clastic debris is reduced to a minimum, and non-clastic rocks are most likely to be produced in times of shallow, widespread seas and slow erosion. Whilst the clastic rocks may be of the greatest importance where quantity is concerned, it is the sediments composed of organic and chemical materials that are the most important economically.

The limestones embrace a large group of rocks of chemical and organic origin, but most of them have a percentage of clay or sand (Fig. 8.14). Most limestones are **bioclastic**, containing transported, organic debris rich in calcium carbonate, and grading into clastic rocks. Some are **biogenic**, representing an ancient reef structure, shell bank or ooze. A third group has a **chemical** origin and is often associated with the formation of evaporite salts. Mixtures are common.

In Britain we have a wide range of limestone types. Four of the most common are described here:

(1) **Chalk** is the purest type of limestone, containing a tiny proportion of sand or mud. It is white and brittle, and is mostly formed of the remains of algae and broken shells. One-eighth of the rock is usually made up of other tiny fossils, such as *Globigerina*, and there are also larger shells of sea

urchins (echinoids), belemnites and lamellibranchs. (Details of all the different fossil groups will be given in Chapter 9.) The chalk of Britain forms escarpments in the south and east of England, and the nature of the rock varies so little throughout the 400 km of its outcrop that the conditions of formation must have been uniform over widespread areas. It is thought that the chalk was formed in relatively shallow waters surrounded by deserts, which would supply very little clastic debris to the sea.

(2) **The harder limestones of Carboniferous age** in Britain include a variety of calcareous (i.e. limy) deposits. There are **shelly limestones** where up to three-quarters of the rock is composed of the remains of sea lilies (crinoids), or corals and brachiopods. The rest of the rock is formed of calcitic mud and ordinary detrital mud. Most of these rocks are grey in colour and weathered surfaces often show the fossils to advantage, since the matrix around them has been destroyed first. Such limestones form thick, massive beds which are well-jointed. They are non-porous, but the joints provide easy routes through the rocks for water, which attacks and dissolves the calcium carbonate. This combination of strength, solubility and well-defined joints leads to the formation of caverns and pot-holes.

Reef limestones are another type commonly found in the Lower Carboniferous rocks. They are formed of the hard skeletons of colonial corals, of algae and of many other fossils which once existed near the living reefs. Such reefs form compact, poorly-bedded and sometimes bulbous masses of hard, nodular limestone in the midst of well-bedded layers, as shown in Fig. 8.15. When erosion takes place they stand out as low hillocks or knolls, which are a feature of parts of the Pennines. The porosity of reef limestones has made them important reservoir rocks for oil fields in many parts of the world: one of the most important is in Alberta, Canada, where small reefs have been dissolved away to provide large voids for oil accumulation.

(3) The Cotswold Hills of Gloucestershire are formed of thinner layers of **oolitic limestone** interbedded with shales. Oolitic limestones are largely of chemical, shallow-water origin: wave action in highly calcareous seas gives rise to the fish-roe-like ooliths. Grains of sand, or small shell fragments are rolled smooth by waves and coated with calcium carbonate (Fig. 8.14(3)). A modern environment where this is taking place is the Bahama islands. The beaches there are formed of oolitic sand. Larger fossils may also be included in these rocks, cemented like the ooliths by more calcium carbonate.

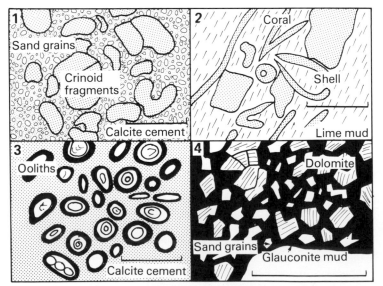

Fig. 8.14 Carbonates. Use the information provided on the diagram to suggest the way in which each rock depicted was formed, and to put forward a name for each. The drawings are taken from magnified thin sections of rocks. The scale line is 1 mm long.

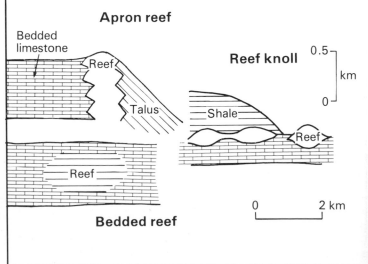

Fig. 8.15 Reef limestones. Notice how these relate to the bedded limestones in each case.

(4) **Magnesium limestone** is an important component of the Permian rocks in north-eastern England. Much of the original calcium carbonate has been replaced by magnesium carbonate, and the rock is a buff-coloured, massive sediment which forms the North Sea cliffs of County Durham, and an escarpment in Yorkshire. When the processes of replacement affect the rock completely all the fossils, reef structures or ooliths disappear, leaving rhomb-shaped crystals of dolomite ($CaCO_3.MgCO_3$).

We thus have many varieties of limestone in Britain. In other parts of the world deposits of **tufa** are found around the mouths of caves from which lime-rich streams emerge. The famous **lithographic stone** of central Germany was formed in the Jurassic age by the hardening of calcitic mud. The deposit is so fine that the tiniest details of fossils entombed in it are preserved, including the imprint of the feathers of the first bird known to man.

Ferruginous rocks: ironstones.

Much of the world's supply of iron ore comes from sedimentary beds of haematite (iron oxide), limonite (hydrated iron oxide), siderite (iron carbonate) and chamosite (hydrated iron silicate), or from metamorphosed sediments. The British supplies of ironstone are now mostly imported, but there is an important band of low quality ore of Jurassic age running northwards from Oxfordshire through Northamptonshire to Scunthorpe near the Humber estuary and thence under the North York Moors. It is a mixture of chamosite and siderite and often has an oolitic texture, suggesting that the iron compounds have been precipitated in chemical conditions which prevented calcium carbonate precipitation. Some haematite is also found, and in places limonite has replaced the original iron compounds. This rock contains between 24 and 38 per cent of iron, and is mined by opencast methods and smelted close to the quarries (Fig. 8.16). Siderite occurs associated with coal in the Carboniferous rocks as blackband ironstone, which in the nineteenth century was the main source of iron ore in Britain.

The famous Lake Superior iron deposits of the USA were all formed from a sedimentary mixture of iron compounds known as taconite. This originally contained only 20 per cent iron, but leaching removed other soluble constituents and left an enriched haematite ore with 50 per cent iron. The recently discovered Labrador deposits have been similarly enriched.

The non-fragmental rocks: salt deposits.

Extensive accumulations of salt are produced by the evaporation of lake and seawaters in very dry climates. Some of the most famous deposits occur at Stassfurt in East Germany (Fig. 8.17), but there

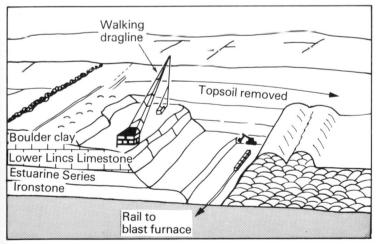

Fig. 8.16 Opencast ironstone mine. Describe the method of operation. Compare the huge Walking Dragline with its boom up to 100 m long with the mechanical shovel tipping the ore into railway trucks. Farming is interrupted for only a few years, and can be restarted by sowing grass as soon as the topsoil is placed on the levelled overburden which has been dumped behind the workings. This is typical of the workings at Corby, Northants.

THE STASSFURT SALT DEPOSITS (PERMIAN AGE)		
Rock-Type	Thickness	Conditions of Formation
Shales, Sandstones, Clays		Wet conditions
Rock Salt (NaCl) Anhydrite ($CaSO_4$)	Variable 30-80 m	Dry conditions, intense evaporation
Salty Clay	5-10 m	Wet conditions
Carnallite ($KCl.MgCl.6H_2O$) + Kainite ($KCl.MgSO_4.3H_2O$) + Sylvinite (KCl) Kieserite ($MgSO_4.H_2O$) Polyhalite ($K_2SO_4.MgSO_4.2CaSO_4.2H_2O$)	>16-40 m	The driest conditions: the most soluble salts were precipitated
Layered Rock Salt and Anhydrite	Hundreds of metres	Seasonal alternations in arid climate
Anhydrite, Gypsum		Dry conditions with evaporation

Fig. 8.17 The Stassfurt salt deposits (Permian age). Relate the succession to climatic conditions and particularly evaporation rates.

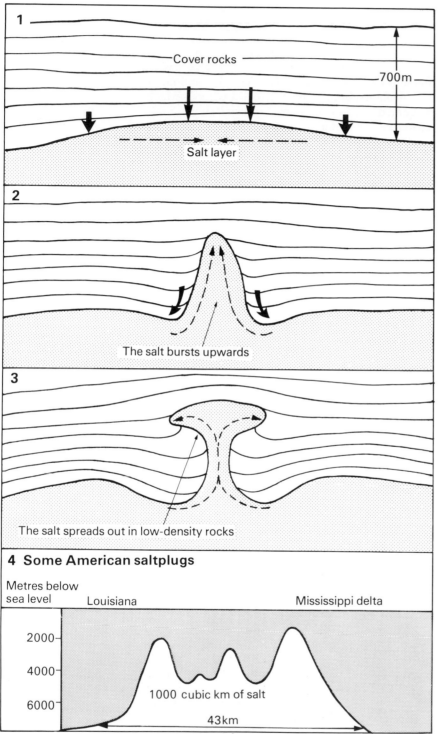

1

Cover rocks

700m

Salt layer

2

The salt bursts upwards

3

The salt spreads out in low-density rocks

4 Some American saltplugs

Metres below
sea level

Louisiana   Mississippi delta

2000

4000

1000 cubic km of salt

6000

43km

Fig. 8.18   Salt domes. Follow the stages 1–3, and write an account of what happens in the formation of a salt dome, remembering that salt has a lower density than the surrounding rocks. The arrows in 1 are proportional to the pressure exerted on the salt layer at different points. Note the vast quantity of salt involved in the American domes.

are also important deposits in Britain near the Tees valley (anhydrite and rock salt), and the Weaver valley of Cheshire (rock salt). Potash deposits have been discovered beneath the North Yorkshire Moors, and are mined near Whitby (Fig. 12.28).

(1) **Rock salt** (NaCl) forms massive layers coloured with yellow or reddish tints by impurities. It is soft and light, and is usually mined by pumping water into the layer and extracting the brine solution. This often leads to subsidence of the landscape.

(2) **Anhydrite** ($CaSO_4$) often occurs as fibrous, granular or more compact masses with a grey, blue or reddish tint. It is harder and less soluble than rock salt, and is usually mined. The ICI mine at Billingham (Cleveland) cut into a layer of anhydrite with an average thickness of 6 m. Passages 16 m wide are cut out in a grid pattern, leaving 16 m square pillars in between: no other support is necessary. The mine is now closed.

(3) **Gypsum** ($CaSO_4.2H_2O$) is rather similar to anhydrite, but is much softer and is used in the making of plaster of Paris.

These are the three main salt deposits, and all are used to a great extent in the chemical industry, where they are made into a variety of products from fertilisers to plastics. Smaller quantities of other salts, including those mentioned in Fig. 8.17, are also utilised.

The weakness of many of the salt deposits compared with the other sedimentary rocks causes them to flow under continuous pressures, even if these are relatively slight, instead of folding or fracturing. Salt is thus often concentrated into large, dome-shaped masses, which are common round the Gulf of Mexico and beneath the North Sea where they have become oil traps. Fig. 8.18 shows how such salt domes form.

There is one great unanswered question concerning the salt deposits: how did they accumulate to such great thicknesses in the past? If you evaporated seawater a thousand metres deep you would only leave a few metres of salt behind. Yet if you look at the diagram of the Stassfurt deposits (Fig. 8.17) you will see that they are too thick to have been formed in a simple phase of evaporation. The Wieliczlic mines in south-western Poland have become a tourist attraction because a whole industrial town has been carved underground in the salt layer which is 400 m thick. The best answer so far is that the area of salt deposition was subsiding causing a continual renewal of inflowing seawater, and that flowage of the mineral led to further concentration of the deposit. It is known that the

Gulf Coast of the USA has subsided by many thousands of metres, and that some of the great salt domes in this area have been calculated to contain 1000 km³ of salt: the concentration of this great quantity probably took place in a combination of the two methods suggested.

Carbonaceous rocks: coal. Coal was the energy source which supported the first 100 years of the industrial revolution. Since 1900 oil has been replacing it to an increasing extent as a source of energy for motive power and heating. Coal is still in demand in the chemical and steel industries, and the main problem of the industry is to discover resources which are economically worth mining. There has been a great increase in open-cast working since 1950, since this requires much less investment in terms of shaft construction and extractive equipment. There is sufficient coal in the rocks of the earth to last for several hundred years at the present rate of exploitation, but this rate is likely to increase towards the end of the century as further supplies of coal are required for conversion to gaseous liquid fuels as the natural resources of oil and gas begin to run out. Coal is likely to have a key role as a buffer-fuel in the future to tide us over the period between extensive dependence on oil and the full development of the technology required for the use of renewable energy resources. This is likely to be even more important if the main energy-consuming nations decide on only limited development of nuclear energy.

Coal is also interesting as an example of a rock which contains conclusive and detailed evidence concerning its conditions of formation. There are several lines of approach to the evidence provided.
(1) The **chemical composition** of coal varies considerably. The most important constituent, carbon, makes up 60 per cent of peat and over 90 per cent of anthracite (Fig. 8.19). Other constituents include oxygen, hydrogen and some nitrogen, and are known as the volatiles, and there may also be variable proportions of materials such as sulphur and mud. The coals with the higher carbon contents have the greater heating values, and are known as coals of higher **rank**.
(2) The **physical characteristic** of coal varies according to the internal make-up: coal may be shiny or dull; dirty and dusty or smooth and clean; it may break into rectangular blocks or may have a more irregular fracture. All these are determined by the microscopic physical constituents known as macerals (cf. the minerals composing the other

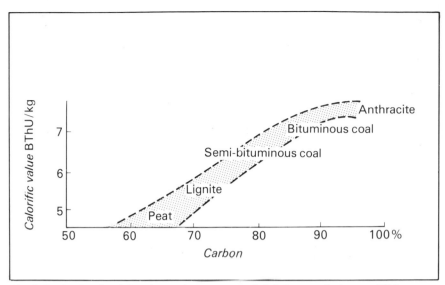

Fig. 8.19 Ranks of coal, based on carbon content and their heat-producing capabilities.

rocks), which are related to the plant matter from which they originated. The most common macerals are vitrinite (shiny, rich in oxygen and possibly formed from woody matter decaying in water), resinite and sporinite (hard, dull material rich in hydrogen, spores and tars) and fusinite (powdery charcoal). The proportion of each affects the appearance and properties of the coal: a high proportion of resinite and vitrinite with a little fusinite gives a shiny coal excellent for heating since it produces little smoke; more sporinite makes a good gas coal. The combinations of these macerals give rise to **coal types**, several of which may occur together in one seam. The soft, bright coal, which is very brittle and easily breaks up into almost rectangular blocks, is vitrain if it is like structureless black glass, and clarain if it is finely layered. Duller coal is durain, largely composed of sporinite and resinite. Larger blocks of coal are often bounded by dusty fusain.
(3) **Coal occurs in layers, or seams**, like most sedimentary rocks. These occasionally reach over 3 m in thickness, but are mostly less than 2 m, and may be mined if less than 1 m if the quality is good. If a seam is traced very far in one direction it may split into two thinner seams separated by sandstone or shale, or it may get thicker. In places seams are interrupted by sandstones occupying a channel through the coal: these are termed wash-outs and are thought to be the courses of old rivers (Fig. 8.20).

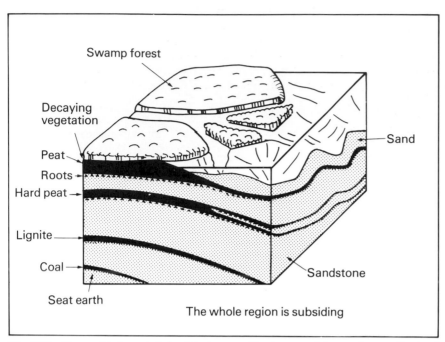

Fig. 8.20 How coal is formed. A flat delta is covered by dense, rapidly-growing woodland and swampy, cut-off lakes. How does deeper burial affect the peat layers? Where will the thickest peat accumulate? (Notice how the peat layers become thinner and split towards the sea, whilst the marine sands become thinner as they pass inland.)

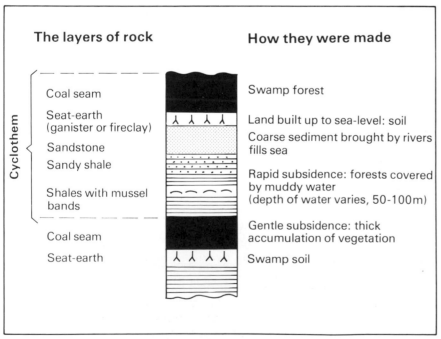

Fig. 8.21 Coal seams and associated rocks. This pattern, with variations, is repeated again and again in the Coal Measure rocks of Carboniferous age. Each unit is known as a cyclotherm.

(4) Coal seams are **associated with certain types of other rocks**. They are often underlain by a seat-earth (the original soil, from which the nutrients have been removed by plants) which may contain the remains of roots: ganisters are sand-grade; fireclays are shales or muds. A shale, or even lime-rich rock, containing marine fossils often lies on top of the coal seam: the fossils in this band may vary from marine animals to freshwater creatures in their environmental indication. On top of the shale the sediment gets progressively, or suddenly, coarser, and thus largely unfossiliferous, and passes into another seat-earth and coal seam. This sequence (Fig. 8.21) is not always repeated in the same form: any unit including the coal seam may be missing. Thus in the East Pennine coalfield of England the lowest group of coal-bearing rocks contains thick, escarpment-forming sandstones between the coal seams; above these are a series of more closely spaced coal seams with only shales between and a decreasing number of horizons containing marine fossils; and at the top there are almost unbroken sands with no coals or marine bands. Coals may occur where there is no evidence of any such succession, and in these circumstances they may contain unusual fossils like fish teeth.

(5) Seams of bituminous coal occur in rocks over a wide **range of geological time**, but only those of Carboniferous age are worked extensively on an economic basis in this country. Nearly all the British coal seams are found in the Coal Measure rocks of Upper Carboniferous age. The cyclothem successions are repeated again and again, and in South Wales the total thickness of rocks formed at this time is over 3000 m. The oldest worked seams are of Lower Carboniferous age in central Scotland, and the only late Carboniferous seams occur in South Wales and east Kent: it seems that there was a southward movement of the coal-forming conditions.

There is also an isolated case of a 1 m coal seam of Jurassic age being worked in northern Scotland, and thinner seams occur in the Middle Jurassic rocks of Yorkshire, but the coal-forming processes were not sufficiently developed at this time to give rise to thicker seams. Most Mesozoic and Cainozoic coals are of the lignite (brown coal) variety.

We can now begin to draw together these various threads of evidence and interpret the way in which coal is formed. The carbon content, plant-like elements and underlying root bed show that coal is formed from massive accumulations of land plants. Plants and trees are largely formed of the carbohydrate plant tissues like cellulose (50 per cent carbon; 44 per cent oxygen; 6 per cent hydrogen) plus various fats, resins, oils and waxes, which are all richer in hydrogen (i.e. the hydrocarbons). The carbohydrates, which form the main body of the plant tissues, are produced, under the action of the sun's rays in the process known as photosynthesis, from the water absorbed through the plant's roots and the carbon dioxide breathed in from the air through the leaves. The resins and fatty substances require in addition mineral salts from the soil for their synthesis: they often serve a protective function.

When a plant dies in your garden it will soon decay completely owing to the action of bacteria and fungi. Their activity is stopped, or reduced considerably, when the supplies of oxygen are cut down. **Peat** is formed in wet, boggy conditions where a very limited amount of oxygen is admitted, and decomposition is confined to the tissue-binding substance and some fermentation of the sugars, forming a carbon-rich deposit with dark brown liquids and jellies permeating the mass together with various gases which escape to the atmosphere:

$$2C_6H_{10}O_5 \rightarrow C_8H_{10}O_5 + 2CO_2 + 2CH_4 + H_2O$$

cellulose   residue   carbon   methane
                           dioxide

If dying plants fall into deeper, stagnant waters, where there is scarcely any air, only the least resistant plant matter decomposes. The organic remains are mixed with fine, slimy mud, and the whole deposit is known as **sapropel**, which has a higher percentage of hydrocarbons like bitumen.

Thus, besides the original composition of the plants, the type of coal will depend on the conditions in which it was formed. The largest quantities of peat are found on lowlying, flat swamps, such as occur on the surface of deltas (Fig. 8.20), where there are also patches of stagnant water in which sapropel forms. The sapropel deposits are usually covered by peat as the pool fills in, or by sands and mud as the streams crossing the delta change course. The weight of all these sediments building up rapidly on the delta causes the whole area to subside, and this allows the individual layers of peat and sapropel to become thicker. They are continuously building up towards the surface of the swamp, but never reach it. The process is interrupted when there is a greater degree of subsidence and the whole area is drowned to become the site for typical marine deposits. The areas in which plants grow most rapidly, and therefore their debris accumulates most swiftly, are those with plenty of sunshine,

high temperatures and heavy rainfall. We thus have to look at the tropical parts of the world to find areas where coal is forming today. Near the equator there is too much rain, and the plant debris is washed away, but the Ganges delta is composed of alternating layers of thick peat and sand such as we might expect.

Bacterial activity ends with the formation and burial of layers of peat and sapropel, but further changes take place in the sediments as they are buried under an increasing thickness of other deposits. The rising pressure and temperature caused by the weight of the overlying matter leads to compaction of the peat, driving off the water and oxygen compounds. This leads to a further increase in the carbon percentage, and the colour changes to black. The whole series of processes finally leads to a hard, brittle coal.

Most of our coals began as peat, but by the time they can be called true coals few of the original plant structures remain obvious for us to see:

(1) **Brown coals**, or **lignites**, are the first in the series of increasing coalification with a higher percentage of carbon than peat. They still retain some moisture and many of the original plant structures. They are only mined where the higher ranks of coal are absent, as in East Germany, or where it is cheap and easy to use opencast methods, as in the Cologne area.

(2) **Bituminous coals** are the most widely distributed and certainly the most valuable. House coals, gas coals, coking coals and steam coals are all in this group.

(3) **Anthracite** may have as much as 95 per cent carbon and burns slowly without a flame, but gives off great heat. It is harder than bituminous coal and is often greyish and shiny. Anthracite is usually only found in the deepest parts of the coalfields, or in highly compressed regions, and it can be thought of as a metamorphic coal.

The sapropel coals are less common and are mixed with inorganic sedimentary matter. **Cannel coal** is mostly formed of the macerals sporinite and resinite, and has a high proportion of volatile hydrocarbon compounds. An unusual feature of cannel coals is that they sometimes contain the remains of fish and other marine life, which are unknown in ordinary coals and suggest transport before deposition. **Boghead coal** is another similar variety, but with more algal remains, giving an even higher proportion of volatile matter: these deposits often grade into oil shales. Coals of this type may be used as a source of gas or oil.

The pattern of sedimentation associated with coal formation and its repetition over and over again is probably due to a combination of the periodic subsidence of the delta area under the weight of accumulating sediment, and the changing positions of the distributary channels. The coal swamp would be drowned and covered by mud and/or sand until the depth of water was filled in and a new swamp established above sea level, whilst the distributaries would shift position horizontally, as in a modern delta (Chapter 4).

Another fascinating point concerns the formation of so much coal at a particular phase of geological history: many variables need to coincide for such an event. Land plants are known first from the Upper Silurian rocks, and the first forests existed in the Devonian period, limited to lowlying, swampy areas by the fact that they reproduced by means of spores. The Carboniferous period saw a full development of these early plants. The lowlying deltaic swamps with a tendency to subside beneath increasing sediments were particularly extensive across northern Europe and central North America during the later part of the Carboniferous period, and were supplied with plentiful rock debris by the erosion of surrounding uplifted areas. There was thus a combination of plant development and geographical environment, which produced a situation where the energy produced by the sun and converted into plant matter could be stored in the rocks of the earth's crust: coal has been well-named as a 'fossil fuel'.

Hydrocarbons in the rocks: bitumens, oils and natural gas. Coal is a deposit which has stored the effects of the sun's energy 250 million years ago and preserved it for use today. Petroleum, whilst not strictly a 'sedimentary rock', was formed by similar processes, and is another vital 'fossil fuel'. One of the great problems facing us at the present moment is that we are using up these fuels at a tremendous rate: it is over 100 000 times the speed at which they are forming today, so there is no hope of enough accumulating to replace those that we are using (p. 138).

The term petroleum includes a vast array of different substances known as the **hydrocarbons**, which include natural gas (methane, CH_4) on the one hand, and solid asphalt on the other. The alkanes (C_nH_{2n+2}) are the most common group found in light-coloured **crude oils** (mixtures of petroleum substances occurring naturally) they include methane, propane and butane, and were formerly known as the 'paraffin series'; the alkenes (C_nH_{2n}) are darker and heavier: they include ethene/ethylene and propane/propylene (formerly naphthene series). These minerals occur in oil shales, where heat is needed to distil them, or in liquid form trapped in porous reservoir rocks.

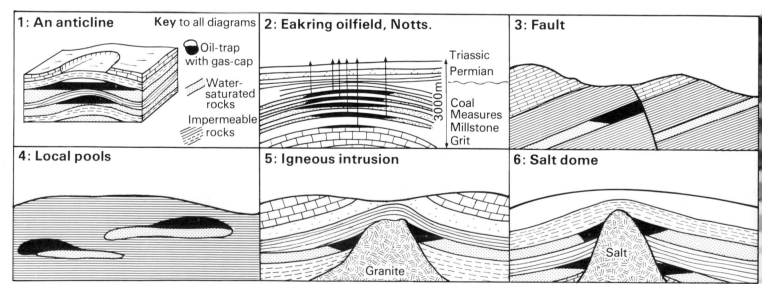

Fig. 8.22 Some oil-traps. Various structures affect groups of alternating permeable and impermeable rock layers and help to trap oil and gas, which rise to the top of any water contained in the rocks.

We can tell a lot about coal and its origin by examining fragments of it, but the crude oils found in the rocks have migrated there, and give few clues as to their origin. These deposits are seldom found near coal, or near areas of volcanic action, and it seems most likely that they were formed from minute marine organisms which decayed over a long period in deep, muddy seas. When the muds were compacted into shales the oil was either trapped (oil shale) in the sediment, or was squeezed out with any water and moved into more porous rocks like sandstones and limestones. The oil separated from the water, and accumulated on top of it because it is less dense; if any gas was present that would form a cap on top of the oil. The oil could be trapped in a number of ways, as shown in Fig. 8.22.

Siliceous, phosphatic and aluminous rocks.

The organic, silica-rich rocks are rare. **Diatomite**, composed of millions of the skeletons of the microscopic diatom plants, and **radiolarite**, formed of equally small animals, are both found as soft creamy-coloured deposits, often formed in old lakes near volcanoes or geysers. **Siliceous sinter** is built up in bulbous layers round the mouths of volcanic geysers. Nodules of silica occur in many rocks. Where these hard masses break unevenly they are known as **chert**, and this type is common in the Lower Carboniferous limestones of northern England. Some chert layers are now known to be composed largely of radiolaria. **Flint** occurs in more compact nodules, varying from grey to black in colour, having a brittle nature and fracturing conchoidally. These nodules are mostly found in the white chalk rocks of south and eastern England, occurring in lines along the bedding planes, but also across them, indicating that some may have been formed after the main body of the rock (cf. Fig. 12.8). Many flints have a silicified fossil as a central core. The source of the silica in these lime-rich rocks is a matter for controversy. The silica may have originated as globules which were attracted together into larger masses, often round the shells of dead animals, in a lime-rich sea. Alternatively, it may have come from percolating, silica-rich, groundwater at a much later date.

Phosphate is derived organically from the bones and teeth of animals, or from their excreta. Thus **bone-beds** and **coprolites** (formed of droppings) are common phosphate-rich deposits. Coprolites are forming today off the west coast of Peru, where the guano of bird droppings is very thick. **Phosphorite** is the main rock containing phosphate,

being a mixture of calcium phosphate and calcium carbonate: it may have been formed in deep, stagnant seas, or by the replacement of limestone as guano-type deposits seeped downwards.

Aluminium is found in all clays, but **bauxite** has been enriched, either by tropical weathering, or by percolating volcanic waters. It is the only rock mined economically for aluminium, and is usually a reddish, mottled deposit of earthy character. The main deposits are in the tropics, but some occur in the USSR, the USA and in southern France (near Les Baux which gave its name to the ore).

Sedimentary rocks and time. As layer after layer of sedimentary rocks are piled up on top of each other they preserve a historical record of the different conditions which affected a particular area. We have seen how this works in our interpretation of the Coal Measure rhythmic successions (Fig. 8.21). The rocks at the base of a series are the oldest, those at the top the youngest. It is unusual, however, for the same type of rock to extend far in any direction. The oolitic limestones of the Gloucestershire Cotswolds, for instance, pass into deep-water marine clays farther south, and into reefs and deltaic deposits to the north (Fig. 8.23).

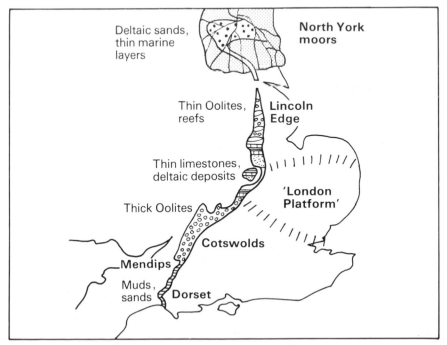

Fig. 8.23 The main Middle Jurassic rock outcrop in England. Note the changing facies from the Dorset coast, through the Cotswold Hills to the Yorkshire coast.

An unconformity at Oban, Argyllshire. The conglomerate lies on top of ancient, highly-folded slates and limestones. Write a short account of the history of this area. (Crown Copyright)

It is thus difficult to extend our dating of rocks from one part of the country to another based on the recognition of the different rock types alone, since the facies may have changed. We are able to compare sedimentary rocks in different areas because they contain similar fossils. This is where we shall begin the next chapter.

When geologists examine very old rocks, which have been subjected to intense mountain-building movements and overturning, they have to begin by determining which is the top and the bottom of a group of sediments. Such rocks do not contain many fossils, and so sedimentary details, like graded bedding and false bedding, ripple-marks and mud-cracks, are used (cf. Fig. 10.1). When they have decided which are the oldest and youngest rocks, the geologists are then faced with the even more difficult task of working out the age of the group of rocks as a whole when compared with other rocks some miles distant. Without fossils this was almost impossible before the development of radiometric dating techniques.

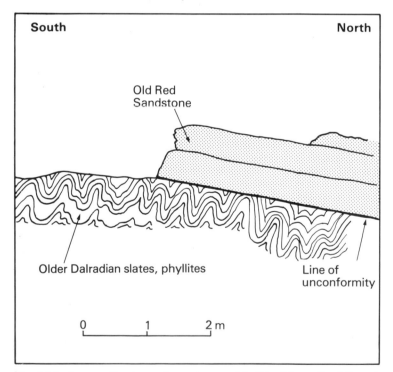

Fig. 8.24   An unconformity on the north coast of the Isle of Arran (Loch Ranza). There is a marked difference between the folded and metamorphosed rocks below the line of unconformity and the gently tilted sandstone above. What happened between the formation of the Dalradian and Old Red Sandstone rocks?

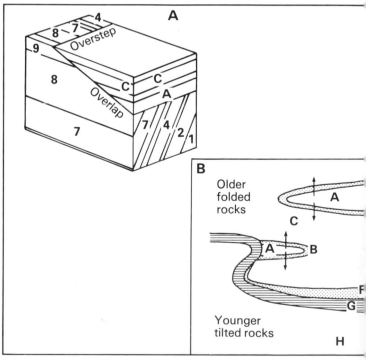

Fig. 8.25   Unconformities: overstep and overlap. Use the block diagram (**A**) and the sketch-map (**B**) to devise definitions of overstep and overlap, which are special aspects of unconformities. What is the difference?

No one area has a complete succession of sedimentary rocks with the most ancient layer at the base and new layers forming today at the top. There have been interruptions in every part of the world because after a certain period of rock formation earth movements have taken place. The bending and breaking, or folding and faulting of the rocks, plus their uplift, lead to a break in deposition followed by erosion until the landscape of hills and mountains is worn right down and the sea comes in over it. New layers of rock are formed on top of the erosion surface planed across the older tilted or folded rocks. Such fundamental breaks between two phases of deposition are recorded in the rocks as **unconformities**, as illustrated in Fig. 8.24.

Figs. 8.25 and 8.26 show some of the features of unconformities and successions of sedimentary rocks as they are represented on geological maps. Answering the questions posed under the maps will help you to understand more about the nature of unconformities.

Sedimentary rocks and plate theory

Most sedimentary rocks have been formed in environments akin to present-day continental shelf and rise conditions. The shelf sediments include most sandstones, clays, shales, limestones and conglomerates. They tend to be formed in relatively thin (a few cm to a few m) layers with a maximum of 100–500 m of the same rock type in a single sequence. Their major characteristic is thus variety. They also contain the major part of the fossil record. The deposits of the rise are more monotonous, but thicker in total (Fig. 5.18).

Other sedimentary rocks are formed on the land, particularly in situations of inland drainage in uplands and beneath the margins of ice sheets. Deep oceans also experience very slow accumulation of fine oozes and red clay.

These situations are related closely to plate movements. Thus the most extensive development of continental shelves and rises is along the trailing edges of continents, and the major sedimentary environments can be related to the features of plates and plate margins (Fig. 8.27).

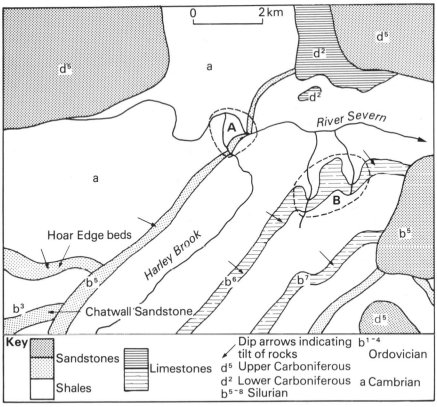

Fig. 8.26 Map exercise. Identify two unconformities, and an overlap and overstep situation. How are the valley outcrops related to the dip directions at **A** and **B**? (Sketch-map based on the Institute of Geological Sciences 1:50 000 Sheet 152, Shrewsbury. *Crown Copyright Reserved*)

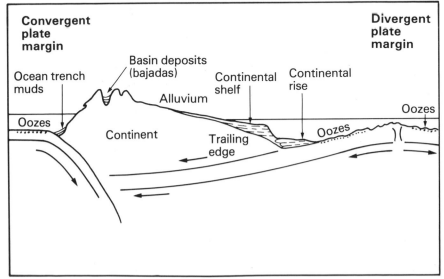

Fig. 8.27 Plate margins and sedimentary rock formation. Most sediments which later become rocks on the continental surface are formed in the continental shelf and rise areas.

Igneous rocks: products of molten, mobile material

Forms taken by igneous rocks

The igneous rocks are those associated with fiery eruptions (*ignis* is Latin for fire), and thus include the volcanic rocks described in Chapter 6. Although we see only what is going on at the surface, a large proportion of molten igneous material is trapped in the rocks of the earth's crust long before it reaches the surface, and solidifies there. Such rocks are known as intrusive, in contrast to the extrusive types which form the volcanoes and lava flows (Fig. 8.28).

One important group of intrusions occurs in large sheets, only a few metres thick, but extending over many kilometres and often found together in great numbers, or swarms. Many of these sheets push their way along bedding planes in between layers of rock, but others take any cracks in the rocks, like faults, and it is thought that sometimes the molten rock forces its way through the earth's crust by permeating and replacing the rocks in its way. **Dykes** are the sheets which cut across the other layers (Fig. 8.29), whilst **sills** are those which push in along the bedding planes and are more or less parallel to the other layers of rock in the area (Fig. 8.30). A **laccolith** is a thickened, lens-shaped sill

mass, often formed of more viscous rock. Dyke swarms are common in the islands along the west coast of Scotland, and are sometimes found in cone-shaped sheets along conical fractures, and sometimes in parallel swarms. The Isle of Arran is crossed by over 600 dykes which have widened the island by 2 km. Sill sheets are also common in this island, but one of the most continuous is the Whin Sill of northern England, which causes a series of waterfalls in the upper Tees Valley, is the foundation for part of Hadrian's Wall, and eventually reaches the Northumberland coast at Bamburgh Castle and the Farne Islands. Most sheet intrusions have solidified more slowly and have larger crystals than the rocks of lava flows, except on the margins, where chilling took place most rapidly in contact with the country rocks (Fig. 8.29). Whereas the lava flows have a lower chilled margin and upper slaggy layer, the minor sheet intrusions have chilled margins on both sides, an important point for distinguishing between lava flows and sills in a group of rocks. The rocks of sheet intrusions are often porphyries, having some crystals larger than the majority: this reflects a two-phase history of cooling. The larger crystals formed more slowly first, and then the rest of the rock crystallised rapidly on injection to its present position.

The more massive intrusions tend to cool even more slowly because they do so at greater depths, and because the retention of the gases trapped in a

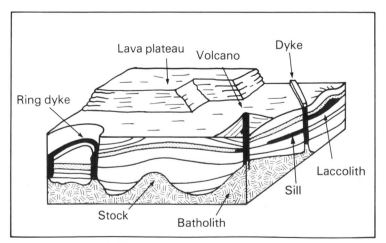

Fig. 8.28   Igneous landforms. The main associations in which igneous rocks are found, and some of the surface features they produce.

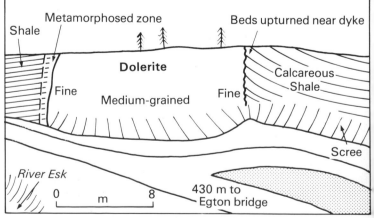

Fig. 8.29   The Cleveland Dyke. The outcrop of the dyke near Egton Bridge, North Yorkshire. This exposure illustrates the variation of grain-size within a dyke, and the effects on the surrounding rocks. Write up your own account of this feature, including the terms: country rock, chilled margin, baked zone, wall of intrusion, discordant junction.

large mass of rock material helps to lower the rate of heat loss. They are characterised by very large crystalline grains, easily seen and distinguished. Because these intrusions solidify at depth they are known as **plutons**. The largest **batholiths** may be several hundreds of kilometres long and up to 150 km wide, and usually push the surrounding rocks aside to such an extent that they are folded, and altered considerably by the heat of the igneous mass. Large fragments of country rock may be broken off in the course of the intrusion, and are incorporated within the magma to form definite patches (**xenoliths**) in the igneous rock, or may be completely 'digested' by melting. Dartmoor, and most of the granite masses in Britain, are parts of batholiths which have been exposed at the surface after millions of years of erosion of the overlying rocks. As the weight of the overlying rocks is gradually released in this way joint cracks open up in the granite in addition to those formed as the rock cooled: the later group is usually parallel to the margins of the intrusion. It has recently been shown that the series of granite masses in the South-West Peninsula are connected at depth, and it is probable that Dartmoor, Bodmin Moor, Land's End and the Scilly Isles are smaller offshoots (**stocks**) of a large, still uncovered batholith.

In some places molten rock has been forced upwards from its deep-seated reservoir by the collapse of a section of the crust, especially in the vicinity of volcanic action. The process is known as **cauldron subsidence**, and happened on the Isles of Mull and Arran. A great slice of crustal rocks subsided, bounded by almost circular cracks, up which the molten material was forced to solidify as **ring-dykes**.

Igneous rocks and time

Igneous rocks become incorporated with a whole series of sediments and metamorphic rocks once they have solidified, and an important task for the geologist is to examine their relationships in time with the surrounding rocks.

The **extrusive lavas and tuffs** can be treated like sedimentary rocks, since they were formed on top of an older layer, and were covered by a younger bed. Lava flows are usually chilled at the base (resulting in smaller crystals), and have a slaggy or weathered upper surface (Fig. 8.30A).

Intrusive rocks were formed after all the rocks which they affect by contact (Fig. 8.30B). They have chilled margins on both sides, and may also

Two lava flows in the cliffs above the Giant's Causeway, Northern Ireland. The columns are at the base of each flow. Notice the difference in the degree of weathering between the upper and lower set of columns. (Crown Copyright)

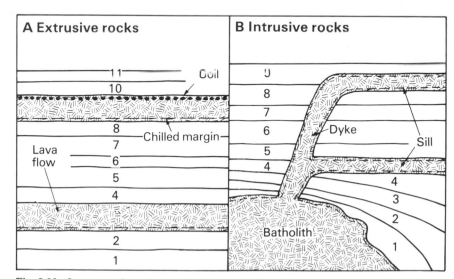

Fig. 8.30 Igneous rocks and time. The igneous rocks can be dated by comparison with the surrounding sediments, numbered according to age (1 is oldest, 11 youngest). The sedimentary rocks would have been heated and altered at the point where they meet the hot igneous intrusions or lavas; the igneous rocks are cooled at this contact – the lavas mainly at the base, but along both margins of the intrusions.

have baked the surrounding rocks. We can thus discover the earliest date for the period of intrusion, but the latest date at which it could have taken place may be less easy to place. In the situation depicted in Fig. 8.30B the upper sill is younger than rock layer 9, but it may be equal in age to layer 10, or to any layer formed at a later date. It is most helpful if a dyke is cut across by erosion and then covered by later rocks: then we can say that the dyke is older than the rocks which cover it.

The origin of volcanic activity and igneous rocks: the magma

The impressive and often catastrophic results of volcanic activity have occupied many scientists in a consideration of their causes. Unfortunately the means for studying the origins of such activity are not within our grasp, as are those for studying the origins of the sedimentary rocks, for all igneous phenomena are the result of deep-seated processes which we cannot observe.

When we look at a map of the world (Fig. 6.12)

we can see that volcanic igneous activity is at present confined to certain areas closely associated with earthquakes, and that large areas of the world are free from both. Even within these major zones there are differences amongst the lavas erupted from the volcanoes. It seems therefore that there is no worldwide reservoir of igneous material, but that each area is subject to local melting of the rocks near the base of the earth's crust, and that there is a certain amount of modification of the molten material as it migrates upwards through the crust. If this is so, we do not really understand how the melting takes place, since temperatures at the base of the crust are probably 500° or 600°C, which is well below the melting point of igneous rocks even at the surface. Thus experiments in the laboratory have shown that a mixture of quartz and feldspar will become liquid at 610°C in the presence of water. We have to assume that some abnormal heating takes place locally in order to make the solid rock into a molten and gaseous **magma**. This magma migrates towards the surface along cracks in the rocks, and is sometimes trapped on the way.

Magmas are silicate melts which vary a great deal in composition.

(1) **Acid magmas** contain a great deal of silica (65–75 per cent SiO_2) and are rich in alkalis (potash and soda), but deficient in iron, lime and magnesia.

(2) **Basic magmas** have much less silica (45–55 per cent), small proportions of alkali elements, but a good deal of iron, lime and magnesia.

(3) **Alkaline magmas**, besides containing a lot of soda and potash, have about 60–65 per cent silica. Much more basic magma than acid or alkaline reaches the surface. It does so at higher temperatures (1000–1200°C) and is much more fluid, whereas acid varieties are cooler (800–900°C), and more viscous and explosive, since they solidify earlier from lower temperatures, trapping the gases beneath the surface until they explode: this factor accounts for many of the features of volcanic activity (Fig. 6.3). The gases contained in a magma are extremely important, and may be equal in volume to the liquid constituents. The gases supply the motive force for moving the magma through the crust, and build up pressure to the point where an explosive eruption takes place. Cotopaxi is an andesitic volcanic peak on top of the Andean ranges in South America, but the pressure of gases ascending 7000 m above sea level was still enough to hurl a 200-tonne block of rock 16 km in 1929. The most intensive volcanic explosions like those of Mont

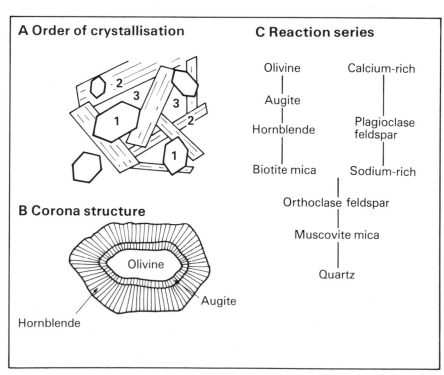

Fig. 8.31 Minerals in igneous rocks. Some of the relationships are shown in **A** and **B**, and the general order of cooling reactions in **C**. The order of crystallisation shown (**A**) is very generalised: compare this simple drawing with Fig. 8.32. The minerals in Corona structure (**B**) increase in silica content towards the outside.

Pelée, Katmai, Krakatoa and Bezimiannyi are accompanied by such great volumes of gas that the magma becomes a froth and eventually disintegrates into a spray as the gases expand in their upward rush (rather like a bottle of pop when you open it after a shaking).

As the magma cools, be it in a surface lava flow, or in an intruded mass, it solidifies by the crystallisation of the rock-forming minerals at different temperatures. Those minerals that are rich in magnesium, iron, calcium and titanium tend to have high melting points and so tend to crystallise before minerals rich in silicon, potassium, sodium and aluminium. Using the principles shown in Fig. 8.31, work out the order of crystallisation in the four igneous rocks shown in Fig. 8.32.

The first minerals to crystallise may float freely in the remaining melt, and the resulting rock will have an even composition throughout. Alternatively, the early crystals may be dense and sink, forming a zone near the base of the igneous body rich in that mineral, or lighter and float to the top. This process is known as **differentiation**. The molten magma remaining after the first minerals have crystallised may react chemically with the first crystals, and if there is sufficient time these crystals will be completely replaced. **Zoned crystals** are formed when the process is not completed and involve mineral groups where some change is possible in their composition without affecting the basic structure: the best examples are the plagioclase feldspars, in which each zone has slightly different proportions of sodium, calcium, aluminium, silicon and oxygen. In other cases reaction with the residual magma may lead to the formation of overgrowths of other minerals, as in the example of **Corona structure** (Fig. 8.31). The order in which minerals replace each other in this way is known as the **reaction series**. The two main branches involve the ferromagnesian and feldspar minerals. As cooling proceeds further new crystals form round the first group, giving rise to a mesh of almost solid rock, and finally the last minerals fill in the interstices that are left. We can usually tell the order of crystallisation by examining the shapes of the minerals which make up an igneous rock. The first crystals to form have good crystal faces (**euhedral**); the second group are formed round the first (**subhedral**); and the last fill in the gaps and have no recognisable shape (**anhedral**). When all the main minerals have crystallised there may still be fluids and gases circulating through the rock, and the last stages in the cooling of a magma involve the

reactions of these fluids and gases with the minerals that have already formed, and their eventual crystallisation in cracks to form distinctive **mineral veins**. These latter stages are most common in the deeply buried plutonic rocks, because the gases escape easily from lavas.

The cooling processes may be interrupted. A magma may have been cooling slowly, forming large crystals, as it moved upwards towards the surface, but may finish the series of events rapidly as it comes into contact with the atmosphere. This will result in a matrix of very fine crystals between the larger ones formed in the early stages of cooling: the example of the lavas on Tristan da Cunha has already been quoted. A rock of this type is known as a **porphyry**, and the larger crystals it contains as **phenocrysts**. Porphyritic texture is also common in the rocks of sheet intrusions. On the other hand two magmas may meet and mix, or some country rock may be absorbed and alter the composition of the magma. This process of **contamination** is important, like that of differentiation, in giving rise to the many varieties of igneous rocks.

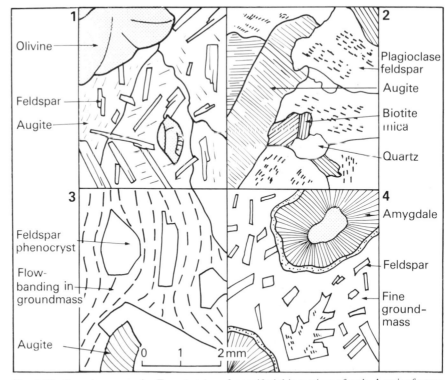

Fig. 8.32 Some igneous rocks. Four sketches of magnified thin sections of rock. A major feature of igneous rocks is that they have an interlocking crystalline structure: note the varying sizes of the crystals in each; which ones have a 'better shape'; and attempt to give names to the four rocks depicted.

The general characteristics of igneous rocks

Before we begin describing the different types of igneous rocks we are likely to find in the field, it will be a help to summarise the main grounds of comparison and contrast. Each rock will have a distinctive chemical composition, mineral composition and textures, and all of these aspects have a special significance, since they are closely linked with the way in which the rock was formed.

Chemical composition. The chemical composition of igneous rocks is dominated by **silica**, which makes up between 40 and 75 per cent of the total in its important position as the basis of the silicate group of minerals. Igneous rocks have been divided according to the quantity of silica they contain, but this is not obvious when examining hard specimens and so we do not recommend such a procedure. It is important to remember that the silica may occur on its own as quartz, and also in combination with other substances in the major group of rock-forming minerals, the silicates.

(1) If igneous rocks contain over 65 per cent of silica they are **acid** and will contain up to 20–25 per cent of quartz.
(2) If igneous rocks contain 55–65 per cent of silica, they are **intermediate** and will contain a small proportion (5 per cent) of quartz.
(3) If igneous rocks contain 45–55 per cent of silica, they are **basic**, and will contain little or no quartz.
(4) Igneous rocks with less than 45 per cent of silica are **ultrabasic**.

Other important chemicals are the oxides of aluminium, iron, magnesium, calcium, potassium and sodium. Water is common in magmas, but usually escapes from lavas as steam, though it may be trapped in deep intrusions and be absorbed in micas and other special minerals there.

Mineral composition. This is largely dependent on the type of magma and on its history of cooling. Each rock has a few minerals, upon which we base its name, and a host of others in very small quantities. The important group are the **essential** minerals, and those of less significance are the **accessory** group.

Rocks rich in silica (acid) are also rich in potassium, sodium and aluminium, and the minerals forming will be rich in these elements, i.e., quartz, alkali feldspar (orthoclase and sodium plagioclase) and the micas. Hence the rock will be light in colour and weight.

As the proportions of the oxides of silicon, potassium, sodium and aluminium decrease those of calcium, magnesium and iron oxides increase, and in the basic rocks (low silica) the minerals to form will be rich in calcium, magnesium and iron, and poorer in silica. These are the heavier and darker minerals (the ferro-magnesian group – olivine, augite, hornblende), plus calcium plagioclase, and so the rocks will be heavier and darker.

You should also realise that the colour of a rock is affected by the size of grains or crystals: in a finer-grained rock the darker minerals predominate in the overall colour, whilst the lighter-coloured minerals show up best in a coarse-grained rock.

Textures. The textures of igneous rocks also reflect the ways and places in which they were formed. The grain size, for instance, is indicative of whether the rock cooled on the surface (glassy or fine-grained), as a minor sheet intrusion (fine or medium-grained), or as a major plutonic intrusion (medium or coarse-grained). Other textures, such

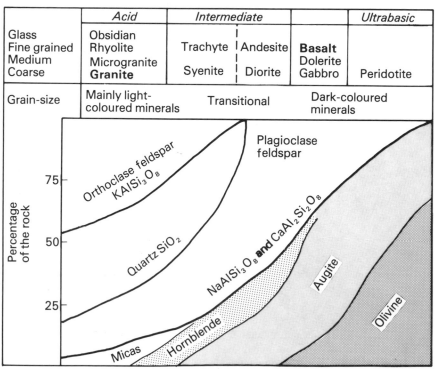

Fig. 8.33 Igneous rock classification, based on grain-size and composition. The darker-coloured minerals are shown in the bottom right-hand corner of the chart.

as porphyritic crystallisation, flow structures and the shapes of the individual crystals, give us further knowledge concerning the cooling history of the rock. Flow structures are where the minerals have a common orientation (their longest dimensions are in similar directions): this is caused when the minerals crystallise as the molten rock is flowing, and is common in porphyritic lavas.

These are the main factors we use to 'pigeonhole' the different types of igneous rock. Over 600 varieties have been named, but these can be placed in the larger groups shown on the chart in Fig. 8.33. As we examine the main groups of igneous rocks you will find that a very important group, the granites, are not mentioned in this chapter. We cannot discuss them until we have studied the characteristics of metamorphic rocks.

The basic igneous rocks: basalts, dolerites and gabbros

Eighty per cent of the lavas poured out on the earth's surface are basalts. Dolerites are the most common rocks occurring in minor sheet intrusions, but gabbros are relatively rare plutonic rocks.

Basalts. These are black, heavy rocks found in lava flows, and sometimes on the margins of minor intrusions. They are the most common type of lava rock, and it has been estimated that more than 5 million km³ of basalt has been extruded from the earth's interior in the last 180 million years. The crystals in basalts are too small to be seen in the hand specimen of rock, which has a dull, broken surface. The only minerals which stand out are the phenocrysts of porphyritic varieties, and amygdales may also be conspicuous. Basalts are formed essentially of the calcium-rich plagioclase feldspar (labradorite) and augite, but they are usually divided into two major groups depending on the proportion of another ferromagnesian mineral, olivine. The first consists of rocks which contain **essential olivine** (up to 20 per cent); these **olivine basalts** are not very common but do occur in the Pacific ocean islands such as the more recent lavas on Hawaii. Some varieties in these areas contain even more olivine because of differentiation which took place in the magma during cooling. The plateau-forming basalts of continental margin regions and many in ocean basins are known as **tholeiitic basalts**, have very little or no olivine, and sometimes even contain a small amount of quartz. A third group of basalts is formed beneath the sea as pillow-lavas, and the rocks are known as **spilites**. The plagioclases have reacted with the seawater, and have been altered to varieties richer in sodium.

Dolerites. Dolerites are similar in composition to the basalts, but have larger crystals, which can just be distinguished in hand specimens and give the rock a small-scale mottled appearance. These rocks occur in sills and dykes. Like basalts they are dominantly dark in colour, and also have a related variation in composition – some are **olivine-dolerites** and others are **quartz-dolerites**. Porphyritic dolerites are common with phenocrysts of olivine, augite or feldspar. The dykes and sills formed of dolerite are very common in areas dominated by basic igneous rocks, and their total volume adds up to thousands of cubic kilometres. Most of our British examples of minor intrusions (e.g. Whin Sill, Cleveland Dyke) are formed of dolerite, and the southern half of the Isle of Arran is riddled with dykes and sills of many varieties of dolerite. One of the most famous sills in the world is the Palisade Sill of New Jersey, USA, which forms steep cliffs looking across the Hudson River to New York. It is an interesting example of the process of differentiation, and the result is illustrated in Fig. 8.34.

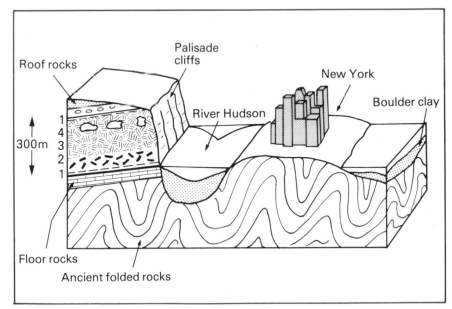

Fig. 8.34 The Palisade Sill, New Jersey. This sill forms cliffs facing New York across the Hudson river. Zone A represents the fine-grained, chilled margins; Zone B is olivine-dolerite; Zone C is coarse-grained dolerite; Zone D has patches of coarse-grained rock rich in feldspar and quartz. Which part is a cumulate rock (i.e. has been formed by differentiation and the sinking of denser olivines and some plagioclase feldspar through the remaining melt)?

Gabbros. Gabbros are the coarse-grained equivalents of basalts and dolerites, and like the latter sometimes have a mottled appearance due to the pale feldspars. The large black crystals of augite show up, even in a completely dark variety, because of their shiny cleavage faces. The gabbros show even greater variety than the dolerites and basalts. Differentiation has led to a layered structure in many of these plutonic intrusions. The Skaergaard intrusion of gabbro on the east coast of Greenland has layers of **olivine-gabbro** at the base, whilst the topmost horizons are formed of **quartz-gabbro**.

The Bushveld Complex of South Africa is also layered, as Fig. 8.35 shows in a simple summary, but is on a much larger scale: it is the same size as Scotland, and several kilometres thick. For some time it was regarded as a saucer-shaped sill type of intrusion, known as a **lopolith**, but this is now disputed. The weight of the rock has caused the area to sag, and the complex is regarded as a whole series of intrusions emplaced whilst earlier intrusions were still molten (there is little evidence of marginal features). The alternative interpretation is that it is a body in which differentiation took place

to enrich the lower layers in heavier minerals and in which contamination took place in the upper layers.

British gabbros are mostly to be found in northeast Scotland and in Skye (the Cuillin Hills), where there is also layering and the composition varies because of differentiation and contamination: in places large xenoliths of shales have been caught up into the igneous mass. This latter process results in patchy rocks containing unusual minerals like garnet. One of the best-known British layered gabbros is Carrock Fell, on the eastern edge of the Lake District.

The ultrabasic igneous rocks

There is a graded group of rocks from the gabbros, which contain a high proportion of feldspar, to the ultrabasic rocks with a very low percentage of silica and scarcely any feldspars. They are composed almost entirely of the heavy, dark, ferromagnesian minerals like olivine, augite and hornblende, and are always coarse-grained. Rocks rich in olivine are known as **dunites**, and those dominated by a mixture of olivine and augite as **peridotites**.

Ultrabasic igneous rocks are a small group occurring sometimes as differentiates at the base of gabbro intrusions, and containing mineral ores such as chrome, platinum and nickel. At other times they are found in larger masses in mountainous areas, and it is thought that the intense movements during mountain-formation led to the incorporation of rock material from beneath 'Moho'. The investigations which have been made of earthquake shock records indicate that the upper Mantle is formed of a substance like peridotite. Rocks of this type occur on Cyprus, and the serpentinites (metamorphosed peridotites) of the Lizard area in Cornwall may have been formed in a similar way.

Other lava-forming rocks: andesites, rhyolites and trachytes

Although basalts form most of the lava rocks, some of the less important types are very distinctive and must have a place in any consideration of igneous activity.

Andesites. The andesites are closely related to basalts. The main differences are due to a higher percentage of silica: the plagioclase feldspar has slightly more sodium in it (andesine), hornblende may be as important as augite, and quartz is more often present in small quantities. Most andesites are almost as dark as basalt and equally fine-grained;

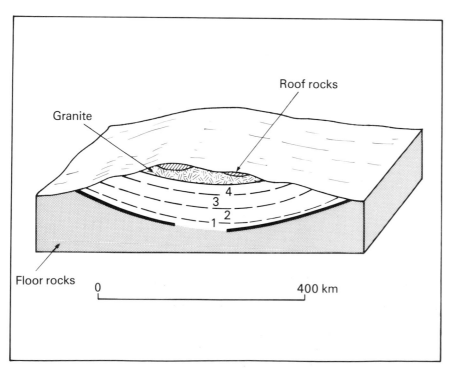

Fig. 8.35   The Bushveld complex. Zone 1: Chilled basic igneous rocks. Zone 2: Critical Zone, up to 2 km thick – gabbro with veins of chrome-rich ores. Zone 3: Main Zone, over 5 km thick – gabbro with veins of titanium-iron ore. Zone 4: Upper Zone, 3 km thick – red rocks including syenite, diorite.

they may be porphyritic and often contain amygdales. The two rocks are difficult to tell apart in hand specimen. Andesite is second in importance to basalt as a lava rock, and is especially characteristic of the mountain-building regions and island arcs around the margins of the Pacific Ocean: its name comes from the Andes Mountains of South America. It occurs with both basalt and rhyolite lavas in these areas. The mountain-building movements must have caused large-scale contamination of the original basic magma with crustal material, because the intermediate and acid varieties are more common than the basic rocks.

Rhyolites. Rhyolites are the most acid group of lavas, and come to the surface in a viscous state. Rapid cooling leads to the formation of a glass, or even to pumice, and these acid lavas are commonly associated with tuffs. Snowdon, the highest mountain in Wales, is built of a pile of largely acid lavas and tuffs of Ordovician age over two kilometres thick. It is now recognised that many acid lavas and tuffs were formed in great Peléan-type explosions, and they are known as **ignimbrites**, formed by solidification from nuée ardente type flows. The main minerals in the acid lavas are quartz, orthoclase feldspar, sodium-rich plagioclase feldspar, and tiny quantities of mica and hornblende. The glassy varieties are **obsidian**, which is shiny black with a conchoidal fracture and common flow-banding, and **pitchstone**, which contains more water and has a dark greenish, resinous appearance. Both of these varieties may contain phenocrysts and the tiny beginnings of crystals, which were 'frozen' as they formed and are known as crystallites. The acid lava rocks occur in small quantities in a variety of conditions, and are most probably formed from the local melting of rocks in the upper layers of the earth's crust as they are affected by large basic intrusions. Pitchstone often forms minor sheet intrusions, such as the sills along the coast of Arran just south of Brodick.

Trachytes. These are alkaline lavas, moderately rich in silica and characterised by their richness in the alkalis so that orthoclase and soda-rich plagioclase are abundant. They are often grey in colour and commonly porphyritic. Such rocks are found around the East African rift valleys, where their formation must have been influenced by the earth's movements. The discovery of unique carbonatite lavas rich in minerals like calcite, which do not normally occur in igneous rocks, in this area, has led to the conclusion that these rocks have been formed from a special magma ascending from great depths. In other parts of the world it is evident that there has been contamination of such rocks by limestone, with a consequent lowering of the silica percentage.

The **medium-grained equivalents** of andesites, rhyolites and trachytes, which sometimes occur as lavas, and more often in minor sheet intrusions, or on the margins of plutonic intrusions, are of small importance compared with the dolerites. They are often porphyritic, and may be known by the main phenocryst mineral (e.g. **quartz-porphyry**). Such rocks are known as 'elvans' in Cornwall, and are used widely for roadstone. In other cases they are related to the coarse-grained members of similar composition (e.g. **microgranite**). Igneous rocks of similar mineral composition, with only differing grain size to separate them, are often grouped together in 'clans', named after the coarse-grained variety: thus granites, rhyolites and microgranites are part of the **granite clan** of rocks.

Patterns of igneous activity

We come to the point where we can summarise the evidence produced by our studies of igneous rocks and their associated landforms. Our conclusions concern the origin of the igneous rocks, their involvement in the major processes affecting the evolution of the earth's surface and the part they play in working out more detailed, local earth history.

(1) The most fundamental conclusion is that igneous activity transfers molten rock, known as magma, from a position at greater depth into the crustal rocks, or even to the surface. Surface activity is manifested in volcanic outbursts.

(2) The scale of volcanic activity varies: there are local cones, e.g. Vesuvius near Naples in Italy, which involve a relatively small quantity of material; there are the vast volcanic domes, e.g. the Hawaiian Island group, which include thousands of cubic kilometres of volcanic rock; there are the extensive basalt plateaus, such as the Deccan plateau in the north-west of the Indian peninsula and the Snake river plateau of north-western USA, which are composed of hundreds of thousands of cubic kilometres of igneous material; and there are the ocean floors, which are now seen as the products of igneous activity and cover two-thirds of the earth's surface. When all these forms of activity are added together it is seen that even the surface volcanic

effects alone contribute approximately 3·5 km³ per year to the crustal rocks; taken over the millions of years of geological time this adds up to immense quantities. When we consider that we live on a planet which is gradually using up its own internal sources of energy, we must at least allow for the fact that volcanic activity has been of greater importance in the past.

(3) The distribution of igneous activity at the present moment, as shown in Fig. 6.12, is confined largely to narrow zones which are also characterised by earthquakes: these are the mobile plate margin zones of the earth's crust, which are separated by much larger areas where disturbances of this nature are few. Compare the map with Fig. 2.4 showing the distribution of ocean-floor features, and the folded mountain belts of most recent origin. Most of the volcanic action is connected with particular sites: the oceanic ridges, the island arcs on the continental sides of the ocean trenches and the ranges of young folded mountains. Other sites of volcanic activity include a more widespread distribution over the ocean floors, rift valleys on the continents and the major basalt plateaus – often near the continental margins.

The ocean ridges seem to be the site of continuous upwelling of basic magmas from the upper mantle: these push apart the ocean floor on either side as they are intruded between walls of rock which have just cooled and solidified. The many ocean volcanic islands and submerged volcanic peaks (seamounts) are sometimes associated with the ocean ridges (e.g. Iceland, which straddles the ridge in the North Atlantic, its new neighbour Surtsey, and the Azores farther south), but are also more widely scattered. These scattered ocean volcanoes include the Hawaiian mass, rising over 10 000 m from the ocean floor. The Hawaiian islands are the south-eastern part of a line of old and now extinct volcanic peaks and seamounts stretching from the central Pacific Ocean to the coast of Siberia (Fig. 5.22). Only the south-eastern part of Hawaii is now active (Fig. 6.1), and it seems that a 'hot spot' (source of magma) has pumped out molten rock for 100 million years, whilst movements of the Pacific plate towards the north-west have carried away the volcanic peaks formed above this spot from time to time. Whilst the lower horizons are normally tholeiitic basalt lavas, upper layers, erupted later, may include olivine basalt or trachyte and rhyolite.

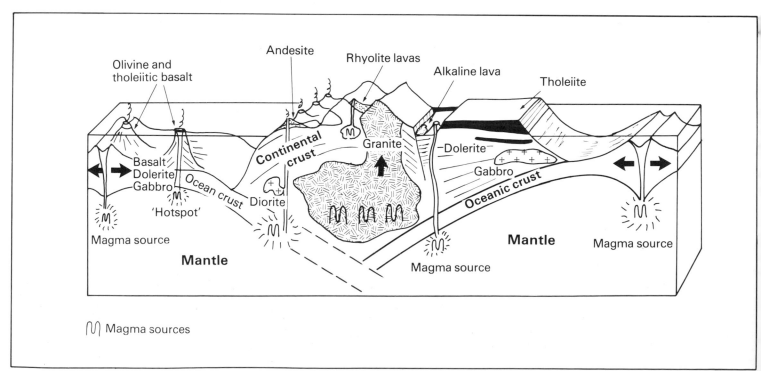

Fig. 8.36 A pattern of igneous activity. Relate the types of igneous rocks to the situations in which they occur. Contrast the igneous rocks produced in the ocean areas with those along the continental margins and in the continental interiors.

Another distinctive volcanic association includes the young folded mountains and island arcs. Volcanic rocks are found amongst those involved in folded mountains: basalt pillow lavas erupted on the sea floor where sedimentation was taking place, piles of andesite lavas like those associated today with the island arcs, and ignimbrites resulting from highly explosive activity taking place into the atmosphere. Since uplift these regions have been the continuing scene of a variety of volcanic activity, involving rock-types which range from fluid basalts to highly explosive, silica-rich varieties. In addition these regions contain larger bodies of igneous material, trapped as major plutons before they could reach the surface. Granites are particularly important in volume, and their relationship with the deeper seated mountain-building activities is discussed more fully in Chapter 10. There are also masses of gabbro, and smaller intrusions, ranging from dolerite to microgranite and silica-rich porphyries.

Other continental regions subject to volcanic activity are the rift valleys, like those in East Africa, where an unusual variety of rocks is produced. Large volcanoes, like Mount Kilimanjaro, and piles of alkaline basalt lava (Ethiopia) are interspersed with smaller volcanoes built of rare carbonatite rock, rich in carbonate minerals. The large basalt plateaus, erupted from elongated fissures in the crustal rocks, form a contrast with the rift valley areas, and show closer relationships with the oceanic lavas.

(4) These distributional features, and the rock-types involved, suggest that the igneous rocks are formed as follows. Almost all magmas must originate near the base of the earth's crust, where the rocks are of basaltic type, rich in olivine. Rocks which have solidified from this magma without any modification are normally only found in oceanic areas, where they have not had to flow through great thicknesses of crustal rocks. The tholeiites may have originated from a magma of that composition, but some have probably suffered a small amount of contamination during their progress through the crust, and andesites, involved in intense mountain-building movements, are contaminated to a greater extent by the crustal rocks, which tend to increase the silica-content of the magma. The reasons for alkaline lavas being associated with the rift valleys is not clear, but there is evidently some relationship with an unusual, deep-seated magma. Acidic lavas are produced from local melting and the formation of a migrating magma from crustal

rocks rich in silica, like sandstones. On rare occasions melting takes place at exceptional depths, incorporating ultrabasic rock material into magmatic fluids. All these phases of igneous activity are illustrated in Fig. 8.36.

Differentiation is most common in the larger intrusive bodies of basic magma, and leads to several distinctive layers within a single intrusion. Most of the large plutonic intrusions, however, are of acidic granite.

Metamorphic rocks

Metamorphism means 'change'. Many rocks have been subjected to different conditions of pressure and temperature from those in which they were first formed, and this has led to the formation of new minerals, new rock textures, and even completely new rocks with no traces of the original. Any previously existing rock may be affected – sedimentary, igneous and even rocks which have been metamorphosed already.

Both metamorphic and igneous rocks are **crystalline** in nature, but the fact that distinguishes them is that whilst the igneous rocks are formed from molten, flowing magmatic material, the metamorphic varieties are recrystallised in place, and whilst still largely solid: they do not migrate at all. Some rocks are mixed in the sense that they are partly altered by reaction with migratory fluids. Granites are found to have relationships with surrounding rocks which suggest that they can be either metamorphic or igneous in origin. In Finland the granites commonly pass gradually into intensely

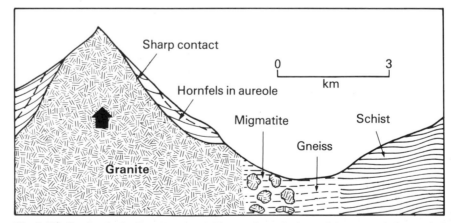

Fig. 8.37 Granites in the Bergallo Massif. The granite at the base of the mountain has typical metamorphic relationships, but that at the top is obviously intruded, and is therefore an igneous rock. This is a schematic diagram of the Bergallo Massif in the Alps.

metamorphosed rocks, and were obviously the result of the same phase of alteration. The Dartmoor massif in Devon, however, is equally clearly an intrusive body of granite, which has pushed aside the surrounding rocks but only produced a narrow zone of alteration caused by the original heat of this igneous mass. There is a striking case in the Bergallo massif in the Alps, where both aspects are combined (Fig. 8.37). The deep-seated fusion of crustal rocks to form the granite was followed by upward injection of the relatively light rocks. We shall return to this subject at the end of the chapter: it is mentioned here to illustrate the essential differences between igneous and metamorphic rocks.

Metamorphism takes place because of changes in temperature, because of increasing pressure and because of chemical reactions caused by migrating fluids. **Temperature changes** may be the most important factor locally, and thermal metamorphism is particularly effective when rocks are in contact with large masses of molten igneous rock, or when they are buried deeply. Rocks take millions of years for any visible changes to take place below 200°C, but as the temperature rises above this figure the reactions become more and more vigorous until parts of the rock become fluid at some stage over 700°C. The minerals contained in the final metamorphic rock give a guide to the temperature at which the alteration took place. The other main agent of metamorphism is **increasing pressure**. As the confining, or hydrostatic pressure increases due to sideways compression, or deep burial, reactions within the rock are encouraged and minerals having their atoms 'packed' more densely are favoured. Garnet is the most common example of such new minerals formed under extreme pressure. Rocks affected by directed stresses as in dynamic metamorphism have their minerals reorientated with their long axes at right angles to the directions of pressure. Recent laboratory experiments have shown that certain rocks under pressure adapt themselves to more confined spaces, whilst others shatter. Finally the **fluids** percolating through and permeating the rocks near igneous intrusions and deeper in the earth's crust may affect the rocks and provide the medium for chemical exchange between the fluids and surrounding rocks (metasomatism).

Some features of metamorphic rocks

Banding, or **lineation**, is a common feature of many metamorphic rocks, in which **cleavage**, or **foliation**, has developed as the minerals are reorientated with their long axes parallel to each other. This is best developed in **schistose** rocks, with their narrow bands of flaky micas. Other metamorphic rocks are non-foliated and **granular**, having almost equal-sized grains interlocking, but without any sign of being arranged in bands or in parallel fashion. The difference between foliated and non-foliated metamorphic rocks is caused by the relative absence or importance of directed stresses in the process of metamorphism.

The grain sizes in metamorphic rocks depend at first on the sizes of grains in the original rocks; thus there are fine-grained slates as well as coarse-grained marbles. As the processes act over a longer period, or react more rapidly, the coarser textures become dominant. **Porphyroblasts** are large crystals which have grown in the solid as metamorphic processes have intensified. They often retain some of the structural features of the surrounding minerals from which they were formed, such as the fine laminations continued in quartz inclusions across a garnet porphyroblast (Fig. 8.40).

The type of metamorphic rock that results from the various changes is related very closely to the geological circumstances in which it is produced, such as nearness to a large intrusion of hot, molten, igneous rock, or involvement in a zone of mountain-building, or deep burial. We shall consider the different rock types in these connections.

Contact or thermal metamorphism

The heat generated by an igneous intrusion affects a band of country rock of varying width. Metamorphic changes take place in this zone, known as the **aureole**. It may vary from a few centimetres wide near minor intrusions to several kilometres around major plutons, and the amount of alteration decreases as the distance from the intrusion increases. The heat from the molten body of rock effects many changes, but the fluids and gases invading the rocks around the intrusion also cause widespread alterations. Pressure plays a very minor role, except locally along the line of contact.

The most common type of rock produced under these circumstances of intense heat and little pressure is **hornfels**, a coarsely granular (i.e. little or no mineral alignment) type of rock in which almost all trace of the original structure is lost. It grades outwards into partially altered, 'spotty' rocks containing clots of new crystals. The fine-grained sedimentary rocks are altered most easily by these processes, and demonstrate most clearly the progression from fine- to coarse-grain size, the increase in hardness, and the higher temperature minerals as the intrusion is approached. As Fig. 8.38 shows, the gabbro intrusion at Insch in north-east Scotland has affected rocks which had already been subject to mild metamorphism in which many of the original clay minerals had been altered to chlorite and white mica. The spotted zone of the new aureole contains tiny masses of dark biotite mica, chiastolite and cordierite crystals, whilst the inner-most zone, next to the gabbro, is a coarse hornfels composed of quartz, feldspar, cordierite and anda-lusite crystals. If the metamorphism had been at higher temperatures another mineral, sillimanite, would have been formed; if lower temperatures had prevailed, the 'cordierite zone' might have been next to the igneous mass. A series of such mineral zones has been recognised, ranging from the low temperature grade chiastolite and cordierite zones, through andalusite to high temperature grade silli-minate: these minerals have the same chemistry, but different forms related to the temperature conditions of alteration. One has to examine thin sections of the rocks under the microscope to determine which of these minerals are present.

Orthoquartzite sandstones and acid igneous rocks are probably affected least by contact metamorph-ism. The sandstones, largely composed of quartz with a little feldspar, may be recrystallised to form a compact and resistant **metaquartzite** with inter-locked quartz crystals, which will have lost most of the original structural details like current-bedding or ripple-marks, though these are sometimes pre-served by lines of mica flakes. Less pure sandstones have new minerals after metamorphism: thus clay minerals are altered to mica. Acid igneous rocks are not affected much, because the temperatures attained in contact metamorphism resemble closely those at which they were formed. Basic igneous rocks, however, show more evidence of change because the temperatures at which metamorphism takes place are usually lower than those at which they were formed and new minerals, stable at the lower temperatures, come into being. Thus at Pitts Cleave quarry, near Tavistock in Devon, older dolerite intrusions have been affected by the Dartmoor granite intrusion and augite has been changed to hornblende.

Limestones recrystallise to white, sugary looking **marbles** if they are almost pure calcium carbonate. Limestones often contain fossils, and these may persist in ghostlike relics with decreasing detail if the metamorphic grade is low. Tiny amounts of

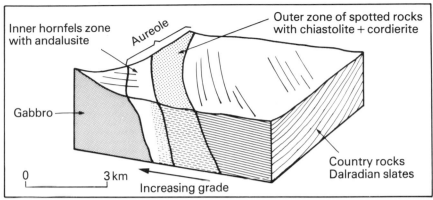

Fig. 8.38 Contact metamorphism. A generalised diagram showing the metamorphic aureole around the Insch gabbro intrusion in north-east Scotland.

impurities may cause red or green streaks, but these white rocks are the only true marbles in spite of the fact that many people give that name to any polished rock. Impure limestones often contain some silica, alumina, iron and magnesia and when metamor-phosed develop a whole group of new minerals such as diopside (one of the pyroxenes) and tremolite (a fibrous amphibole).

Dislocation or dynamic metamorphism

The enormous stresses set up by the directed pressures in earth movements cause rocks to be fractured along fault lines, or to be compressed into folds (cf. Chapter 10). The breaking of rocks often causes a wide zone to be shattered, and under moderate pressures this forms a breccia. Increasing pressure leads to the rock being broken down into fine microscopic particles, which may become mol-ten in extreme cases. These 'crush-rocks' are known as **mylonites**, and involve considerable recrystallis-ation, as well as the distortion and flattening of the more fragile minerals. As in the case of contact metamorphism, however, rocks formed exclusively by high pressure stresses are limited to relatively narrow belts with a maximum of a few kilometres width.

Regional or dynamothermal metamorphism

Whilst contact and dislocation metamorphism are caused by intense heat or pressure affecting the rocks of a relatively small local area, regional metamorphism involves the deep burial of large masses of rock, and its results can be seen over widespread areas. Regional metamorphism is usually associated with mountain-building, where the rocks are buried, and subjected to both high temperatures and intense pressures. The products vary according to the depth of burial, and various intensities of action can be distinguished mainly on the basis of temperature.

(1) **Low-grade regional metamorphism.** This involves considerable pressures, but relatively low temperatures. The resistant rocks, like sandstones, are altered gradually to more compact rocks with an interlocking granular structure. Many meta-quartzites or marbles formed in this way show evidence of the stresses in deformed crystals or fossil relicts, and a degree of flakiness or preferred orientation of quartz crystals. As in contact metamorphism the fine-grained shales and mudstones, containing minerals stable at low temperatures (e.g. the clay minerals), are most affected. They may be recrystallised completely to phyllites containing new flaky minerals (mostly chlorite and mica). The rock is brittle and has a surface sheen. Organic matter is broken down to specks of graphite or cubes of iron pyrites.

Cleavage planes at right angles to the direction of compression are formed when fine-grained rocks are subjected to intense pressure (Fig. 8.39). **Slates** are dull grey rocks which may have a greenish or mauvish tint, and they cleave along fine parallel planes. This made them in great demand for roofing in the past, because they are also tough. Most of the roofing slates came from great quarries in the

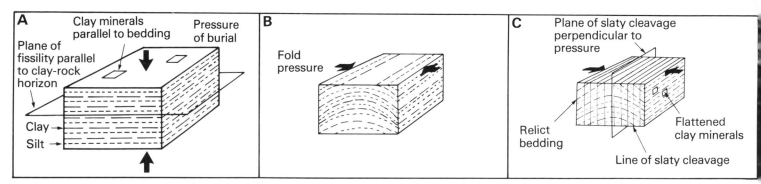

Fig. 8.39   The formation of cleavage in slates under the effect of increasing pressure stresses. **A** A fine-grained rock develops fossility due to the pressure of overlying sediments – a shale (sedimentary rock). **B** Directed pressures cause folding of the sediments, but the clay minerals are still parallel to the bedding planes. **C** Increased directed pressures lead to a re-orientation of the clay minerals with their flattened at right-angles to the pressure direction. This forms the slaty cleavage in rocks and may include recrystallisation.

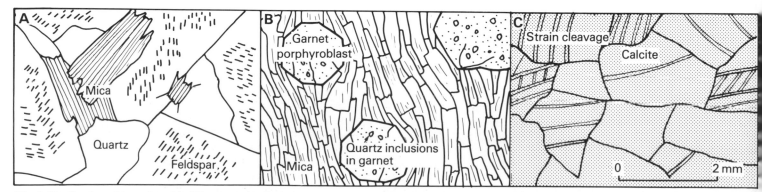

Fig. 8.40   Granite, schist and marble. Three magnified thin sections of the rocks: **A** granite, **B** garnet-mica-schist, **C** marble. Can you explain the differences between the shapes of the mica crystals in **A** and **B**? What are the differences between **C** and a sedimentary limestone (Fig. 8.14)?

Snowdonia area of north Wales, where the sediments and lavas, which had otherwise been unmetamorphosed by the mountain-building movements at the end of the Silurian period, were squeezed up against the resistant block of Anglesey. The fine Cambrian shales were altered to slates in this process.

(2) **Medium-grade regional metamorphism.** Rising temperatures cause the increasing loss of relicts of sedimentary and igneous structures as the rocks become more plastic and develop strong foliated structures (Fig. 8.40). The fine-grained sediments form **mica-schists** with very thin alternating layers rich and poor in mica. As the degree of alteration increases, porphyroblasts become more common, with the garnets 'growing' in the rock, and garnet-mica-schists, having large red crystals in the shiny micaceous rock, are often found. Even at this stage the resistant sandstones take on only a very rough schistose texture: the quartz grains of these rocks recrystallise in an interlocking mesh and quartz-schists or granulites are formed. British examples of these rocks can be seen in Anglesey and in the Scottish Highlands. Basic igneous rocks are also recrystallised, resulting in **hornblende-schists** (greenschists) or **amphibolites**.

(3) **High-grade regional metamorphism.** The most intense metamorphism causes the rocks to become plastic with free recrystallisation. In the earlier stages the recrystallisation took place within the framework of an almost solid rock, but under the most extreme conditions the rock is completely transformed. **Gneisses** are coarse-grained, roughly banded rocks and often have alternating streaks of dark- and light-coloured minerals. The darker layers have a greater degree of schistosity, being composed of hornblende and mica, and the lighter part is more massive quartz and feldspar. Gneisses often grade into granites, and are associated with migmatites (described below).

When regional metamorphism takes place large areas of many square kilometres are affected, but there are usually patches of more intense metamorphism in the centre and a gradation outwards to low-grade rocks. Thus there may be a change from fine slates on the margins to schists containing garnet porphyroblasts, and then to coarse gneisses in the centre. As in contact metamorphism there is a series of recognisable zones.

Regional metamorphism is associated with mountain-building, but some areas show evidence of more intensive alteration than others. The Snowdonia area of north Wales has already been referred

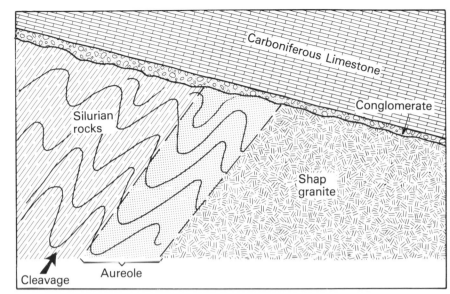

Fig. 8.41 The age of metamorphism. Two phases are shown here. In the first the Silurian rocks were folded and the fine-grained rocks were altered to form slaty cleavage. A little later the Shap granite was intruded and produced the zone of contact metamorphism, or aureole. Suggest a date for each phase and compare this situation with Dartmoor (Fig. 8.1).

to as an area of low-grade action, and a similar result was produced by folding at a later date in Devon and Cornwall. Schists and granulites predominate in the Scottish Highlands. It is only the very ancient 'shield' areas of Scandinavia, northern Canada and Africa that are largely made up of gneiss and granite. The features of two metamorphic rocks can be examined in detail in Fig. 8.40.

Metamorphic rocks and time

The date of a period of metamorphism is normally assessed by working out the relative ages of the rocks affected and overlying unaltered sediments (Fig. 8.41). The separating unconformity provides a major line of reference in such studies. In north-west Scotland the most ancient rocks are highly metamorphosed schists and gneisses, and are covered, above an unconformity, by Precambrian sediments. The metamorphic rocks are thus of early Precambrian age. The alteration of the aureole around the Dartmoor granite must have taken place after the Upper Carboniferous period (the age of the rocks affected by metamorphism), but as there are no rocks of later date in contact with the granite, or lying near it without any alteration, it is difficult to give a final date for the ending of the metamorphism without using the dates provided by radiometric dating (Chapter 11).

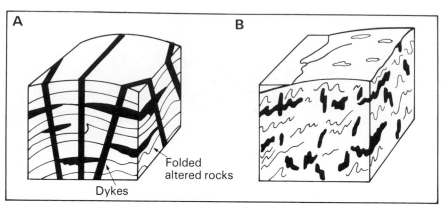

Dykes

Folded
altered rocks

Fig. 8.42 Polymetamorphism. **A** is a reconstruction of a series of ancient Lewisian rocks in north-western Scotland before the last phase of metamorphism – they were already folded and altered. **B** The last phase of alteration and folding led to the disruption of many of the older features of the rocks.

Polymetamorphism means that an area has been subjected to more than one phase of alteration, and that these can be detected from a study of the minerals and other structures contained in the rocks. Geologists who make studies of ancient, altered rocks are dependent on the preservation of distinctive minerals, and there are often cases where an area is affected by a second phase of metamorphism similar in grade to the first and this results in few, if any, changes. Some parts of the extreme north-west of Scotland show evidence of several phases of regional metamorphism and it is supposed that there have been as many periods of mountain-building. One example of this is shown in Fig. 8.42, which is a good illustration of how the structures of complex rocks can be sorted out, and can give us a lot of information about the history of an area.

If a region of gneiss rocks is affected by low-grade alteration the rock will be modified, but usually some of the high-temperature minerals like andalusite or sillimanite will remain to record the earlier phase of high-grade alteration, and the process is known as **retrograde metamorphism**. Fig. 8.43 provides an example of dating different rock types.

Migmatites ('mixed rocks')

Migmatites are highly metamorphosed rocks which have an added component. This has resulted from the migration of magmatic fluids circulating through the rock pores in an extremely fine state. Most of these rocks have a streaky or patchy appearance, and are characterised by fold structures which testify to highly plastic conditions at the time of formation. The actual rock-type depends to a large extent on the original 'host-rock'. Thus former sandstones and shales will be richer in quartz and mica than the basic igneous rocks, which will have more darker patches. All these rocks, however, with the addition of the acid quartz–feldspar mixture, become richer in potassium feldspars.

The migmatites occur mostly in the centre of the most intensely metamorphosed regions, where the effect of the migrating fluids is obvious, but has not dominated and completely altered the rock by being involved in the highly plastic deformations such as produce gneisses or granites.

Granites

Granites and the related granodiorites are twenty times more important than all the other plutonic rocks. They have always been regarded as igneous rocks, and are included in the classification of the igneous rocks adopted in this book (Fig. 8.33). The fact that granites are nearly always associated with areas of mountain formation, and are often an integral part of a group of regionally metamorphosed rocks, has caused lengthy disputes to arise concerning their true position and origin.

All granites are similar so far as rock characteristics are concerned. They are all largely formed of quartz and potash feldspar, with some plagioclase, biotite and muscovite. When the plagioclase is a little more abundant and a little more calcic, and the amount of biotite increases, the rocks should strictly be called granodiorites – which are actually commoner than true granites. Granites are mostly coarse-grained and granular, the grains being of roughly the same size and having irregular outlines: it is often difficult to distinguish which cooled and crystallised first. Occasionally a complex interlocking of some crystals shows that they became solid during the same phase of crystallisation. In hand specimen granites are amongst the easiest rocks to identify, as the crystals are large and stand out plainly. The dominating white or pink feldspar (up to 75 per cent of the volume of some varieties), the greyish-looking but glassy quartz (usually about 20 per cent of the rock volume), and the shiny, black mica grains, are all distinctive. The appearance of a granite under the microscope can be seen in Fig. 8.40. Some granites are porphyritic, like the variety from Shap on the eastern edge of the Lake District. This rock has large pink phenocrysts of orthoclase feldspar set in a mass of small crystals of white plagioclase feldspar, grey quartz and black mica, and is quite unique: it is of particular use when found as a glacial erratic.

When most of the granitic material has crystallised there are still mineral-rich liquids and gases circulating through the deeply buried mass, and these finally solidify in a distinctive series of deposits which form veins across the granite or in the surrounding rocks. **Aplites** are medium- or fine-grained granitic veins, and **pegmatites** are very coarse-grained. Exceptionally large crystals of mica and feldspar have been obtained from pegmatites, which must have continued to be mobile for a long time. Crystals of orthoclase the size of a house have been found in Norway, and micas up to 3 and 5 m across are reasonably common in South Africa. The reasons why aplites and pegmatites differ are not fully understood, especially as they can occur together (e.g. in the Tregonning granite of west Cornwall). It is most likely that the pegmatites crystallise from a magma rich in volatile gases, whilst the aplites crystallise from magma fractions poor in volatiles. Other liquids permeate the surrounding rocks, and often form mineralised veins containing copper, tin, tungsten and other metal ores (Fig. 7.2). The final stages of magmatic activity may be accompanied by the alteration of the feldspar minerals by gases. The kaolin (china clay) of the granite moors near St Austell in east Cornwall was formed in this way.

The main distinctions between granites are not so much in their composition and rock characteristics, as in the circumstances in which they are found. Some are closely associated with metamorphic rocks, since they grade into migmatites and banded gneisses; others are just as certainly intruded into a group of other rocks which they have pushed aside and altered by heat action; and in many cases the granites have made room for themselves by replacing the original rock (stoping). Nearly all the granites, however, are associated with areas of mountain-building in the recent or more distant past.

It seems that the **metamorphic granites** were the original granites, formed in such conditions of migmatic injection, heat and pressure, that a series of crustal rocks was completely broken down to a state where they became virtually molten. Such rocks have diffuse margins with surrounding migmatites and gneisses, and show no evidence of intrusion. They mingled together to form a rock composed of quartz, alkaline feldspar and mica, having a granular crystalline texture, but no banding, as a result of flowage, pressure or partial injection. We cannot tell what the previous rocks were like. It is possible for any type of rock to be

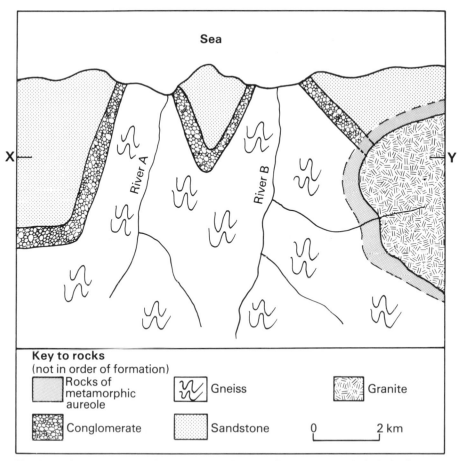

Fig. 8.43 Give an account of the geological history of this area, which includes igneous, sedimentary and metamorphic rocks. Draw a sketch section from X to Y. What rocks would you expect to find in the metamorphic aureole?

altered to a granite after millions of years of additions of granitic material until the host is eventually unrecognisable. As granitisation takes place the incoming fluids react with the rock, leaving the acid, granitic material and removing any basic constituents. These processes take place in rocks which have been dragged down to great depths during mountain formation (cf. Chapter 10). This phase is followed by the uplift of the rocks into great ranges of folded mountains, and as this happens there are large-scale migrations of the granitic material, which is relatively light, into the rising mountains where they are emplaced as **forceful intrusions**. Such granites have discordant margins with the country rocks, and also metamorphic aureoles. They are often emplaced in rocks which have been subjected to low/medium grade metamorphism (Figs. 8.1 and 8.41). Some granites have been found outside areas of mountain-building: they also have discordant margins, but narrow aureoles and are intruded into unaltered rocks. Some are associated with **ring-dyke structures** (e.g. on the Isles of Mull). Still other granites may be associated with a series of volcanic rocks resulting from basic magmas, and are probably formed as the residual magma (after differentiation has removed all the basic constituents) solidifies (e.g. Isle of Skye granite).

Diorites and **syenites** are two other rocks found in plutonic associations, but only occur in minor quantities compared with the granites, or even the gabbros. Diorites are akin to the andesites in composition; syenites to the trachytes. They are most probably formed by the contamination of granitic or basic magmas, and many of these rocks can actually be traced into large masses of granite from which they are offshoots. They are coarse-grained, and can be distinguished from granite only by the absence of quartz in hand specimen. Syenites often have pinkish feldspars, whereas diorites usually have white plagioclases and a large proportion of dark minerals (e.g. hornblende or augite).

Metamorphic rocks and plate theory

Metamorphic rocks are formed in response to compression and locally excessive supplies of heat energy. These characteristics are found mostly in zones of plate convergence in association with fold mountains (Chapter 10) and major igneous intrusions. Attempt to produce your own diagram related to the plate theory model (Fig. 2.7), which summarises the main areas at which the three major groups of rock are formed.

Energy and fuels

Energy is the key to all activity, not least to what man is able to accomplish on this planet. The earth has three major sources of energy – the sun, the oceanic tides, and the earth's own interior heat flow. Most of the energy used by man to date has come from various forms of stored energy, or fuel (e.g. timber, coal, oil, natural gas and uranium). The source of energy stored in many of these is the sun, since the energy trapped in the process of photosynthesis (wood and coal) is also passed on through the food chain (oil). Fuels like coal, oil and natural gas, which have been stored up over millions of years in the earth's crustal rocks, are often known as 'fossil fuels': at present 95 per cent of the world's energy production comes from such resources.

World demand for energy is increasing rapidly (Fig. 8.44), and it is estimated that the world will be running out of oil and natural gas by the end of the twentieth century. Coal reserves are greater (several hundred years), but there is going to be a need for other sources of energy by the year AD 2000. The use of oil and gas exemplify some of the problems to be met. By the 1970s the world has become dependent on the largest producers in the Middle East, and this has had important political implications. New technology has continued to make new sources available, as in the North Sea area. It is expensive to extract oil from beneath the North Sea, but the high prices demanded by the producing countries elsewhere have made it a profitable venture. It is to be expected that other offshore deposits (e.g. off the eastern coasts of North America) will also be exploited. Still more costly sources of oil are available. The tar sands of Alberta, Canada, are loosely cemented sands saturated by asphaltic hydrocarbons, and oil is already being extracted from these deposits. It would require an extremely heavy investment, however, for these deposits to make a major contribution to supplies. Similarly the oil shale deposits of western USA contain more oil than all the underground reservoirs in that country added together. The major problems in their exploitation lie in the cost of damping down the waste dust produced in the extraction process, and in returning the landscape of the area to something acceptable.

In addition to further sources of conventional fuels, other sources of energy must be considered. The use of nuclear power is increasing, but there are still nagging safety doubts as well as environmental concerns over the use of water for cooling, and a number of national programmes have been slowed in the 1970s. Water power has been used over the centuries, first directly to turn wheels and later to generate electricity, but such facilities are limited, and constitute a small contribution to the total energy demands of most countries.

Further possibilities are still in the research stage. It would seem most sensible to use the energy direct from the sun, and this may eventually be possible. Tidal differences are used to generate electricity in a few places, but the number of suitable sites is limited, and the capital costs are high. Geothermal power has been developed in specially favoured areas like California and New Zealand, and there are imaginative plans to tap the earth's heat where is comes close to the surface in areas like the granite batholiths of south-west England. Another favoured source is the energy which can be generated from wave movements, but that would probably be limited to the most active storm wave environments.

The major problem in the use of many of these forms is in the difficulty of storing electrical energy. We still do not have adequate storage facilities, and strikes of electricity workers demonstrate our vulnerability in this area. Reliance on natural forces, like the wind, waves, tides, volcanic activity, or even sunshine, which all fluctuate in intensity, could only be possible in conjunction with other means, or when better means of electricity storage are available.

One of the side-effects of burning so much fossil fuel at the present time is atmospheric pollution: car exhaust gases are turning the atmosphere of central parts of American and Japanese cities yellow. Another effect is the increase of carbon dioxide gas, which threatens to raise the average atmospheric temperature around the world. At one stage it was calculated that by the time all fossil fuels were burnt up the temperature could increase by 2°C,

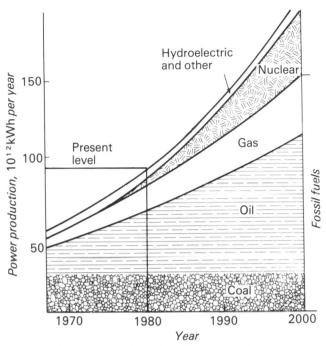

Fig. 8.44 The predicted world energy demand to AD 2000, based on the maximum growth in demand. This is just one view of the possible future, and coal might increase in place of oil. Discuss other possibilities.

Alternative energy sources. The oil shales of the Green river valley, Utah (left), and the geothermal steam fields at Wairakei, New Zealand. (USGS)

causing major climatic changes including the melting of ice caps and the consequent raising of sea level. At the same time, however, increasing dust particles in the atmosphere seem to have balanced this effect, since they reflect the sun's rays and the rises in temperature to 1940 have not been continued.

Rock resources

In addition to the mineral resources (Chapter 7), and the energy resources, the earth is the source of many other raw materials used by man, such as building materials and industrial minerals. Although less spectacular in terms of value and demand, these are often important in the development of an economy. Due to their low value and bulkiness they cannot, in many instances, stand transport costs, and so local supplies are of importance. In developing countries, for instance, it is vital to have materials for house-building and supplying basic industry. A country may have its own iron ore, but if it has no moulding sand it will have to export the ore and lose the benefit of manufacturing it at home.

Investigation 1. Examine specimens of sedimentary rocks, and make notes in your practical books as follows:

Size of grains:
Colour:
Textures/structures present:
Any fossils?:
Type of rock (a conclusion from the above information):

Locality where found
   (be accurate):   } You can only answer these
Possible age   } if you found the rock yourself
   (giving evidence):

You should be able to identify most of the rocks mentioned in this chapter.

Do not restrict your observations to individual specimens, but examine local quarries and record the associations of rock types (e.g. alternations of sandstone and shale) and the large-scale features of their arrangement (e.g. cross-bedding; whether layers maintain an even thickness).

Investigation 2. There are a number of simple experiments which illustrate sedimentation. (a) Shake a mixture of sand, gravel and clay in water in a jam jar and then let it settle. This will result in a graded deposit. (b) Colour salty or very cold water, and pour it into the end of a trough containing warmer water to see how turbidity currents move.

Investigation 3: igneous rocks. Examine specimens of igneous rock (basalt, dolerite, gabbro, andesite, obsidian, pitchstone – and granite) and ask yourself similar questions to those suggested in connection with the sedimentary rocks. If possible have a look at some of these rocks under the microscope in thin section and try to recognise the minerals contained. Make neat notes of your descriptions in your practical books.

Investigation 4: the chemical composition of igneous rocks. Plot the following figures on a graph, and answer the questions:

Oxide	Peridotite	Gabbro	Diorite	Syenite	Granodiorite	Granite
SiO_2	43.54	48.36	51.86	59.41	66.88	72.08
Al_2O_3	3.99	16.84	16.40	17.12	15.66	13.86
Iron oxides	12.35	10.47	9.70	5.02	3.92	2.53
MgO	34.02	8.06	6.12	2.02	1.57	0.52
CaO	3.46	11.07	8.40	4.06	3.56	1.33
Na_2O	0.56	2.26	3.36	3.92	3.84	3.08
K_2O	0.25	0.56	1.33	6.53	3.07	5.46
Others	1.83	2.38	2.83	1.92	1.50	1.14

(1) Describe the variation in silica (SiO_2), iron oxides, and the oxides of calcium, sodium and potassium.
(2) Compare the chemical compositions of the rocks with the mineral compositions as shown in Fig. 8.33.
(3) The rocks mentioned here are all coarse-grained varieties. Add the names of the medium- and fine-grained varieties to the graph.

Investigation 5: metamorphic rocks. Examine specimens of metamorphic rocks, including hornfels, slate, marble, schist and gneiss. Write up the results and description in your practical books in terms of mineralogy, texture, structures and possible conditions of origin.

Building and roadstones

A wide variety of rock-types have been used as **building stones**. For many ages the nearest compact, load-bearing stone which could be quarried in squarish blocks was used for normal buildings. Today little natural stone is used anywhere in Britain, since cheap bricks and concrete are easily transported anywhere. Even where natural stone is used it is largely as a decorative, facing material, rather than as the load-bearing part of the building.

A study of older buildings suggests that almost every locality had its source of building stone. The best are often known as freestones – i.e. they could be cut or split in any direction: some of the Permian red sandstones and the Jurassic oolitic limestones are of this nature. Granites have been used extensively where hard-wearing properties are valued: kerbstones in the older sectors of our towns and the old London Bridge were built of Dartmoor granite. At times the best building stones were not available, and a second best had to suffice – or else the inhabitants of the area used local clays, not of any use as a building stone, for brick-making. The villages along the northern margins of the Vale of Pickering in Yorkshire were close to Jurassic oolites and the older houses were built of this material, but the villages along the southern margin, separated by marshes from the oolites, used the local chalk for their buildings.

Roofing materials have also changed with the years: the slate quarries of north Wales have been put out of business by the manufacture of clay and concrete tiles. Before the slate boom in the nineteenth century houses were roofed with the nearest thin-bedded rock material: this may have been a sandstone several centimetres thick (e.g. in Yorkshire and Lancashire the tilestones of Carboniferous sandstone were often used). Failing the presence of such materials – which also necessitated the construction of strong walls and cross-beams – thatch was the answer.

Roadstones require a number of properties: they should be resistant so that they will not wear away rapidly and will not dissolve; and they are required to break into cuboidal fragments as well as being able to bind easily with the tar. The most important stones used today for road surfaces are the Palaeozoic limestones (e.g. Carboniferous limestones of the Peak District and North Wales) and metamorphosed silty sandstones, together with medium- and fine-grained basic igneous rocks. Shales are excavated for road bedding. At the Cliff Ridge Wood quarry, near Great Ayton, Yorkshire, the Cleveland dyke was excavated many years ago for roadstone, leaving a great trench across the landscape, and now the hardened shales on either side are being removed for road hard core and foundations.

Bricks, cement and industrial rocks

Limestones are the most-quarried rocks in Britain today. Their most important use is in the production of **cement**: two-thirds limestone and one-third clay are mixed in the manufacture. Some rocks, like the Lower Lias at Rugby, are natural cementstones because the lime and clay are already mixed before extraction. Chalk is the basis of 75 per cent of British cement, and the largest works in the country line the banks of the lower Thames, located where the Chalk outcrop comes close to the river muds (with which the chalk is mixed) and the transport artery. Other large works, such as that near Castleton in northern Derbyshire, also thrive on a site near the junction of limestone and shale.

Bricks and **tiles** are produced from clays, mixed with a small proportion of sand to reduce shrinking in the firing process. The best brick-making clays are those containing natural fluxing materials like soda and lime, iron to give the red colour, and organic matter to help the firing. The Oxford Clay (Jurassic age) is one of the major sources in Britain, and is quarried extensively near Peterborough and Bedford. Most clay horizons have been used for brick-making at some stage, but the many abandoned brickpits testify to the centralisation of the modern industry. There are still small brickworks, such as the one at Hambledon in Surrey, producing bricks directly from the Weald Clay (Cretaceous) for special uses (e.g. to bear heavy loads, or for special facing effects). Locate a brickworks (used or disused) near your home. Tiles are made in a similar way, but increasing numbers are being made of concrete in moulds and being given a surface dressing for colour and appearance.

Refractory clays and **moulding sands** are important rocks which are used in industrial processes. Moulding sands need to be even-grained and to allow the gases to escape whilst holding the shape to be moulded. Refractory clays are often taken from the 'fireclay' of coal-seam associations since these deposits contain aluminium.

9
Studying fossils

No alteration	Organic compounds	Soft parts	Frozen, mummified	Extremely rare
		Some hard parts	eg chitin beetle wings	Very rare
	Inorganic compounds	Most hard parts	Calcium carbonate most common in shells; bones of calcium phosphate; noncrystalline silica	Only in recent rocks
Alteration during fossilisation	Organic compounds	Soft parts	Carbon films, eg plants, worms	Unusual
		Hard parts	Carbonised, eg graptolites	Common only in this group
	Inorganic compounds	Hard parts	Shells pitted; impregnated with minerals in pores; recrystallisation; replacement of original shell material	Very common
Moulds and casts	Organic compounds	Soft, hard parts	Imprints in fine-grained sediment, eg bird feathers	Unusual
	Inorganic compounds	Internal or external moulds after shell solution. Cavity refilled as cast		Common in older rocks
Trace fossils		Tracks and trails — signs of movement. Burrows and borings — homes. Coprolites, castings. Tooth-marks		Quite common

DECREASING USEFULNESS →

Fig. 9.1 Fossilisation. This chart summarises the ways in which animal and plant remains may be fossilised. Notice the relationship between the most common modes of fossilisation and the degree of information supplied.

Fossils in rocks

A fossil is any piece of evidence in the rocks which tells us something about ancient life. Fossils include the unaltered remains of animals and plants, shell and leaf impressions, and even burrows and footprints. The study of fossils is called **Palaeontology** ('the study of ancient life'). Leonardo da Vinci (1452–1519) was amongst the earliest to recognise the true nature of fossil markings in the rocks:

When the floods from rivers turbid with fine mud deposited this mud over the creatures which live under the water near sea shores, these animals remained pressed into the mud, and being under a great mass of this mud, had to die for lack of the animals on which they used to feed. And in the course of time, the sea sank, and the mud, being drained of the salt water, was eventually turned into stone. And the valves of such molluscs, their soft parts having already been consumed, were filled with the mud; and as the surrounding mud became petrified, the mud which was within the shells, in contact with the former through their apertures, also became turned into stone. And so all the tests (shells) of such molluscs remained between the two stones, that is, the stone in which they were and that which covered them; and these are still found in many places. And nearly all the molluscs petrified in the stones of the mountains still have their natural shell, especially those which were old enough, which would be preserved by their hardness; and the young ones, already calcined, were penetrated in great part by the viscous and petrifying liquid.

Fossils include a wide variety of remains and marks in the rocks. Like living forms today they range in size from the skeletons of enormous animals to a wealth of microscopic forms. Fig. 9.1 summarises the ways in which they occur. What strikes you about the parts of animals and plants which are most commonly preserved?

The fossils we find in the rocks have undergone a long series of processes since the death of the animal or plant which they represent, and each of these processes removes some of the evidence for life in the past. When an organism dies it will decay, assisted by bacteria and atmospheric weathering, unless it is rapidly removed from their influence by burial in sediment. Soft parts disappear most quickly, but even hard skeletal features are soon lost unless they are buried. When the remains of a dead organism are covered by sediment in this fashion they then become part of the sediment and are subjected to the processes of sedimentary rock formation which we studied at the beginning of Chapter 8. Compaction may crush the skeleton if it is fragile (Fig. 8.5); circulating groundwaters commonly cause the skeletal material to be dissolved away, to be replaced by another substance with some loss of detail, or they may cause the pore spaces in the original shell to be filled by the precipitation of chemical salts. Earth movements, incorporating folding, faulting and even alteration of the rocks by heat and pressure, add to the likelihood of distortion or the complete removal of the fossil evidence from the rocks. The longer the fossil is in the rock, the longer this combination of processes will have to act, and the less chance we shall have of examining the record of life for a particular point in geological time.

We shall thus expect to find fossil remains in sedimentary rocks, and we shall expect to find fewer in older rocks or in rocks formed on the land, where atmospheric oxygen and predators soon destroy any remains. There are certain animal groups which have an impressive and detailed fossil record: we have a good idea of their past history. On the other hand some of the most important and varied groups alive today have hardly any fossil record: the most obvious examples are the insects and worms (Fig. 9.2). The reason for this is straightforward when we understand something of the processes of fossilisation. Animals which have hard (often calcified) internal or external skeletons, bones or shells, are more likely to leave a record of their existence. If you study biology as well as geology you will find that the animal groups you investigate are quite different in the two courses. Some of the most important groups of fossils are extinct (trilobites, graptolites, ammonites, dinosaurs), or almost extinct (brachiopods). Another consequence is that whilst the biologist bases his classifications on the soft parts of animals and plants these are rarely preserved in the fossil state, and so classifications used by palaeontologists are often necessarily quite different.

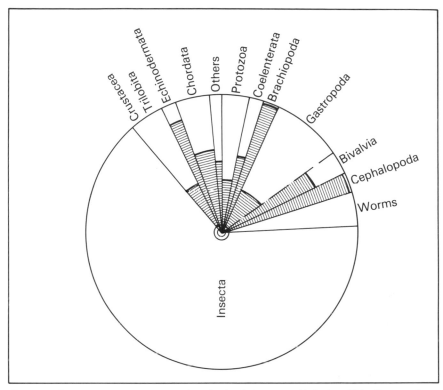

Fig. 9.2 There are over a million species of animals known to biologists. This diagram shows how the proportions are grouped in the major phyla. The shaded areas of each zone represent the fossil species known. What does this tell you about the differences between the animals you are most likely to see today and those you are most likely to find as fossils in the rocks?

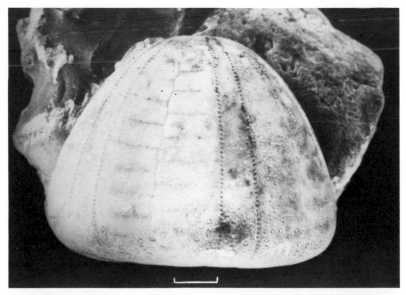

An irregular echinoid preserved in a nodule of flint from the Chalk. Notice the degree of detail preserved (e.g. plate and pore structures). The scale line always represents 1 cm. (JS)

There is thus a built-in bias to the fossil record, and we should always be aware of this. When we examine a group of fossils extracted from a layer of rock we must realise that there is not likely to be any record of the worms, insects or plants which lived in the area of sedimentation represented by that rock layer. We must also be aware of this 'missing factor' when we devise theories based on the fossil record.

The variety of living and fossil animals and plants

The earth is the only planet within the Solar System on which life can exist in a wide diversity of forms. That diversity becomes apparent as soon as we start to make a detailed study of the living organisms within a few miles of our homes. Even those who live in towns must be aware of the range of living creatures from human beings, through the domestic animals, birds and fishes, to the insects and disease-causing microscopic forms of life. There is also a wealth of plant life, ranging from the tallest trees down to the shrubs, grasses, mosses and microscopic algal forms.

We normally make a major distinction between the animals and plants, because of their fundamental differences in feeding and growth mechanisms. Whilst plants draw inorganic matter from the soil, and convert it into organic substances by reaction with carbon dioxide and water in the presence of sunlight, animals are unable to gain nourishment from inorganic matter and have to feed upon plants or other animals. Such distinctions tend to break down amongst the microscopic, single-celled creatures, and many biologists would place them in a separate group, the protists.

When we extend our studies from our own neighbourhood to the entire world, the number of

Collecting fossils

Fossil-collecting is a fascinating hobby. See how many different varieties you can collect from any exposures of rock near your home (always remembering to obtain permission of the owner of the land). You may be fortunate and find a rock containing many fossils: the longer you are able to spend on one thin layer the more you will find out about it.

This hobby needs few special tools. Fossils in soft rocks will be prised out easily with a penknife; in harder rocks you will need a hammer and chisel. Many fossils are fragile and must be wrapped carefully in newspaper or sometimes in cotton wool and carried in tin boxes. A notebook is essential to record the exact locality where each one was found, and a small hand lens often helps in the examination of the tinier details.

When you find a layer of rock containing a number of fossils, concentrate on a detailed study of it. Collect samples of each type of fossil, noting their importance in the group as a whole and the place where you found them. It often takes a little time and careful examination before you find the first few fossils in a rock exposure, but the longer you persevere the greater will be your chance of success: you will soon get to know the best places.

You will need a special place to keep your collection tidy for easy reference. When you get home clean each specimen as far as possible so that the details show up, and make neat drawings of all of them. Add your notes on the place where you found each fossil, plotting them all on a diagram or map. If more than one bed is exposed at a place, indicate from which each fossil came. Then look up the name of the fossil in a reference book, which will also give you some assistance in determining its age.

Fossils are found by looking at the surface of fallen blocks of sedimentary rocks. This photograph was taken near Lyme Regis in Dorset.

different varieties, or species, of organisms multiplies. A **species** can be defined most simply as a potentially interbreeding group of organisms which is able to reproduce a further generation capable of carrying on this process (thus horses and donkeys, two distinct species, can breed, but their offspring, the mule, cannot). Over a million animal species and a third of a million plant species have been described as existing on the earth. Our knowledge of the life on our planet is still far from perfect: over 10 000 new species of animals are added to the list each year, and scientists have intimated that the total number of species on the earth may lie somewhere between 2 million and 4 million.

Two questions strike the person faced with a study of this tremendous diversity: how can it be reduced to manageable units for investigation? and what is the explanation for the existence of so many varieties of life on the earth?

Throughout the history of the systematic observation of nature men have attempted to classify what they have seen into groups. The differences between the animal, vegetable and mineral worlds were noted by some of the most ancient writers, but our modern system for classifying living creatures dates from the mid-eighteenth century and a Swedish botanist, Linnaeus. He suggested a more natural type of classification than had hitherto been in use: species showing close similarities could be grouped together as a **genus**. Names were Latinised to give them a world-wide application, and each species is now referred to by its generic (with a capital letter) and specific names. Thus modern man (a potentially interbreeding group which includes every variety of mankind on the earth today) is known as *Homo sapiens*. He is related by the features of his bodily anatomy to other animals. There are ancient forms of man which are also included in the genus *Homo*. Several genera of manlike creatures form the **family** Hominidae. Together with the families of monkeys, apes, lemurs and tarsiers the Hominidae are members of the **order** Primates. Another step in the hierarchy joins the primates to the lions, elephants, cattle, kangaroos and rodents in the **class** Mammalia. The mammals and other groups of animals with backbones (fishes, frogs and reptiles) are included in the **phylum** Chordata, and eventually linked with all the other animal phyla in one **kingdom**. As soon as a new species is discovered and named it is slotted into this system, a method having the great virtue of placing it amongst its near relatives. Fig. 9.3 lists the major phyla of the plant and animal kingdoms.

But why are all these species represented? How did they come into existence? You can arrive at a part of the answer by examining the habits of a particular animal more closely. It is limited to a certain area in which certain conditions operate; it will eat a restricted variety of foods and may well confine its eating habits to the members of a single animal or plant species. You will notice that it is very well suited to living in such conditions: it is adapted to its particular mode of life in that environment, and the make-up of its body will be such that it is able to move and capture its food supply in the most efficient manner. Study different types of creature in one area, and they will each be found to be specialised for a particular mode of life: birds, for instance, will go for worms or insects or nuts and will have beaks suited to the consumption of their diet. In widely separated parts of the world the same type of environment may be occupied by different groups of animals. Thus in Australia we find that there are a variety of marsupials (feeding their young in pouches), which not only take over the ways of life enjoyed by the more advanced mammals in the rest of the world, but often look like them as well.

Thus one part of the answer to our question is found in the fact that the bodily features and modes of life of organisms are closely related to the different conditions of temperature, soil, water salinity and light in the places where they live.

For the full answer, so far as biologists understand it today, we need to investigate what life was like in the past. What evidence do we have? How reliable is it? To what uses can it be put? This is our study in the present chapter. The animal- and plant-like markings on the rocks are the remains of creatures which were once alive. As our knowledge of the past has been extended it has been suggested that only 1 per cent of the total number of species which have lived on the earth are alive today. It may well have been less than this figure, but in any event there is a vast realm of life in the past to be studied in addition to the one living today.

The major fossil groups

The major groups of fossils occurring in sedimentary rocks thus provide us with important sources of information concerning the environments of the times in which they lived. Many, especially those which lived fixed to the sea floor, will reflect the conditions at the time when they lived. In addition,

	PHYLUM	Approx. number of species now	Important fossil groups (class or order)	Geological occurrence
ANIMAL KINGDOM	*Protozoa*	30 000	Foraminifera Radiolaria	Ordovician-Recent Cambrian-Recent
	Parazoa (Porifera) (sponges)	4 200		Cambrian-Recent
	Coelenterata (Cnidaria)	9 600	Anthozoa (corals)	Ordovician-Recent
	(Several phyla of segmented worms)	27 000	(fossil record scarce)	
	Brachiopoda	260	(important fossils)	Cambrian-Recent
	Bryozoa (Polyzoa)	4 000		Ordovician-Recent
	Mollusca	100 000	Gastropoda Bivalvia (Lamellibranchia) Cephalopoda	Cambrian-Recent Cambrian-Recent Ordovician-Recent
	Annelida (Segmented worms)	7 000		Precambrian-Recent
	Arthropoda	765 000	Trilobita Crustacea Insecta	Cambrian-Permian Cambrian-Recent Devonian-Recent
	Echinodermata	5 700	Crinoidea Echinoidea	Cambrian-Recent Cambrian-Recent
	Chordata	45 000	Hemichordata (Graptolites) Vertebrata (fish, amphibia, reptiles, mammals, birds)	Cambrian-Carboniferous Ordovician-Recent
PLANT KINGDOM	*Cyanophyta* *Euglenophyta* *Chlorophyta* Algae *Pyrrophyta* *Phaeophyta* *Rhodophyta*	20 000	Diatoms	Precambrian-Recent
	Schozomycophyta *Myxomycophyta* Fungi *Eumycophyta*	48 000		
	Bryophyta	25 000		
	Tracheophyta (land plants)	250 000	Psilopsida Lycopsida Sphenopsida Pteropsida (ferns, conifers, flowering plants)	Devonian-Recent

Fig. 9.3   The major groups of animals and plants.

the progressive change with time, or evolution, of living forms enables geologists to give a general date to a layer of rock containing particular fossils. When studying each group of fossils, therefore, it is important to have a knowledge of the main functioning parts and their relationships to the animal's or plant's mode of life. It is also necessary to know which periods of time were important for each group of fossils. This knowledge can then be used in the interpretation of new situations.

The following list of major groups can be compared with Fig. 9.3.

Animals

1 **Corals:** forms within the phylum Coelenterata which secrete calcareous skeletons.
2 **Brachiopods:** a phylum of almost extinct bivalves (i.e. they have two shells), mostly calcareous.
3 **Lamellibranchs** or
  Bivalves (bivalved molluscs e.g. clams, oysters, mussels, razor shells)
4 **Gastropods** (e.g. snails)
5 **Cephalopods:** extinct, shelled forms (ammonites, belemnites)

} the three main classes in the phylum Mollusca: calcareous shelled forms are preserved.

6 **Trilobites:** extinct group in the phylum Arthropoda, having a jointed, calcified skeleton.
7 **Crinoids:** sea lilies, with skeleton of calcareous plates supported on stem.
  Echinoids: sea urchins – mobile animals with skeleton of calcareous plates.

} two major groups within the phylum Echinodermata having rigid skeletons.

8 **Graptolites:** an extinct colonial group of uncertain affinities in terms of classification.
9 **Vertebrates:** fishes, amphibians, reptiles, birds, mammals – larger creatures with strong internal skeletons.

Plants

10 Leaf and stem remains, usually carbonised.

This list leaves out a large proportion of today's living creatures. Many of these have a fossil record, but it is sparse, or not obvious to the observer without special equipment. Thus the vast realm of microscopic life gives rise to plentiful **microfossils**, which are becoming more and more important to geologists in such specialised fields as the search for oil, but they require advanced extraction techniques. Many groups of animals which are less numerous, such as the sponges, bryozoans and other groups of molluscs and echinoderms, also secrete skeletons and are often fossilised, but make up a tiny proportion of the fossil record. There are also the soft-bodied animals which are scarcely represented at all in the fossil record: coelenterates, e.g. the hydra and jellyfish, many phyla of segmented and unsegmented worms, and most of the huge groups of insects and crustacea with their uncalcified skeletons, as well the greatest part of the plant kingdom. Once again we have emphasised the bias of the fossil record.

1 The corals (class Anthozoa of phylum Coelenterata)

Corals are the simplest animals included in our list of the commonly found fossils. Coelenterate cells are arranged in two layers separated by a jelly-like substance (mesogloea), but cell specialisation and organ development are very restricted. Each animal has a large central cavity (the coelenteron) which acts as a stomach, and has a central mouth opening, encircled by tentacles bearing stinging cells. In corals the coelenteron is partitioned by fleshy walls (mesenteries) which increase the digestive powers of the animal by enlarging the area of digestive cells. Some coelenterates, e.g. the sea anemone, are just soft-bodied, but many corals secrete skeletons forming a rigid base for the animals. The hard parts are related to the soft parts in life, and are preserved as fossils: the geologist is particularly interested in these hard parts. There are approximately 6000 Anthozoan species today (two-thirds of all coelenterates), plus a further 6000 species of fossil corals.

A colonial coral from Lower Carboniferous rocks: *Lonsdaleia*. Note how the corallites are joined together: the smaller examples are juveniles growing in the spreading colony. Draw a portion of this rugose coral, labelling the parts shown. (JS)

At the present time there are two distinct groups of skeleton-forming corals, and these are so different that they are placed in subclasses. The **Octocorallia** have eight tentacles and mesenteries and always occur in colonies. They secrete internal skeletons, formed of unconnected calcareous spicules or horny gorgonin: when the animal dies such a skeleton soon disintegrates. These skeletons are often colourful and are used ornamentally. Their structures are simple, including the calcareous tubes and transverse plates of the organ-pipe coral, and the fused spicules of the sea fans.

On the other hand the stony corals produce more massive and resistant calcareous skeletons. These belong to the subclass **Zoantharia** and have similar bodies to the sea anemone with six pairs of complete mesenteries and shorter incomplete fleshy walls in multiples of six between (this group is sometimes known as the hexacorals). The stony corals, or **scleractinians**, secrete septa formed of calcium carbonate crystals between the mesenteries, and these give rise to a pattern of radiating walls. The whole skeleton has the shape of an elongated cup (corallite) (Fig. 9.4). Both colonial and solitary forms occur: the solitary forms are often larger (with a corallite diameter of up to 250 mm), but the majority are colonial, where the smaller individuals average only 1–3 mm in diameter.

The extensive **coral reefs** (cf. Fig. 5.5 and Fig. 5.6) of the present day are largely formed of the stony corals, but include some octocorals. The colonial stony corals live in close symbiosis (i.e. living together for mutual benefit) with algae, and so are limited to shallow, well-lit waters where the algae can photosynthesise. They only flourish in tropical areas with temperatures of between 25°C and 29°C and need copious oxygen. These reefs are associations of a remarkable variety of creatures and are amongst the richest living communities in this sense on the earth, but their existence is threatened by the plague of 'Crown of thorns' starfish, which is stripping the stony coral areas of the Australian Great Barrier Reef and the many Pacific and Indian ocean islands. The delicate balance of the coral reef food chain has been upset, possibly by the human demand for attractive *Triton* shells: this large marine snail eats the starfish and previously restricted its effect. We wait to see whether the starfish plague will cause the stony corals to become extinct, or whether this is part of the predator–prey cycle whereby the predator will increase in numbers until its food supply is almost wiped out; the predators then starve to death and the prey has an opportunity to re-establish itself. Our answer should not be long delayed, since already 25 per cent of the Great Barrier Reef coral and up to 90 per cent of the coral around some Pacific islands is destroyed.

The **solitary corals** are not limited in depth by an algal association: they occur over a much wider climatic zone and in water of low temperatures down to depths of 7000 m.

Fossil corals. The skeletons of corals, like other fossils, are filled with mud or mineral matter after death; they may be bored through and through by sponges and bivalves; and in older rocks the whole mass may be recrystallised, destroying the organically secreted structures.

The stony corals (**scleractinians**) are the only modern, skeleton-forming corals which are important in the fossil record: they form the reefs and solitary fossil corals which occur in the Mesozoic and Cainozoic rocks (Fig. 9.5). The **octocorals** have few fossil ancestors.

Two other groups of fossil corals are found in the Palaeozoic rocks: both became extinct before the development of the modern corals. The **tabulate** corals are the first to occur in the geological record and were important in the late Ordovician, Silurian and Devonian periods, but were later overshadowed by the **rugose** corals. The tabulate skeleton is simpler (Fig. 9.4), with few internal structures,

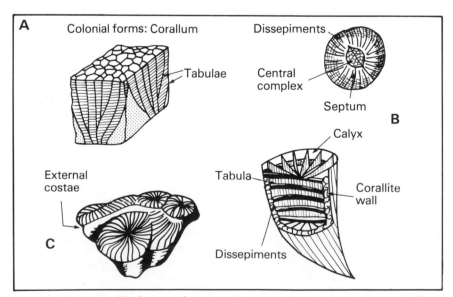

Fig. 9.4 The corals. The features of tabulate (**A**), rugose (**B**) and scleractinian corals (**C**) are drawn. Relate these to specimens which you have drawn yourself. What are the differences between the main groups?

and these corals were always colonial. The rugose skeletons are more complex, with a variety of internal features more akin to the scleractinian group: they are common in rocks of middle Silurian, Middle Devonian and Lower Carboniferous age, but then gradually became less important and died out. Many of these rugose corals are used as zone fossils: this means that they are used in dating the rocks in which they occur (the principle is further explained in the section 'Fossils and time' at the end of this chapter).

Reef corals of these distant periods may have demanded similar conditions to the reef corals today, and if this is assumed we can obtain an indication of the ancient environment in which they were formed. Fossil reefs, however, have a different distribution from those today: they occur in places like Yorkshire and Devon well removed from the tropics. This suggests that either tropical conditions were more widespread in the past, or the land masses must have moved their positions relative to the climatic belts, or that the distribution of reefs is controlled by special conditions necessary for the symbiotic algae, and that the reef corals have become increasingly dependent on these with time. The solitary coral fossils may indicate deeper water conditions for sedimentation of the rocks in which they are found.

2 The Brachiopods (phylum Brachiopoda)

Brachiopods form a small, rare group today: they are sometimes called the 'lamp-shells' because the shape of the shells of one group resembles the old oil lamp. The most distinctive feature of the brachiopod body is its food-catching organ, the lophophore (Fig. 9.6). This is a fold in the body wall which is covered in fine hairs, or cilia: as these move they direct a current of water towards the mouth where microscopic food particles can be extracted. The animal is enclosed between two shells, or valves, but most of the space is occupied by the lophophore and by the complex system of muscles used for opening and closing the valves. The bodily organisation is more complex than the coelenterates, but is still at a low level: there are few specialised sense organs, digestion is by means of cells in the gut wall, the blood circulation is very weak and there are no special gas exchange organs (i.e. gills).

Externally the brachiopod shell is similar in some ways to an important group of living molluscs, the lamellibranchs: palaeontologists must learn to dis-

tinguish the features of the two groups. Biologists did not separate these two bivalve groups until the nineteenth century. (The features of the two groups are compared in the next section.)

The brachiopods living today are the remnant of a much more important past: 260 living species compared with 30 000 fossil species. *Lingula* burrows in the sands or muds of warm, shallow seas close to the shore and can survive less salty estuarine conditions ('brackish' water). Its valves are not hinged, but are held together and adjusted by complex muscles. Similar shells are found as fossils

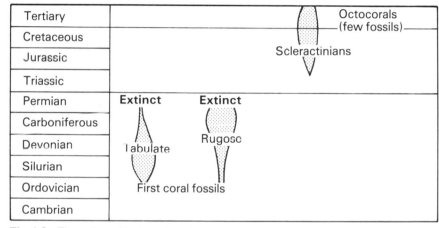

Fig. 9.5 The geological history of corals.

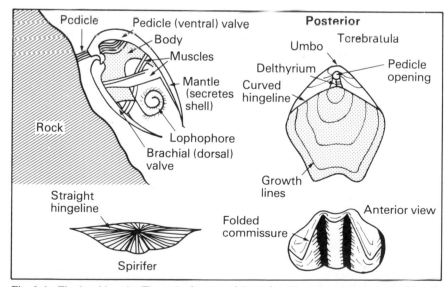

Fig. 9.6 The brachiopods. The main features of living brachiopods and their shells. Identify similar features on the brachiopod fossils you find.

in rocks since the Cambrian system (i.e. 600 million years old), showing little change in shape and generally occurring in sediments interpreted as being of coastal origin. All other living brachiopods are also bottom-living (benthonic) creatures. Forms like *Terebratula* attach themselves to the rocky sea floor by a fleshy pedicle and have a sedentary existence; their shells are hinged and vary from shorter, heavier and more rounded shapes in shallower and rougher water to more fragile, elongated forms in quieter conditions. All brachiopods are marine, but hardly any live beyond the edge of the continental shelf (i.e. 100–200 m deep).

Fossil Brachiopods. The Palaeozoic brachiopod fossils do not normally have the shell preserved as such, but only the **moulds** and **casts** of the internal or external features (Fig. 9.7). Mesozoic shells have either been replaced or recrystallised, and only the most recent fossils retain the original shell materials. This illustrates the effect of the passage of time on the preservation of fossils.

Palaeozoic brachiopods are particularly important in the fossil record (Fig. 9.8). The earliest known forms, the Cambrian, did not have hinged shells and include *Lingulae* like those we find today: the group without hinged shells is known as the **inarticulates**. Hinged (**articulate**) brachiopods became really important in the Ordovician and Silurian: genera like *Orthis* with a long, straight hingeline, and the pentamerid *Conchidium* with its bulbous shape and rounded hingeline, were particularly common. Spiriferids, with a straight hingeline at the widest part of the shell, were important in the Devonian, and the Lower Carboniferous rocks contain many productids. At the end of the Palaeozoic most of these groups died out, leaving the small, ribbed rhynchonellids, which tended to repeat certain varieties of shell form at intervals in the fossil record, and the smoother-shelled terebratulids: these two groups were at their most important during the Mesozoic, since when there has been a further decline to the present situation.

It is probable that brachiopods have always inhabited shallow seas, and therefore they can be used as a criterion for such an environment. Many of them existed over a very short time range, being rapidly replaced by newly developing forms. Some brachiopods are thus useful as zone fossils, and in spite of their sedentary adult habit the larval forms are free-swimming and ensure a widespread distribution of the group. Brachiopods are most import-

Some brachiopods. **A** and **B** show the features of the outsides of shells: what are the main differences between the two? **C** has been distorted as the rock in which it was entombed was affected by earth movements. **D** shows the cast of internal features following solution of the shell—muscle scars and even blood vessels. (JS)

ant in this way in the lime-rich rocks of the Devonian, Lower Carboniferous, Middle Jurassic and Upper Cretaceous.

3 The bivalve molluscs (class Bivalvia phylum Mollusca)

The molluscs include a wide range of animals, and the phylum is the second largest today, including over 150,000 living species. Oysters, snails, squids and the octopus are common members, and many molluscs secrete hard calcareous shells: this means that they are amongst the most numerous forms in the fossil record, since it seems that the molluscs have always been a large and successful group. Common fossils, e.g. the ammonites and belemnites, are extinct molluscs.

Even the simplest molluscs show an advance in bodily development compared with the corals and brachiopods. All molluscs with the exception of the Bivalvia have a mouth region equipped with a scraper-like tongue (the radula), respiration is by means of complex gills or even lungs, the blood circulation and excretory system become well-organised, and movement in the search for food is more efficient. The most advanced molluscs are amongst the most complex and highly organised of all invertebrate animals: squids, for instance, can swim as fast as fishes, digest their food as rapidly as advanced mammals, and their nervous system enables them to change their body colours over a wide array within seconds.

Bivalve molluscs: lamellibranchs or pelecypods. This group of filter- and deposit-feeders is the most sedentary of the three major mollusc classes: few move far, and a great number are fixed to the rocks or pebbles beneath the water in which they live, or plough into the soft sediment to varying depths. They secrete bivalve shells round their bodies, and these act as a support and protection. Food enters the shell, is sorted on the gills and swept into the mouth. The animal only extends a muscular foot outside the shell to pull itself through the sediment, or, if it is burrowing, a pipe-like pair of siphons to connect it with the clear water and the supply of food above the sediment. Those forms which are cemented to the surface, like the oysters and mussels, have hardly any foot and no siphons. No lamellibranch possesses a head or radula: the group is highly specialised for its particular mode of life and feeding.

Fossil record. The shells of lamellibranchs are common Mesozoic and Cainozoic fossils, although

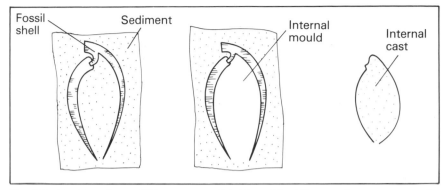

Fig. 9.7 Moulds and casts. Write your own definition of the differences between a cast and a mould.

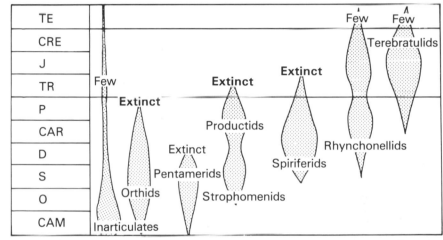

Fig. 9.8 The geological history of brachiopods. What age would the rocks be if you found productids, spiriferids, rhynchonellids and the occasional terebratulid in them?

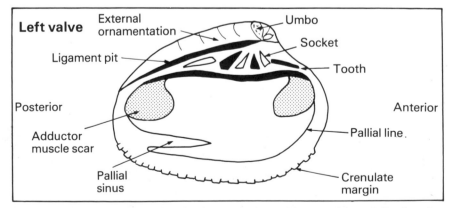

Fig. 9.9 The bivalve mollusc. You can often see the internal features of these shells, since they fall apart on the death of the animal. External features such as radiating or concentric ribbing, and the shape of the shells are also distinctive.

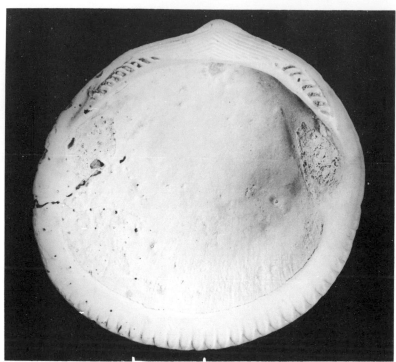

A bivalve mollusc: *Glycimeris glycimeris*, of Pleistocene age. (JS)

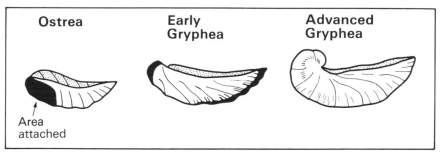

Fig. 9.10 *Gryphea* evolution. This series of stages from an oyster to the coiled *Gryphea* took place on several occasions. It is possible that the excessive coiling led to extinction of each line, but the same trend was repeated again later.

preservation is less common in older rocks, as is the case with other shell creatures: once again the oldest records are merely impressions of the mould or cast variety. Lamellibranch shells are composed of an outer dark organic layer (never fossilised), and a calcareous, layered shell consisting of varying proportions of aragonite and/or calcite: most fossilised forms show a high proportion of the calcite. The features of their shells are shown in Fig. 9.9.

Lamellibranchs are important fossils throughout the geological record from the Cambrian period (Fig. 9.14). The Palaeozoic forms show an increasing range of diversity. At first they were mostly attached in some way, but by the Carboniferous period there were burrowing and swimming varieties, as well as freshwater forms. The most advanced forms, which became more active burrowers, developed through the Mesozoic and into the Cainozoic, and the thick-shelled, sedentary oysters also appeared at this stage. There is a correlation between the thicker-shelled varieties and shallow, warm, lime-rich seas, in contrast to the thinner shells associated with deeper or freshwater conditions. The lamellibranchs outnumbered the brachiopods by the Mesozoic era and became the dominant bivalves. It is probable that the competition between the two groups was settled in favour of the more advanced and better-adapted lamellibranchs.

Lamellibranchs are particularly characteristic of shallow waters and stand a very good chance of fossilisation if their mode of life involves burrowing into the sea-floor sediment. They may also be used as zone fossils in circumstances where other groups are absent. Thus the non-marine lamellibranchs are used for this purpose in the Coal Measures of the Upper Carboniferous; the lowest Jurassic rocks in South Wales are divided according to the Gryphea

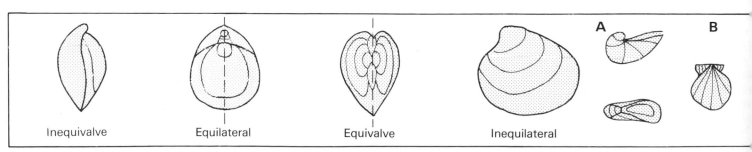

Fig. 9.11 Brachiopod or bivalve mollusc? What do the terms used here mean? Which apply to the brachiopod, and which to the bivalve? In which category would you place the examples **A** and **B**? Why are these two slightly different from the normal rule?

sequence (Fig. 9.10); and the British lower Tertiary rocks are partly zoned by the lamellibranchs because of the great variety and numbers contained. **Lamellibranchs and brachiopods are similar** in many ways, including their modes of life and the shapes of their shells. It is thus important to be able to distinguish between them, and there are several features which make this possible (Fig. 9.11).

(1) In life the lamellibranch valves are to the left and right of the hinge, which faces upwards in the active varieties, whilst the shells open downwards (ventral); sedentary forms, like oysters, rest on their right valve. The brachiopod valves are normally one on top of the other (i.e. the pedicle, or ventral, valve on top of the brachial, or dorsal, valve).

(2) The two brachiopod valves can each be divided into two equal halves (i.e. they are equilateral), but the lamellibranch valves cannot (inequilateral). A little care is necessary, since some lamellibranchs are almost equilateral.

(3) Lamellibranchs normally have two identical valves (i.e. they are equivalved), but the two brachiopod valves are always of different sizes (inequivalved), the pedicle valve being the larger. Some of the attached lamellibranchs (oysters) have unequal valves, but they are also obviously inequilateral. It is common, however, for the valves of lamellibranchs to become separated on death, whereas the brachiopod shells are more commonly held together.

Thus, whilst there is some variation within the forms of lamellibranch shells, the brachiopods are always equilateral and inequivalved.

(4) Brachiopods often have a prominent pedicle opening, which may be in the form of a round opening at the top of the pedicle valve, or a triangular gap. Lamellibranchs do not have this feature.

(5) If the hingeing mechanism can be seen, lamellibranchs will have mirror image patterns of teeth and sockets in each valve, whilst brachiopods have only two teeth in the pedicle valve, and two sockets in the brachial valve.

(6) The interiors of lamellibranch shells have the scars of only one or two muscle attachments, and the same pattern occurs in each valve. Brachiopod shells have more muscle scars, and different patterns in each valve.

(7) Lamellibranch shells are opened by a ligament, housed in a pit outside, or inside the hingeline. Brachiopods do not possess such pits.

(8) Another interior feature, unique to the lamellibranchs, is the pallial line, which marks the outer edge of the mantle attachment to the shell. When siphons are developed for deeper burrowing there is a deep notch in this line, known as the pallial sinus.

4 The gastropods (class Gastropoda, phylum Mollusca)

This mollusc class includes snails, slugs, limpets and several varieties of tiny swimming or planktonic organisms. Being molluscs they have many points in common with the lamellibranch bodily organisation (e.g. a muscular foot used for movement, gills, blood circulation and nervous system), but in addition they have a mouth area with radula and a head in which the sense organs are concentrated, they come out of their shells for moving around and are found in a vastly greater range of environments.

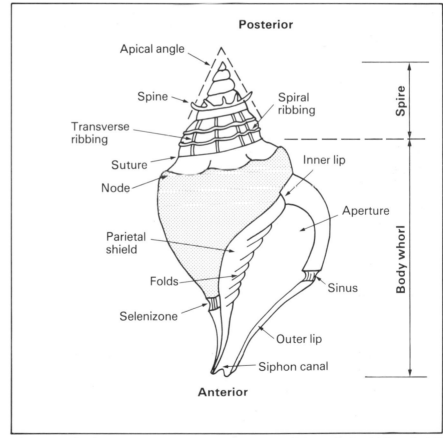

Fig. 9.12 The gastropods. The general features of gastropod shells: no single animal ever had all of these on its shell!

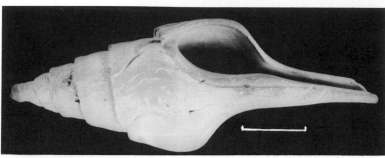

A gastropod: *Clavella longaeva*, of Eocene age. (JS)

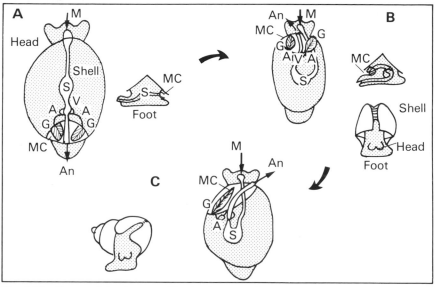

Fig. 9.13 Gastropod development. The development of the gastropods from an ancestral form (**A**), for which there is no evidence, to a typical Palaeozoic form (**B**) with a plain coiled shell, and a modern form (**C**). A process known as torsion took place between (**A**) and (**B**), in which the back end was twisted round and above the front. A later development led to the spiral shell, and to the progressive suppression of one set of any paired features (e.g. gills). A–auricle; V–ventricle; M–mouth; S–stomach; G–gill; An–anus; MC–mantle cavity.

TE	Great expansion		Expansion
CRE		Increasing variety	
J			Many new groups: oysters, active forms
TR			
P			
CAR	Simple and plane-coiled shells		Mussel and scallop-like forms
D			
S			
O			First bivalves
CAM	First gastropods		

Fig. 9.14 The geological history of the gastropods and bivalve molluscs. Compare the history of the bivalves with that of the brachiopods.

The most distinctive feature, however, concerns the arrangement of the body. It seems that, although a primitive, shelled mollusc may have had a head and mouth at one end and a cavity containing the gills and excretory organs at the other, the gastropod larva went through a process known as torsion which brought the rear cavity to a position above the head in the adult gastropod. Gastropod shells, apart from specialised shapes like the conical limpet, are coiled (Fig. 9.12), and at first this took the form of a flat spiral, only found in the early, and now extinct, bellerophontids. As the animal expanded the space available in its shell, it adopted a helicoid spiral form (Fig. 9.13): this necessitated a redistribution of weight, and the shell assumed a position with its axis slanted with respect to that of the body. As a result one side of the animal's bodily development was restricted, and paired organs like the gills have become reduced to one.

The marine gastropods include the shelled bottom-crawlers, which feed in various ways: some scrape rock-encrusting algae from the sea floor; some scavenge through sediment and extract organic matter; others are carnivorous and reach their prey by boring holes into protective shells or by direct attack. In addition many gastropods have lost their shell, or have reduced it so that swimming and floating are possible. On the land there are freshwater snails with thinner shells, and land snails and slugs, which have modified their gill chamber into lung-like organs for breathing air: all these land varieties feed on plant matter.

Amongst this immense variety of forms, adapted to a great range of environments, the shelled gastropods are of most interest to the palaeontologist. The shapes of gastropod shells do not vary greatly through geological time and are in fact often repeated. One of the most distinctive features is the presence of a canal at the anterior margin of the aperture in some forms (Fig. 9.12): this protects an inhalant siphon which projects forward to sample the water ahead of the animal and separates it from the exhalant organ, and is a feature of the more advanced and carnivorous varieties.

Fossil record. Gastropods are more plentiful today than they have ever been (Fig. 9.14), and there are more varieties of gastropod than of any other mollusc group. Palaeozoic and Mesozoic forms are not common, and it is only since the early Cainozoic that they have really become important. In such rocks they are sometimes used as zone fossils and are particularly valuable in distinguishing between marine (thick, well-ornamented shells sometimes having siphon canals) and freshwater (thin, simple shells) conditions in the formation of sediments.

5 The cephalopods (class Cephalopoda, phylum Mollusca)

The squids, octopuses and their close relations are the most highly developed animals without back-bones. Their degree of muscular coordination and sensory development is only rivalled by the higher vertebrates. They are quick-moving, carnivorous predators, and although the octopuses are not quite as formidable as they are portrayed in science fiction (the largest has a body 30 cm long), some of the largest squids are up to 16 m in length, including the tentacles.

These creatures have, however, lost any hard shell they may have possessed, and they have a poor fossil record. And yet there are over 10 000 fossil cephalopods to compare with the 400 existing species. Fossil cephalopods are related most closely to the three species of *Nautilus* living in the Indian and Pacific Oceans. These have coiled shells, like the extinct ammonites and their close relations, and their bodies show some marked differences from the squids: *Nautilus* tentacles are short, the animal has four gills as opposed to two, and the funnel which it uses as a form of jet engine is formed in a different way. It lives only in a part of the coiled shell, the rest being divided into chambers (cam-era) by walls of calcium carbonate (septa). The chambers are filled with gas, which gives the animal buoyancy in the water. The *Nautilus*, however, is slow-moving and clumsy compared with the squid.

Fossil record. The fossil record is thus particu-larly important in this group. The **nautiloids** are the earliest to appear and originally had straight shells: once they became coiled tightly their shells showed little record of further changes (Fig. 9.16). The **ammonoids**, which are nearly all coiled shells in a plane spiral (Fig. 9.15), are found commonly in the rocks of Upper Palaeozoic age, where they are known as **goniatites**, and in the Mesozoic rocks, where they are known as **ammonites**: these separate names reflect two phases of development, each of which gave rise to a rapidly-changing variety of forms and ended in a major wave of extinctions (Fig. 9.16).

The cigar-shaped belemnites (Fig. 9.15) were important fossils only during the Mesozoic, although first reported in Carboniferous rocks, and they

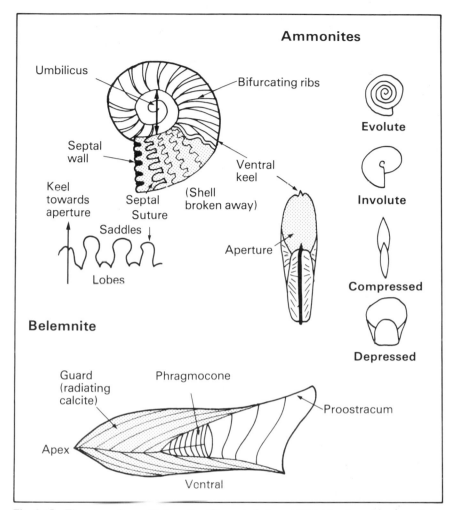

Fig. 9.15 The ammonites and belemnites. The main features of these fossils: neither have living representatives.

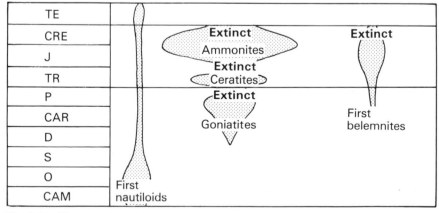

Fig. 9.16 The geological history of the cephalopods.

became extinct with the ammonites in the late Cretaceous. The belemnites are assumed to represent the internal structures of the ancestors of today's squids, although these animals have now lost all but the faintest remnant of their internal skeletons.

Goniatites, ammonites and belemnites are all used as zone fossils, and the first two are the most important of all groups in this respect. Many of them must have been extremely mobile and are widely distributed geographically. New forms rapidly succeed old in the rock succession, and they are extremely common fossils. The Jurassic rocks in particular can be divided into very narrow time zones on the basis of the ammonites they contain.

A cephalopod: the ammonite *Dactylioceras commune* from the Lower Jurassic.

A belemnite guard preserved with part of the proostracum. (JS)

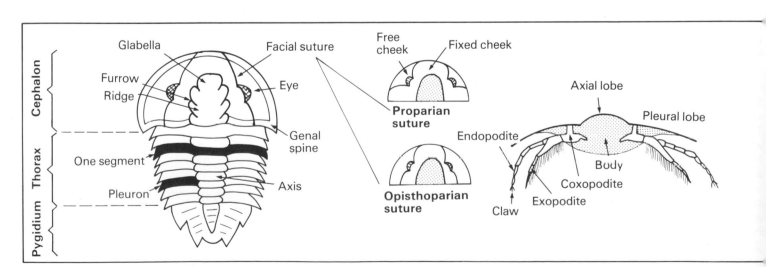

Fig. 9.17   Trilobites. The main features of these fossils are shown on the left, and details are enlarged on the right.

6 The trilobites (class Trilobita of phylum Arthropoda)

This is another group of fossils with no living representatives, like the ammonites, belcmnites, ancient corals and most brachiopods. The segmented nature of the fossilised skeleton (Fig. 9.17), together with the fact that some trilobite fossils indicate that this was an exoskeleton, shed several times as the animal grew, have led to the classification of the trilobites in the vast phylum **Arthropoda**, which today includes three-quarters of a million species of insects, spiders and crustacea (crabs, lobsters). Trilobites were the most primitive of these arthropods and are amongst the oldest animals found with a plentiful fossil record: they occur in the earliest Cambrian rocks. Like the other arthropods the trilobites possessed jointed legs and advanced sense organs, such as the oldest known examples of compound eyes (similar to those in bees). Most crawled on the sea floor, but some probably burrowed in the sediment (streamlined shape, no eyes), floated or swam through the water (large area of spinose skeleton, eyes beneath body), or lived in darker deep zones (large eyes).

Fossil record. All our knowledge of the trilobites is derived from the fossil record, although it is important for us to attempt to establish their relationships with modern animals.

The first trilobites in the Cambrian rocks (Fig. 9.18) are the oldest zone fossils. At this stage they are characterised by forms with spinose skeletons, small tail-shields, and if eyes are present they are attached to the glabella on the headshield. Some of the largest trilobites, up to 45 cm long, occur in Cambrian rocks. The following Ordovician period witnessed an almost complete change and replacement by new families with a much wider adaptation to different marine environments. Silurian forms were similar, but at the end of that period many groups became extinct. The decline continued through the Upper Palaeozoic, as the fishes and ammonoids provided increasing competition, and final extinction came in the Permian period (about 225 million years ago).

The trilobites thus provide us with an almost complete record of the progress of a major group of animals: we miss only the earliest stages of their history (which is true of virtually every major group), and we can study their increasing importance, the height of their development and their decline to extinction.

7 The echinoderms (phylum Echinodermata)

This name means 'spiny-skinned' and is a good description of such members of the group as sea urchins and starfish. It is not so accurate for others such as the sea lilies and soft sea cucumbers. The essential features of the members of this phylum are that they are all marine creatures living largely inside a shell, or test, of interlocking calcite plates (or separated calcite spicules) and have a unique internal circulatory system known as the water vascular system which is particularly concerned

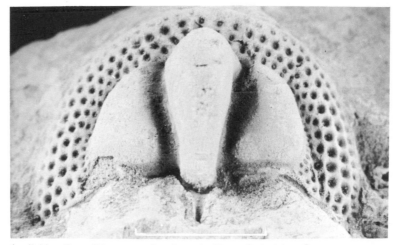

A trilobite: *Cryptolithus concentricus* from the Upper Ordovician. Compare this head shield with Fig. 9.17 and note which features differ from the 'typical' form depicted there. It is common to find headshields and tailshields separated from the rest of the skeleton. (JS)

TE	
CRE	
J	
TR	
P	**Extinct**
CAR	Only one major group left
D	Fewer groups: less important
S	
O	Wide range of new groups
	Major extinction
CAM	(Many trilobites)

Fig. 9.18 The geological history of the trilobites.

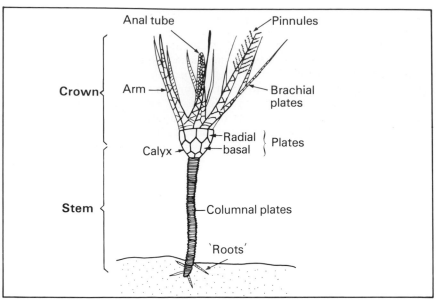

Fig. 9.19   A typical crinoid. These animals have skeletons composed of interlocking plates of calcite. Fossils of anything apart from the columnal plates are rare.

with respiration and locomotion. In many forms the plate arrangements can be divided into five zones (Fig. 9.20), but in others the symmetry of the test is essentially bilateral. To effect growth of the test, new plates are secreted on the margin of the apical system, enlarged as they reach the widest part, and trimmed down near the mouth.

A variety of forms inhabit different marine environments. Some have thicker and stronger shells and live close to the low tide mark, often burrowing into sand or making a home beneath a rock; others live in deeper water, moving across the surface of the sediment in search of prey; and the sea lilies are often attached to the sea floor by stems of varying lengths, using a system of arms to direct the rain of fine organic debris from the upper layers of the ocean into a central mouth.

Fossil record. The possession of a hard, calcite-plate skeleton means that many groups of echinoderms are common fossils: the original shell material is commonly permeated by additional chemical matter and made considerably heavier. In addition,

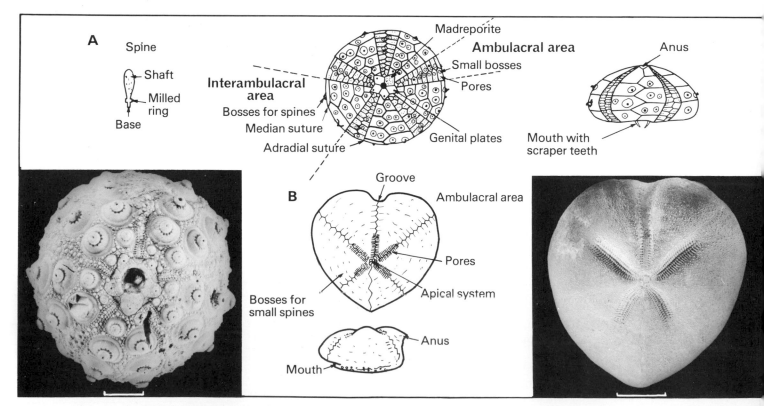

Fig. 9.20   Echinoids. The features of a regular (**A**) and an irregular echinoid (**B**). Compare the two sets of features and relate them to the modes of life of the animals.

the many forms in which living tissue divided the calcite plates tended to disintegrate on death, leaving piles of calcitic detritus. Many of the Carboniferous limestones of Britain are composed largely of such fragments, sorted by marine currents before final deposition.

The greatest variety of the fixed, stem-based sea lilies (**crinoids**, Fig. 9.19), together with several extinct related varieties of echinoderm (e.g. **cystoids, blastoids**) is found in Palaeozoic rocks (Fig. 9.22), but complete specimens are rare, and it is most common to find the separated calcite plates. Only one group of crinoids survived into the Mesozoic and Cainozoic, where stems and cups, including those of free-swimming, stemless varieties, are found in the Middle Jurassic and Upper Cretaceous rocks.

Sea urchins (**echinoids**) are rare as fossils until the Mesozoic, where they are commonly associated with shallow, clear, lime-rich seas including reefs. They are particularly common in the Upper Cretaceous Chalk, where the evolutionary sequence of Micraster has been suggested (Fig. 9.21) and the group is used for zoning. Sea urchins seldom live deeper than 200 m and their presence in the Chalk shows that the rock was formed in relatively shallow water.

8 The graptolites

The graptolites are an extinct group of tiny colonial animals, which leave markings on ancient shales and silts resembling the blades used in fretsaws (Fig. 9.23). They have been extinct for 300 million years and bear such little resemblance to any living

creatures that it has been difficult to associate them with any major group of animals. The colonial form at first suggested that they should be assigned to the coelenterates with the corals and many other small colonial creatures. Modern opinion tends to place them closer to the very primitive members of the phylum Chordata because of the distinctive way in which they build their skeletons: this phylum includes forms which have no skeletal backbone to support the interconnecting notochord, as well as the higher vertebrate animals.

Stages	Zones	Basal view	Side view
Micraster chalk or **Lower Senonian stage**	*Micraster coranguinum*		
	Micraster cortestudinarium		
	Holaster planus		
Upper Turonian	Terebratulina lata		N.B. A: Sulcus B: Labrum C: Posterior

Fig. 9.21 Evolution in echinoids. Notice the changes in particular features as time advances (towards the top of the diagram).

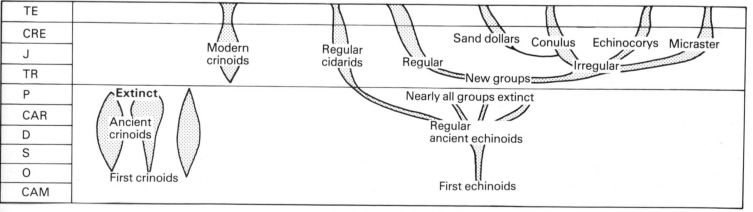

Fig. 9.22 The geological history of the crinoids and echinoids.

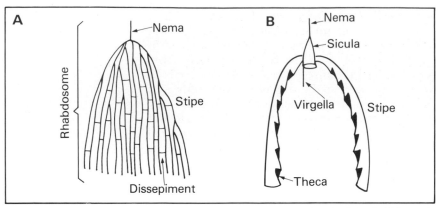

Fig. 9.23 Graptolite features. Simple diagrams showing the many-branched dendroid form (**A**) and the more typical graptoloid form (**B**).

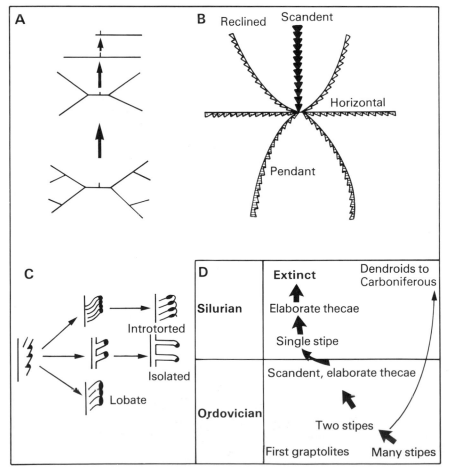

Fig. 9.24 Graptolite development. Graptolite development took place in a variety of ways. The simple features were altered by reducing the number of stipes (A), mostly during the Ordovician; by changing the direction which the thecae faced (B), a process which took place in the later Ordovician; and by elaborating on the basic form of the thecae (C) – in the late Ordovician scandent forms and in the Silurian monograptids. The general progress within the Group is shown in D.

Fossil record. Graptolites form extremely delicate skeletons and their fossils are largely restricted to fine-grained sediments deposited in quiet conditions; some are preserved in silts or sands, but with less detail.

The lowest Ordovician rocks contain many-branched (**dendroid**) forms, which were probably attached to the sea floor by a stem. The branches were soon reduced in numbers, and the two-branched 'tuning-fork' varieties became important. Further modifications appeared rapidly in the Ordovician rock succession, involving the alteration of the orientation of the branches until they coalesced into a single, double-sided branch (Fig. 9.24), and the elaboration of the individual cups in which the animals lived. By the Silurian the graptolites were nearly all single-branched with the cups (thecae) on one side only; a second period of elaboration was experienced with lobate, hooked and isolated forms, but all except for a few dendroids became extinct by the end of the Silurian. It is thought that most of the graptolites with a small number of branches lived in a floating existence in the plankton, since their fossil remains are found with a similar distribution to other planktonic forms, i.e. in sediments formed in conditions which were unfavourable to bottom-living organisms so that the only fossils they contain have dropped in.

The rapid changes in skeletal shape and the widespread geographical distribution of the graptolites make them good zone fossils for the fine-grained rocks of Ordovician and Silurian periods. Much of the history and structure of complex areas such as the Southern Uplands of Scotland, and central Wales, has been worked out after a study of the graptolites contained by the rocks. It is a pity that the fragile graptolite skeleton disintegrated when incorporated in coarser rocks, or when involved in the rougher conditions of sedimentation in which only shelly fossils like brachiopods, trilobites and corals were preserved, but occasionally graptolites do occur in such rocks and enable their age sequence to be tied into the finer, graptolite-rich shales.

9 The vertebrates: from fishes to man

All the groups of animals so far mentioned are invertebrates, having no backbone. The vertebrates (fishes, amphibians, reptiles and mammals, including man) are a very important group today, and their development is particularly interesting to us as human beings. All the vertebrates are members of the phylum Chordata: their vertebral column supports and protects the main nerve chord.

Fossil record. The fossil remains are generally much larger than those of the other groups with which we have been dealing, and are less common. The bones of fishes are often extremely delicate and are preserved only in fine-grained sediments. Many vertebrates live on the land, where weathering processes and scavengers soon cause even the most massive bones to disintegrate. The palaeontologist may discover a huge fossil 'graveyard' with the bones of hundreds of animals, but is often restricted to a piece of bone from a limb, skull or vertebral column, or even just a few teeth. The sharks, for instance, have a non-calcified skeleton which is not preserved, and their history is virtually known only from their fossilised teeth.

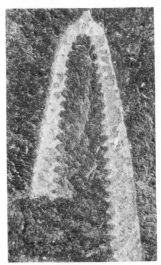

Graptolites: *Didymograptus*, Ordovician (left)
Climacograptus, Upper Ordovician. (JS)

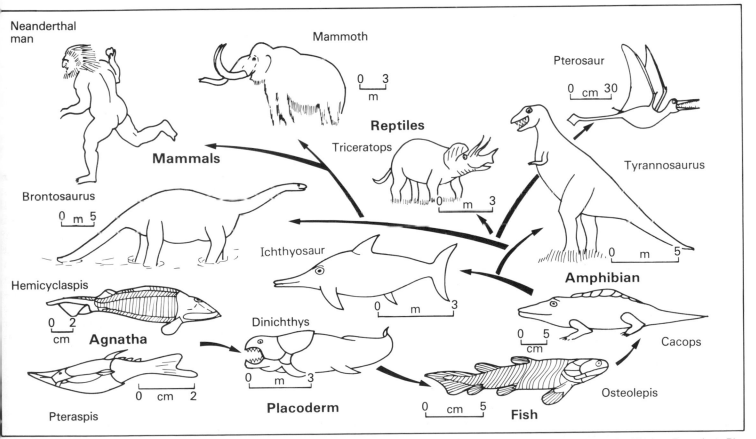

Fig. 9.25 From fishes to man. Agnatha – Hemicyclaspis, Pteraspis (Silurian, Devonian); Placoderm – Dinichthys (Devonian); Bony Fish – Osteolepis (Devonian); Amphibian – Cacops (Permian); Reptiles – Brontosaurus, Ichthyosaur, Pterosaur (Jurassic), Triceratops, Tyrannosaurus (Cretaceous); Mammals – Mammoth, Neanderthal Man (Pleistocene).

A study of the different groups of vertebrate fossils illustrates their development (Fig. 9.25).

(1) The **Agnatha** are the most primitive fish-like animals, and have no jaws or paired fins. They are the oldest fossil vertebrates: scales occur in the Ordovician rocks of Wyoming and their remains become more common in the Upper Silurian and Devonian rocks. These early varieties were mostly ostracoderms, which had bony plate armour over their head regions and lived until the end of the Devonian in mainly freshwater conditions. The lamprey is a living representative which has adopted a parasitic mode of life.

(2) The **placoderms** were another early group of fish-like creatures, but had primitive jaws and paired fins. They contained a variety of forms: some, like the ostracoderms, were small (10–30 cm) bottom-living forms, but others were highly mobile predators, like *Dinichthys*, which was 10 m long. They were particularly important in the Devonian period, but became extinct before the end of the Permian.

(3) Another group, of which we have a fossil record beginning in the Devonian period, includes the **sharks** and **rays**. They have a more complex body including more specialised jaws, and a shape which is well-adapted to their role as fast predators (sharks), or bottom-living (rays). Sharks are known mainly from their teeth fossils, but also from carbonised film impressions on fine-grained rocks. They have altered very little through geological time, surviving many competitors, e.g. the huge marine reptiles of the Mesozoic and the whales of the Cainozoic. It seems that they were at their largest (15 m long) in the Miocene.

(4) The Devonian period is often known as the 'Age of Fishes', for besides the three groups already mentioned we also find the first fossils of the **true** or **'bony' fishes** in rocks of this age. These fishes have strong, calcified skeletons, and have become the most common types: in fact they include more species than all the other groups of vertebrates combined – eels, cod, flat-fish, lung-fish, sturgeon, marine and freshwater types. The bony skeletons are preserved frequently as fossils and sometimes the patterns of scales can be detected.

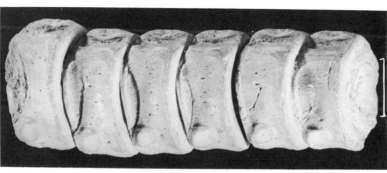

A reptile vertebral column from Jurassic rocks. (JS)

Many of these bony fish have fins with a ray-like pattern of supporting bones (Fig. 9.26), but a small group have a lobe-like pattern which gives a much stronger support. It is thought that the latter feature, combined with other skeletal developments and the ability to breathe air instead of water-filtered oxygen, was part of a series of changes leading towards the amphibian group, which first appeared later in the Devonian.

(5) The **amphibians** are the simplest group of vertebrates able to live on dry land, though they have to return to water to lay eggs: larvae grow in the aquatic conditions and breathe through gills, and the process of metamorphosis is necessary for the change to the adult state. The earliest Devonian varieties had short legs and could only drag themselves slowly along the ground, but by the Permian they were more mobile, and their longer legs could lift them clear of the ground. They then had to meet competition from the reptiles, and now only a few groups, such as the frogs and salamanders, are left.

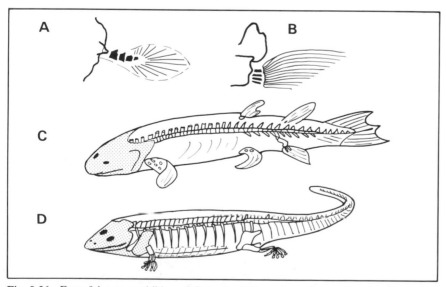

Fig. 9.26 From fishes to amphibians. **A** Lobe-fin, different from the ray-fins (**B**) of fishes: this was the basis for developments which led to land vertebrates. **C** A crossopterygian fish, which had lobe-fins and could breathe air. It is thought that this type may have developed to an amphibian **D**. What features of these animals suggest similarity of bodily form, and what features are related to their modes of life in water (**C**) and on land (**D**)?

(6) The **reptiles** were the first group of vertebrates able to live completely out of the water. The earliest, Upper Carboniferous, varieties looked very much like the amphibians of the time, but had the advantage of being able to lay eggs with a protective covering. From the Permian to the end of the Cretaceous these animals dominated the scene.

The earliest reptiles were small creatures, and although they were freed from the aquatic life by their reproductive capabilities they were still bound to the water areas by the availability of food. It seems possible that it was not until the Permian that meat supplies in the form of large plant-eating reptiles became available over a wide area of land, a development which led to diversification of the group as a whole. Some of the Permian varieties had characteristics which are more reminiscent of the mammals, such as a false palate separating the mouth and nasal passage, and it is thought that this group, or some close relation, gave rise to the mammals.

Many of these varieties are confined to the Permian period, and they were replaced by a wide spectrum of new groups in the Triassic, including the land-based dinosaurs, gliding and flying pterosaurs, and the ichthyosaurs and plesiosaurs which returned to marine conditions. These included the largest creatures on the earth until the end of the Cretaceous, when most became extinct. The reason for this is a great puzzle, especially as many of the groups were at the peak of their development. Suggestions for solutions have included a change in climate, or the results of the changes in vegetation during the early Cretaceous period, but none are satisfactory. Today the snakes, crocodiles, turtles and lizards are the remnants of this once all-powerful group.

(7) The **birds** are feathered and warm-blooded flying animals. The earliest bird-like creature is recorded from the Jurassic rocks of southern Germany. It had feathers, but reptile-like teeth, and is thought to be a link between the groups. By the Tertiary there were large numbers of flightless birds like our present-day ostrich.

(8) The **mammals** are the dominant group of animals throughout the world today, but although the first mammal fossils are found in rocks of Triassic age, the group spent over 100 million years as an insignificant minority overshadowed by the reptiles. The wholesale extinction of the reptiles at the end of the Cretaceous (i.e. just over 70 million years ago) enabled the mammals to spread into many of the environmental 'niches' vacated on land and in the sea. The mammals are warm-blooded, mostly covered with hair or fur, and have the most advanced and adaptable bodies of all the animals living on the earth.

The primitive groups include the egg-laying monotremes (e.g. duck-billed platypus) and the marsupials (e.g. kangaroo and opossum) which protect their immature young in a pouch. Ninety-five per cent of the mammals alive today are of the third, placental group, named after the organ which supplies nourishment to the embryo in the womb: many of these animals are born almost ready to face the rigours of the world around. The placental group includes many hoofed mammals (e.g. odd-toed horse and rhinoceros; more plentiful even-toed cattle, pig, deer, camel, elephant), the edentates with smaller teeth (e.g. armadillo, sloth), the rodents, the flesh-eating group (e.g. dog, cat, bear) and the primates (e.g. lemur, monkey, ape, man). Many of these have well-documented histories in spite of the fact that fossil preservation is unusual on the land where scavengers and the agents of decay are at their busiest (Fig. 9.27).

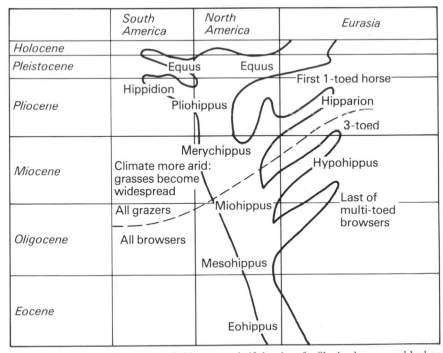

Fig. 9.27 Evolution of the horse. Eohippus was half the size of a Shetland pony, and had an arched back, four hoof toes and low-crowned teeth (for browsing on leaves). Mesohippus was more horse-like, though only 60 cm high; it was three-toed, had a larger brain than Eohippus, and there were sharp crests on its teeth. Merychippus was up to 1 metre high at the shoulder, and developed a deeper jaw to hold the resistant, high crowned teeth suitable for chewing grasses.

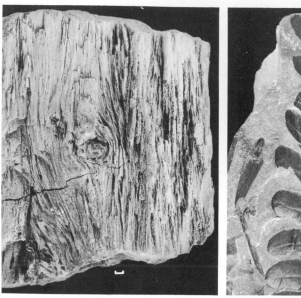

Plants: fossil tree trunk from the Jurassic (left)
Linopteris, Upper Carboniferous (Coal Measures) (JS)

The history of the development of the vertebrates is particularly interesting because it provides us with a record of the diversification and replacement of a wide range of animal types through the geological record. Whereas nearly all the major groups of invertebrate creatures appear almost suddenly in the Cambrian or Ordovician without any record of their earlier connections, a more adequate family tree of the vertebrates can be constructed. Many of the groups of vertebrates can be linked by fossil skeletons showing intermediate characteristics, but we shall probably never be able to answer many of the questions concerning some of the more dramatic steps involved. How did jaws develop in the primitive fishes? Why did some forms move out on the land? How did the reptiles begin to lay protected eggs? When and how did the mammals and birds acquire warm blood together with methods for regulating their body temperatures?

10 Plants

Fossil plants are less common than animals and although they have played a vital part in the development of life on the earth, we can reconstruct only a poor and fragmentary picture of their history. Many plants are soft, or susceptible to rapid decomposition when they die, and it is only rarely that carbonised films of leaves and stems are preserved: rapid burial in fine sediments is necessary, and this means that the high degree of compaction associated with muds squashes plant remains flat. Occasionally tree stems may be preserved in sandstones with the original round cross section (Fig. 8.5). It is common to find the leaves separated from the stems, and they are often transported before entombment in sediment.

The **simplest plants** are those without any circulatory system of their own. Most of them live in water or in very damp conditions. They include the single-celled diatoms and other planktonic forms, which may give rise to thick deposits of their siliceous skeletons; the blue-green algae, which often form large colonial structures and are amongst the earliest fossils known in the Precambrian rocks; the parasitic fungi; and the mosses and liverworts. All are rare as fossils, though the first two groups are occasionally important.

Plants living on the dry land need a more advanced circulatory, or vascular, system and complicated reproductive facilities: roots, stems, leaves, spores, flowers and seeds have developed. The **earliest vascular plants** (Silurian of Australia,

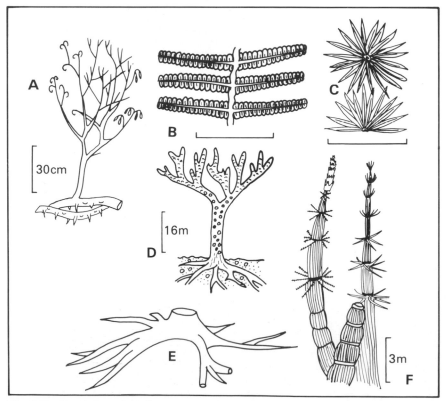

Fig. 9.28 Fossil plants. Scale line represents 2·5 cm unless stated. **A** Psilophyton (Devonian); **B** Pecopteris; **C** Annularia; **D** Lepidodendron; **E** Stigmaria (roots); **F** Calamites (**B–F** all Upper Carboniferous).

Investigating fossils

Investigation 1: Introduction to fossils. It is important to relate the fossils to living organisms of today and to examine the different ways in which fossil material may be preserved in the rocks.
(a) Examine a group of named fossils from the school collection. Which of these are like modern animals or plants?
(b) Which modern organisms are not represented?
(c) Which parts of an animal are preserved as fossils? Is the whole animal preserved, and how much detail can be discovered from the part that is (e.g. shape, structural features, colour)?
(d) Make notes concerning the evidence you can detect relating to the modification of the oganism (or the fragment preserved) from the time of its death up to the time it was extracted from a rock (e.g. What has happened to the original organic features?).

Investigation 2: Examining fossils. Answer the following questions for each of the major groups (corals, brachiopods, lamellibranchs, gastropods, ammonites, belemnites, trilobites, crinoids, echinoids, graptolites, vertebrates, plants – even micro-fossils if you can obtain a slide of some and a microscope). You may find it best to carry out this investigation before reading the relevant section of the text.
(a) What is the basic shape of the fossil (i.e. tubular, flat shells, segmented, globular)? Compare your own description with the technical terms used for each group.
(b) What are the major features (e.g. external ornamentation, internal markings if seen)? Draw one specimen and name the features using the diagrams in the chapter to check that you have picked them all out. Add a scale.
(c) Look at several specimens belonging to the same group. How much variety is evident in the various features you have just been describing?
(d) Discuss the significance and possible functions of the features you have just been drawing and labelling.
(e) In which geological period did the fossils you have been drawing and examining live? Show the range of the group on a simple chart.

Investigation 3: Classifying fossils. Take a group of ten different brachiopods or ammonites and construct a key to their classification in the following way:
(a) Select one feature (e.g. smooth shells or ornamented shells) and divide the group into two parts on this basis. The ratio does not matter: you may have one smooth and nine ornamented.

(b) Select another feature which divides the smooth shells into a further two groups, and repeat this process until each of the individuals has been distinguished from the rest.
(c) Complete the same procedure for the ornamented shells.
(d) Record the information on a chart:

	Feature 2A		Individual 1
Feature 1A	Feature 2B	Feature 4A	Individual 2
		Feature 4B	Individual 3
	Feature 3A	Feature 5A	Individual 4
Feature 1B		Feature 5B	Individual 5
	Feature 3B	Feature 6A	Individual 6
		Feature 6B	Individual 7

(e) Compare your classification with one used in a textbook, and modify it to include any new fossils you discover.
NB This process can also be used as an introductory investigation to distinguish between representatives of each of the major fossil groups listed in Investigation 2.

Investigation 4: Palaeoecology. Examine a large group of fossils obtained from a narrow band of rock.
(a) Do the fossils show signs of abrasion (i.e. have they travelled far before entombment in the rock, or did they live where the rock was formed)? This will enable you to remove forms which could not have lived in the environment where sedimentation took place.
(b) Divide the fossils into groups possessing similar features and modes of life: i.e. which were the bottom-livers and which were the swimmers? Is there a wide range of forms?
(c) Take each group and measure the range of sizes included. These can be plotted on a graph.
(d) Draw your own conclusions concerning the difference between the assemblage of fossils and the possible living situation. Use the evidence from the fossils and the sediment in which they are entombed to work out the sort of environment in which these fossils once lived.

Devonian of Britain) have a simple stem lying along the ground and topped by bare branches carrying spores (Fig. 9.28). By the Carboniferous period there were dense forests of giant horsetails, tall spore-bearing lycopod trees, ferns and early seed-bearing conifers. Other groups of seed-bearing plants, such as the ginkgos, became important in the Mesozoic: all these early seed-bearing varieties are known as **gymnosperms** ('bare seed'). Our present flowering plants, or **angiosperms** ('hidden seed'), which make up 95 per cent of all plant species today, only developed in the early Cretaceous, and grasses only became widespread in the Miocene.

This is the general pattern of plant development. It is interesting to note that the first land plants came into being at a similar date to the land animals (amphibians and insects); that the Cretaceous development of flowering plants was accompanied by that of pollinating insects; and that the spread of grasses in the Miocene caused many mammal groups, e.g. the horse, to change their feeding habits from browsing on leaves to grazing (Fig. 9.27).

11 Microfossils

The fossils we have mentioned so far are those you will find in the rocks you examine with the naked eye. Most of them are between 1 cm and 8 cm in length, although the vertebrate remains in particular may be much larger. None of them, however, are of much use to the palaeontologist who is examining the cores of rock chippings brought up from a depth during drilling. Such a person is dependent on the microfossils, which are present in most rocks in great numbers. Just as there is a wealth of microscopic life on land and in the sea, so there is a vast realm of microscopic fossil evidence. A whole new branch of palaeontology has arisen from the study of these tiny forms, and has contributed important new techniques of investigation. The microfossils can be used in the same way as normal fossils for giving a geological date to the rocks in which they occur and for providing information concerning the development of life on the earth, or the conditions in which the rocks were laid down.

The size of microfossils makes it inevitable that they will be of use only to specialists, since their study requires the use of sophisticated extraction techniques, the possession of expensive equipment and a knowledge of the latest discoveries in the rapidly enlarging field of identification.

Microfossils are drawn from several of the animal groups, including those from which important macrofossils are drawn.

(1) Protozoans of the order **Foraminifera** are widespread and abundant, especially in the Mesozoic and Cainozoic rocks. They are some of the simplest animals, having no cellular division of their body, although the more advanced types may produce calcium carbonate shells with many interconnected chambers. These tiny fossils are used widely in the search for oil.

(2) **Ostracods** are minute relations of crabs and lobsters (i.e. Arthropods), living in the water and related to water-fleas. They have a bivalve shell covering their body, and the wide variety of ornamentation on this shell is amazing for such tiny forms. Different types of ostracod today are very sensitive to changing salinity, so the fossils may be used for interpretations of past salinities.

(3) **Plant spores** are found in many different rocks, including those formed in the sea, as well as those on the land. Coal seams and the overlying shales can be correlated across the country on the basis of the spore assemblages they contain.

(4) New groups are coming into use as the study of microfossils gains ground, although it will be many years before they attain a widespread significance. One such group is known as the **Hystrichospheres**, which were once part of the plankton floating at the surface of the oceans; another is the **Acritarchs** of more uncertain affinities. All marine sediments contain these microscopic forms which have spherical, ellipsoidal or polygonal shapes ornamented with ridges or growths (spines or tubes, branching or interconnected). One gram of shale may contain thousands of specimens, which can be studied after removal from the crushed rock. Because they originated from floating plankton they are widely distributed throughout the world: forms of Jurassic age in England and Australia are identical. Many groups show signs of rapid development into new varieties, another important feature to the geologist. When more is known about these microfossils they will surely help us to gain a fuller picture of the history of our planet.

Fossils and time

Certain facts emerge from a close study of the fossils contained in the rocks. A pattern can be distinguished in the development of life on the earth, even when one bears in mind the bias imparted to the record by the processes of fossilisation. A number of aspects of this pattern are particularly significant.

(1) The record of **the earliest life on the earth** is extremely sparse. There are hardly any fossils in the rocks over a span of 3000 million years: some organic compounds have been recognised from extremely early rocks (e.g. Fig Tree Series of South Africa, *c* 3200 million years old), and there are remains of the most primitive plant life from other rocks of later date (e.g. Gunflint Chert of Lake Superior region, *c* 2000 million years old), but no geologist would pretend that the record is as full as he would like. There is no indication of the actual origin of life: it seems that the most primitive forms were the earliest, but the manner and date of their emergence, or creation, are unknown. A number of theories have suggested answers to this problem, involving life emerging from organic compounds present in the special conditions of a primeval sea. We can never know the complete facts concerning the origin of life. Laboratory studies may indicate how life could have, or might have, emerged, but not how it started.

(2) **The numbers and variety of living things have increased** through geological time: compare the situations in the major groups during the Silurian, Carboniferous, Jurassic and Eocene periods, i.e. at roughly 100 million year intervals. The increase has not been a steady process, but has been punctuated by periods of more spectacular expansion, like the Ordovician and Jurassic, as well as by others which witnessed widespread extinctions (Permian, late Cretaceous). After the very slow Precambrian progress within single-celled creatures, the Cambrian system of rocks presents us with a wide range. This includes not only the first records of metazoan (many-celled) creatures, but the representatives of almost all the major groups of animals. From that time to the present there is a record of increasing complexity, and a diversity which spreads from confinement in the sea to gradual mastery of the land and air.

(3) During the course of this process of expansion of life on the earth **new species** of animals have developed, and others have become extinct. The arrival of the different groups of vertebrate animals illustrates this, and on a small scale fossil groups have been examined closely and show tiny, but definite, changes of shape with the passage of time (Fig. 9.21). Gaps in the records are due to the fact that many of the vital 'links' have not been preserved as fossils, and we saw earlier in this chapter that the scales are weighted against the chances of any animal leaving a record of its existence. As the descent of a particular group of animals is traced through the rocks, modifications clearly happen: this process is known as evolution. Some animals have shown rapid evolutionary changes (e.g. graptolites, ammonites), whilst others have shown little or no change over the whole range of geological time in which fossils are commonly found in the rocks (e.g. brachiopods such as *Lingula*).

(4) Another concept which is useful in the interpretation of the fossil record is that **the environment** clearly affects the features of an animal's anatomy. This is obvious when we compare the shapes of, for instance, fishes living in the buoyant medium of water and their distinctive methods of locomotion and respiration, with those of the land vertebrates, which have limbs to lift their bodies clear of the ground so that they can move efficiently in the less buoyant medium, air. It is also instructive to compare the shapes of the extinct reptile, the ichthyosaur, and the dolphin, a mammal: they are almost identical, and both have lived in similar conditions.

(5) The increasing numbers of animals and plants on the earth's surface eventually led to **competition** between the species. Each environment could support only a limited number, and the entry of a new, better-equipped, or adapted, species would lead to

A palaeontologist comparing mammoth teeth found in glacial deposits in Illinois, U.S.A. (USGS)

the replacement and extinction of the old variety, unless it could readapt itself in another environment. Only one complete phylum has ever become extinct, and that is a rather small group of Cambrian sponge-like creatures, the Archaeocyatha: representatives of all the rest which are first found as fossils in the Cambrian and Ordovician rocks are still with us. As well as competition the extinction of groups of animals must have been due to changing environmental conditions. If an animal became too specialised a slight change would cause its extinction and this could lead to a whole series of extinctions, since the interrelationships in nature tend to be very complicated. Yet many animals survived drastic changes, and this must have been due to a very adaptable body structure and mode of life.

The present animal and plant populations of the world are the result of 'descent with modification' from primitive ancestors of the main phyla. This concept is well-supported by the evidence we have studied. Within each species there is much variability: just look at the human race, all members of the species *Homo sapiens*, but with so many variations in skin colour, height, hair, nose, etc. Such diversity extends to all species of animals and plants, and in many is fostered by nature's bounty in reproduction: salmon lay as many as 28 million eggs each in a year. Some members of each species must be slightly better adapted than the others to their particular environment, with the result that they will survive more easily and have a greater effect on the next generation in reproduction. The study of genetics has helped to show how vital are the hereditary genes in determining the characteristics of the new populations.

Fossils record the general line of evolutionary progress, although the detailed observations concerning the mechanisms of change have to be made on living animals. We can also derive an idea of the slowness of the processes, since geological time is so vast. It has been calculated that the diameter of the horse's tooth changed as it turned from leaf-browsing to grazing – by 1 mm in 5 million years. And this was one of the more rapid changes!

(6) The development of life through geological time is also important in a very practical way for the geologist: it is the basis of nearly all his **time correlations**. Now that radioactive dating is becoming of greater significance we shall be able to fill in some of the gaps in the historical picture and give more accurate numerical dates, but fossils are the only way of comparing the relative age of sedimentary rocks in widely separated areas.

When we have studied a group of rocks in one area, and established their age sequence by such methods as those suggested in the course of Chapter 8, it is vital that we should be able to compare them with other rocks elsewhere. We can often trace a band of distinctive rock across the country: the ridge of Chalk across the centre of the Isle of Wight is an outstanding example. Aerial photographs and geophysical methods help us to do this on a larger scale, but it is dangerous to put complete trust in such methods, for the conditions in which a certain type of rock are formed may migrate across a region with time: the same rock-type may be formed earlier in one place, and later elsewhere.

Fossils provide us with the most reliable method of telling whether two rocks are of the same age. Each rock contains a certain assemblage of fossils, and those which were formed more recently will have more highly developed and modern-looking fossils. This is due to the results of the process of evolution combined with rapid geographical distribution of new species. Each rock, or part of a rock, or group of rocks, with a unique assemblage of fossils is known as a **zone**. Such zones are referred to by an exclusive species of fossil they contain, and are the smallest units of rock-time division. Several zones form a **stage** (cf. Fig. 9.21). Zones vary greatly in thickness since the rate of rock-deposition has not always varied in the same way as the rate of evolution.

The ideal **zone fossils** would be animals which have had a wide geographical sphere of influence; have evolved rapidly and show rapid changes as they are traced upward through the rocks; have been able to live in a variety of environments; are of a relatively small size (i.e. 5–10 cm) so that they can be seen as a whole and easily extracted from the rocks; and are commonly preserved. In practice zone fossils may possess many, if not all, of these qualities, but some rocks have to be zoned by the least likely fossils, since they contain nothing else. The graptolites (Ordovician, Silurian), goniatites (Devonian-Permian), ammonites (Jurassic, Cretaceous) and several groups of microfossils are the best fossils for this purpose. Refer back to your studies of each group and pick out the reasons for this. Free-floating crinoids and some trilobites are also good, and where none of these are present corals, brachiopods, belemnites, freshwater mussels, plant spores and fish have been used (Fig. 9.29). Why are these groups not such good zone fossils?

The main difficulties arise when a rock changes its facies, and it has often been the case that a different set of conditions was not liked by the animals living at the time of formation. Thus graptolites are typical of the fine-grained, deepwater shales of Ordovician and Silurian age, but are seldom preserved in the coarser, shallow-water deposits of the same age which have different fossils (mainly trilobites and brachiopods). It is usually possible, however, to find a rock layer where the graptolites and trilobites occur together.

There have been notable cases where the study of fossil zones has helped geologists to work out the structures affecting the rocks of an area. This was the case in the Southern Uplands of Scotland, where the study of the graptolite zones led to the discovery that the rocks were tightly folded. Palaeontologists assist in the search for oil by tracing oil-bearing horizons across country and determining the structures affecting the rock successions.

Fossils and place

Fossils may also assist us in attempts to reconstruct the environment in which they lived. Ecology is the science which studies the relationships between organisms and their environment today; **palaeoecology** is the study of ancient life environments. Whilst the ecologist is sometimes embarrassed by the wealth of data available to him and the complexity of the relationships he observes, the palaeoecologist is often embarrassed because either his data are too few (owing to the poor fossil record), or because there is a mixture of fossils which once lived in the place of sedimentation and those shells which have been transported to the area from elsewhere, or those which have dropped into the sediment from the plankton or swimming creatures.

Animal and plant shapes are related to their living functions. Thus it is possible to distinguish between lamellibranchs which burrow to different depths by the extent of their pallial sinus; regular sea urchins roam about in any direction on the surface of coral lagoon sediments, whilst irregular echinoids burrow in the sediment; and the features of vertebrate skeletons and jaws reveal whether the owner was a marine or terrestrial beast, how active it was, and what was the main component of its diet. These features are preserved in fossils. The most useful fossils to the palaeoecologist are those which are buried in the place where they lived: sedentary forms like oysters, corals, brachiopods and crinoids, and burrowing forms like many lamellibranchs and some gastropods and echinoids.

Trace fossils are becoming increasingly important, since they are also features formed by animals as the sediment accumulates. They include the tracks (straight-line trails or meandering grazing tracks), burrows, resting marks and subsurface sediment-browsing systems. One snag is that it is seldom possible to recognise the originator of the features, since they leave no remains: many crustacea and annelid worms are responsible for such burrows today. Certain types of trace fossils are associated with particular rock facies, and they may also be used to work out the rates of sedimentation.

Other new techniques have also increased the scope of such studies. The proportions of the different **isotopes** of oxygen in the calcium carbonate of ocean-floor deposits are related to the temperature at which the shells of the microscopic animals were formed near the ocean surface. The **pollens** in thick deposits of peat give a very good idea of the trees living there in the past, and of the temperature and humidity of the climate at the time. A picture of the falling temperature at the end of the Tertiary era, for instance, can be obtained by studying the animal and plant life of the time.

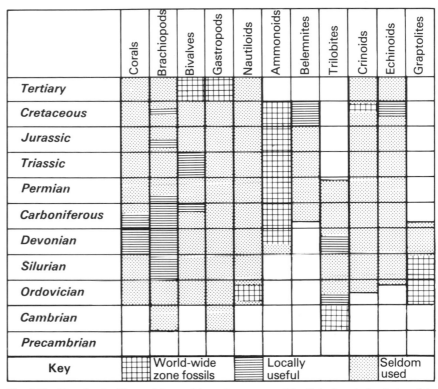

Fig. 9.29 Zone fossils. The usefulness of the various groups of fossils in the relative dating of the rocks in which they occur. Which are the most useful? Notice the outstanding contribution of the various groups of molluscs.

As in the case of the study of zone fossils the study of palaeoecology has an economic application: certain rock facies, identified by their fossil content, are known to be good oil reservoirs. It becomes more and more important to study the complete relationships of the fossils when we find them; not only do we have to recognise the type in order to date the rocks which contain it, but we should note the rock type, the position of the animal in the rock and the group of fossils of which it is a part.

Fossils and plate theory

The development of life on earth has been affected by the changing configurations of the oceans and continents as the system of plate tectonics has led to changes across the earth's surface. The detailed implications are still unclear, but major points are emerging. Thus the fossils occurring in rocks of Cambrian age in northern Scotland, northern Newfoundland and the interior parts of the Appalachian mountains of eastern USA are different from those in southern Britain and southern Newfoundland.

This was a puzzle to earlier geologists, who postulated ancient continents in the middle of the Atlantic ocean, and then allowed them to disappear when they were no longer required. It is now seen that the two distinct zones were on opposite margins of an earlier version of the Atlantic ocean, and that they were brought together along a single convergent plate margin by the end of the Silurian period.

It is also thought that the reptiles were encouraged to spread throughout the world by the existence of the single major continent of Pangaea during the Permian period. After this the continents began to break apart, increasing the isolation and differentiation within the major groups of reptiles. The process does not seem to have been very advanced by the phase of major reptile extinctions at the end of the Cretaceous, in contrast with the extremely rapid evolution of mammals in the early Tertiary. The earlier and more primitive forms of mammals have been preserved by the isolation of continents like Australia and South America for long periods of time, and this helps to account for the immense variety of mammalian forms on the earth at present.

Fossil coral reefs of *Lithostrotion* in position of growth. (Crown Copyright)

10
Rock structures and mountains

Rock structures in the field

So far our examination of the evidence available in the rock record at the earth's surface has been confined to those features which were formed at the time when the rock itself was laid down or solidified. We have studied the composition, grain-size and internal structures associated with the sedimentary, igneous and metamorphic rocks, together with the remains of ancient life trapped in the sedimentary rocks.

Most sediments and lava flows began as more or less horizontal layers, but when we start to investigate rock outcrops in cliff faces and quarries we discover that it is unusual to find horizontal rocks: they are normally tilted, bent or fractured, and often criss-crossed by many cracks. A study of these structures provides us with information concerning the post-formational history of the rocks: their reaction to the stresses and strains which act on them as part of the earth's crust, and to uplift. We refer to such structures as of tectonic origin (Gk, *tecton* = constructor, builder) in order to distinguish them from the depositional or cooling structures.

Studying tectonic structures

Certain observations can be made at rock outcrops.
(1) **Are the rocks the right way up?** Are the oldest at the bottom and the youngest at the top as we would expect, or has the whole group been turned upside down? How can we tell? There are a number of points to consider.

The presence of unconformities is always helpful, because the younger beds cut across the eroded ends of the older (Fig. 8.24). Minor structures in sedimentary rocks, such as false-bedding and graded-bedding (Fig. 8.6), ripple-marks, rain-pits and sun-cracks give us the most useful evidence concerning the 'way-up' of a group of rocks. Fig. 10.1 shows how a knowledge of these facts helps a

geologist to work out the structure of an area. When lava flows are included in a group of rocks we can look for an upper weathered or vesicular zone and lower columnar layer (Fig. 6.4). In coal-bearing rocks we can look for roots going down below the coal seam, but absent from the overlying strata.
(2) **What are the strike and dip of the rocks?** Having determined whether the rocks are the right way up, or have been completely overturned, we shall want to define their position. We must be sure we have an actual bedding surface, and not a cross-bedding lamination, metamorphic cleavage plane, fault or joint surface. We also need a clear surface on which to take measurements: all rock surfaces must be defined in terms of a three-dimensional space, and a flat, vertical quarry face will not give a true reading.

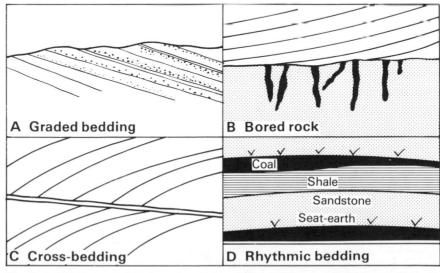

Fig. 10.1 Which way up? Each section of this diagram, **A–D**, contains evidence which will enable you to determine the original order in which the rocks were formed. (The oldest rocks should be at the bottom.)

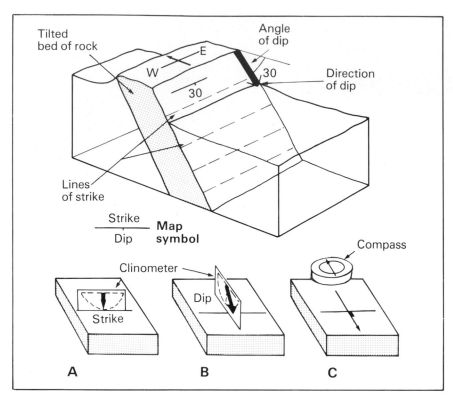

Fig. 10.2 Strike and dip.

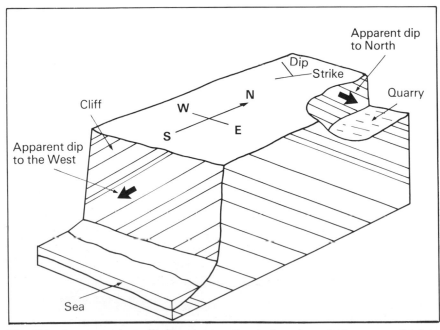

Fig. 10.3 Apparent and true dip. Two apparent dips are shown here – in the quarry wall (where the rocks dip to the north), and in the cliff-face (to the west). The three-dimensional pattern is necessary to find the true dip: the symbol drawn on top of the diagram shows the way in which this is often depicted on maps.

Two directions at right angles to each other are used to define the attitude of a bedding plane. Of these the **strike** is the easiest to determine: it is the bearing of a horizontal line across the surface. The instruments we use to measure the strike include one which will enable us to measure the horizontal (a level or clinometer), and one which will give us a bearing (a compass): the two may be combined in a single instrument. Once the strike bearing has been established a line can be scratched on the rock surface. The *dip*, which is the angle of maximum slope, or inclination, from the horizontal, can then be measured in a direction at right angles to the strike by the clinometer: it may be expressed as, say, 30 degrees towards the south (Fig. 10.2). It is important that such measurements are made, since mistakes can be made by observing just one face of a cliff or quarry (Fig. 10.3).

A similar approach can be adopted when analysing a geological map. Strike lines can be drawn at different heights across a bedding plane: in the case of Fig. 10.2 they would be east–west. They can also be regarded as structural contours – i.e. lines drawn across a bedding plane (or other structural surface such as a fault plane) which join all places of the same height above or below sea level. In this form it is most helpful to relate them to the relief contours, as in Fig. 10.4.

The dip of strata affects the patterns of rocks as they reach the surface – i.e. in their outcrop. If a group of rocks is in the horizontal position their outcrops will run parallel to the contours, whereas if the dip is vertical (90 degrees) the rock will form a narrow, straight outcrop across hills and valleys alike, resembling a wall. When the rocks are tilted at an angle between these extremes, the pattern of outcrops in the valley will depend on the angle of dip and the slope of the valley floor: Fig. 10.4 shows some of these relationships.

(3) **What patterns of folding or fracturing affect the rocks?** As we find out more and more about the rocks in an area we shall see patterns emerging in the structures which affect them. The most important are folds, when the rocks are bent, and faults, when they are broken, but there are also minor breaks in the rocks without displacement on either side, known as joints. Folding and faulting are caused by pressures being applied to rocks. Such pressures sometimes alter the nature of the rock, but always cause some degree of deformation. If the affected rock is still soft and relatively flexible, it will bend into continuous folds, rather like the ones you can make by crumpling the tea-cloth on a

able. If it is brittle, however, the rock will break as stress is applied. Thicker masses of sediment are folded more easily than thin sediments lying on a rigid basement of rocks, as is demonstrated in southern England where the Jurassic and Cretaceous deposits are many thousands of metres thick. When the Alpine folding movements took place the thicker group was compressed into open folds, whilst thinner rocks of the same age north of the river Thames were only tilted and fractured. Occasionally a very weak bed of rock will collapse completely and flow (Fig. 8.18): we have seen how 'plastic flow' occurs in substances like ice and salt, but it can affect a wide range of rocks if a stress is applied for long enough.

A geologist taking a measurement of dip with a compass–clinometer. (USGS)

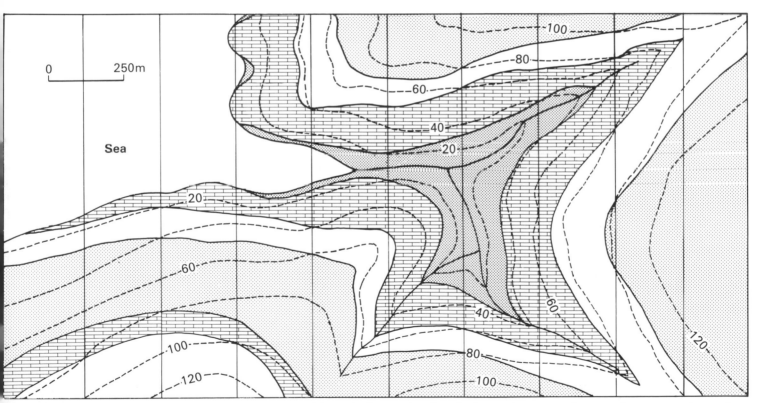

Fig. 10.4 Tilted rocks. The strike lines are drawn across the map so that you can see how to place them. Calculate the dip gradient and direction. (NB Strike lines on geological maps must always pass through an intersection between a relief contour and a bedding plane outcrop line.) Heights in metres.

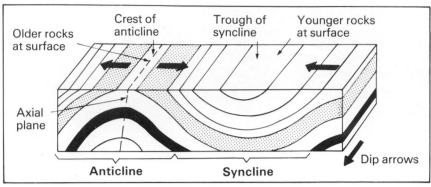

Fig. 10.5   Simple folds. Notice that the rocks at the surface get younger towards the centre (trough) of the syncline, and older towards the centre (crest) of the anticline.

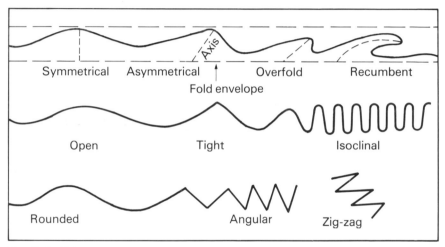

Fig. 10.6   Fold patterns. A few of the patterns which result from the folding of rocks.

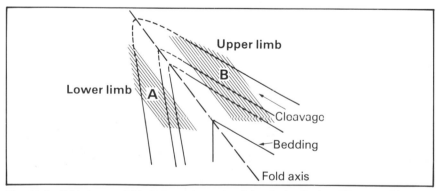

Fig. 10.7   Fold cleavage. This fold was observed and sketched at Bovisand Bay, near Plymouth. Both limbs of the fold are dipping to the south, and this means that the rocks of the lower limb are in reverse order (i.e. the oldest is on top if you draw a vertical line through them). Devise your own rule to determine which way up the rocks are, using the information provided by the dips of the bedding planes and the cleavage.

Folds

There are two types of folds, illustrated in Fig. 10.5. The upfolds are **anticlines**; the downfolds are **synclines**. These basic patterns may vary in a number of ways (Fig. 10.6): folds may be symmetrical, with vertical axial planes, or asymmetrical when the axial plane is inclined to the fold envelope. The term **monocline** is sometimes used for a fold with one very steep limb between horizontal, or nearly horizontal, limbs.

Some are overturned to form recumbent folds, in which the strata on one limb are inverted. They may be open, or tightly pressed together, and they may be rounded or angular in cross-section. All these characteristics are determined by the intensity of the folding pressures, and the softness and flexibility of the rocks when they were folded. A fold may also dip at one or both ends – when it is said to plunge.

The bending of rocks as they are folded often means that the top of an anticline begins to crack open, just as a rubber does when you bend it in half. The joints caused in this way may be filled by percolating solutions, or may form a zone of weakness on the fold crest which is easily attacked by weathering and river action. Although the anticlines form the hills at first, and the synclines the valleys, it is common for this arrangement to be reversed after a prolonged period of erosion. The Vale of Pewsey to the north of the Salisbury Plain is an example of such **inverted relief** since it is a lowlying area along the line of an anticline.

Some of the most spectacular results of folding are to be seen where alternating bands of massive, resistant rock and weak shale are involved. The resistant layers are known as **competent**, and the weak as **incompetent**. Whereas the competent beds resist the stresses affecting the rocks, and form rounded folds, the incompetent rocks crumple easily and are greatly distorted. Some of the results, including overfolds and zigzag folds, can be seen clearly in the cliff rocks of Upper Carboniferous age around Broad Haven, west Pembrokeshire.

The most intense folds are only found in certain regions, and are usually on too large a scale to be traced in a quarry wall, or cliff-face. The Scottish Highlands and the great ranges of fold-mountains such as the Alps, Himalayas and Andes are formed of large-scale folds of the isoclinal and overfold types as well as vast broken slices of rock.

Intensely folded regions may also have experienced metamorphism, which imparts a cleavage to

he rocks. This type of structure enables the geologist to determine which limb is the correct way up
in overturned rocks (Fig. 10.7).

Faults and thrusts

Faults are fractures of the rocks in the earth's crust
where there is displacement of one side relative to
the other. They may be caused when rocks are
crushed together, or when they are pulled apart –
by compression or tension as shown in Fig. 10.8.
The latter seems to be more common, since **normal
faults** are formed in this way, whilst **reverse faults**
and **thrusts** originate when one group of rocks is
pushed over another, and a section of the earth's

crust shortened. Both involve vertical movements
in the rocks, but **tear or wrench faults** affect them
by sideways movements. Faults may also be
described by reference to their relationship to the
directions of dip and strike in the rocks affected.
Thus a fault with an outcrop parallel to the strike of
the rocks it fractures is known as a **strike fault**; one
parallel to the dip as a **dip fault**; and one between
these directions as **oblique**. (NB Another set of
terms for fault patterns is also used. Dip-slip faults
are those involving a movement down the dip of the
fault plane – i.e. what we have termed normal and
reverse faults with a largely vertical movement.
Strike-slip faults involve movements along the
direction of strike on the fault plane, or horizontally:
they include the tear fault.)

Asymmetrical folds in an opencast coalmine, Carmarthenshire. (Crown Copyright)

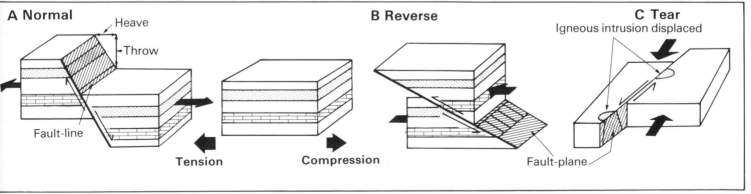

Fig. 10.8 Faults. Some faults involve mainly vertical movements up or down the dip of the fault
plane (normal or reverse), and some mainly horizontal movement along the strike of the fault plane
(tear).

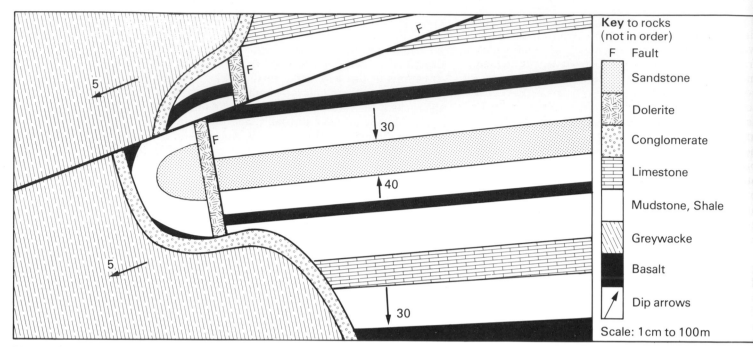

Fig. 10.9 Map exercise. What evidence does the map record which suggests that the rocks in this area have been folded? Write out a list of the geological events which have taken place.

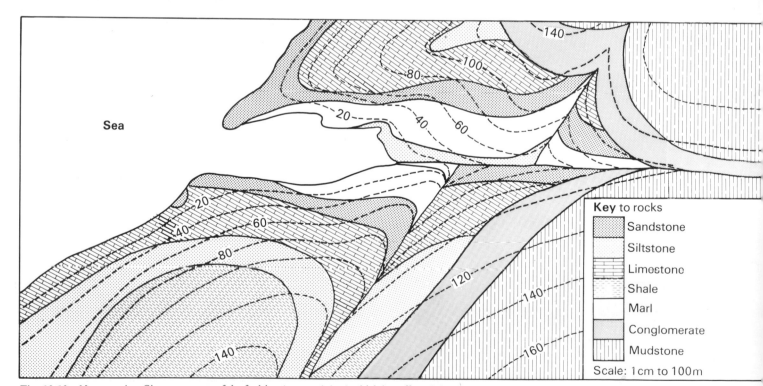

Fig. 10.10 Map exercise. Give an account of the faulting (type and date) which has affected the rocks of this area.

When we examine a small fault closely we often find that there is a zone of fracture a few centimetres wide which is occupied by a mass of small angular fragments and fine clay, produced during the fault movement. If this fault breccia is not present it is common to find the fault planes grooved and scratched by the movement: this feature is known as **slickensides**. In many cases larger faults, extending over hundreds of kilometres, are zones of several near-parallel fractures rather than a single line of breakage. Such faults often do not have a clear outcrop: as zones of weakness they may be occupied by a gully and covered with weathered debris. The outcrop patterns of the rocks on either side of this gully will provide the evidence for the fault's existence, as Fig. 10.11 shows.

Normal faults are often found in two parallel groups where large masses of rock have been let down between them to form **rift valleys** like those of the middle Rhine (illustrated in Fig. 10.12) and East Africa. Both have been caused by local uplift of an elongated dome and the collapse of the central part to form a long trough over 40 km across and bounded by a stepped series of normal faults. Lake Victoria is formed in the wide depression between the two updomed areas which have been rifted in East Africa.

The opposite effect, whereby a faulted mass is uplifted to form a **block mountain, or horst**, is also common. Areas like the Harz Mountains of Germany, and the Longmynd area in Shropshire, have been raised in this way (Fig. 10.12).

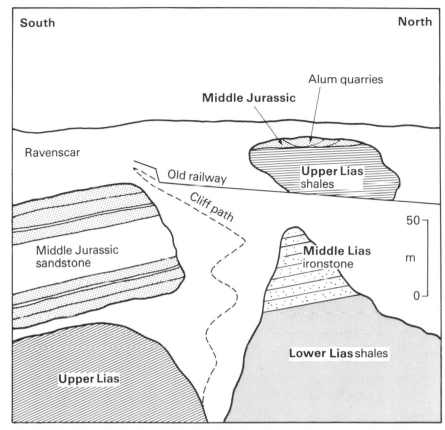

Fig. 10.11 Robin Hood's Bay, Yorkshire. This sketch shows the rocks in the cliffs at the southern end of the bay: the beach level is at the bottom of the diagram. Draw your own diagram with a key to the succession of rocks in the area, and their approximate thicknesses; then place a vertical fault line and calculate the throw of the fault.

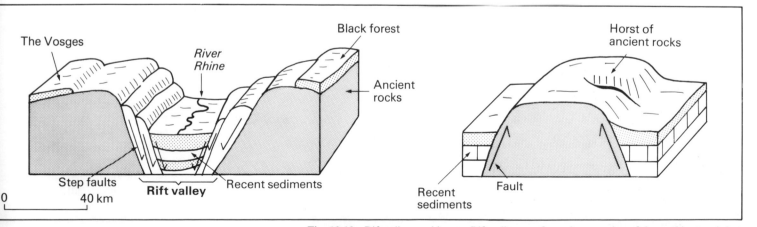

Fig. 10.12 Rift valleys and horsts. Rift valleys are formed as a section of the earth's crust is let down by the collapse of the centre of a mass of updomed rocks. Horsts often result from the effects of earth movements on hard, ancient rocks: these lead to fracturing and the uplift of some sections relative to others.

Reverse faults are most commonly found as low-angle thrusts, especially in highly folded areas where the degree of deformation is greater than that to which the rocks can adjust in simple folds. Thrusts often replace the inverted limbs of recumbent folds (Fig. 10.13). In the Alps there are piles of thrust slices of rock known as **nappes**, which have been carried over the top of younger rocks and unaffected resistant basement rocks. An important thrust zone is found in the Highlands of Scotland, where the folded and metamorphosed Precambrian Moine Series of rocks are thrust over unmetamorphosed Torridonian and Cambrian strata. The

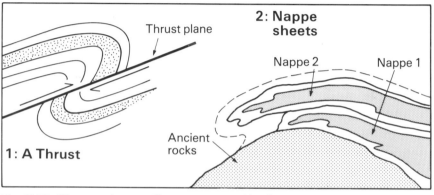

Fig. 10.13   Thrusts and nappes. Two structures resulting from extreme deformation of rocks.

A tear fault: the Great Glen in Scotland. There is evidence to show that the northern side (on the left) moved 100 km to the south-west. (Aerofilms)

section in Fig. 12.16 illustrates the results, and shows the extent of the broken zone of dislocated rocks often altered to mylonite. It has been shown that some of these rocks have travelled at least 16 km from where they were originally deposited.

Tear faults involve horizontal movements on either side of the fault plane. The Great Glen fault in northern Scotland is a broad belt of shattered rocks where there has been a sideways movement of 100 km as indicated by the displacement of granite on either side of the fault. This is similar to the effect shown in Fig. 10.8C; compare the positions of the Foyers and Strontian granites on a geological map. The northern part of Scotland slid 100 km to the south-west, mostly during the late Caledonian movements at the end of the Devonian period, but minor earthquakes are still common in the area. The best-known tear fault is the San Andreas fault in California (Fig. 10.14), which is 1300 km long and has nearly 2 km of brecciated rock in the fault zone. It has been moving since the late Jurassic, and the Pacific side has been moving northwards, wrenching open the Gulf of California and causing some areas to move over 500 km in that time. The movement has taken place in small stages, including the disastrous San Francisco earthquakes of 1906, when displacements of a few metres were enough to break gas mains and cause many fires. The average movement of this fault is to separate the land on either side by 4 km every million years, but at present, in an era of more intensive earth activity the average rate is 50 km per million years.

This fault zone tends to move in sudden jerks after a period of years when stresses build up. San Francisco today, once again, stands in danger of a major calamity, although its inhabitants seem to be extremely casual about it: massive office blocks, schools, hospitals and even the disaster headquarters, have all been built across the line of the San Andreas fault itself, where the greatest damage would occur. Observations in the 1960s have suggested that a fault which is lubricated by pumping a liquid into it moves in more frequent, shorter stages: it will give earthquakes more often, but their effect is very slight. American scientists are now trying to apply this treatment to the San Andreas fault, so that they might avert the impending disaster. They are also monitoring with great precision, using laser beams reflected across the fault, all actual movements which take place. It is hoped that such information may enable adequate warning to be given of an impending earthquake.

Folds, faults and time

How can we tell the age of a fold or fault? We usually have at least two sets of criteria. The folding or faulting movements took place **after** the youngest rocks they affect, but **before** the oldest undisturbed sediments lying on top. The Carboniferous rocks of the Pennines are folded and faulted, but the over-lying Permian rocks are not affected in the same way. We date the movements which caused the Pennine folding and faulting as late Carboniferous because the uppermost Carboniferous rocks found elsewhere are missing. An unconformity usually marks the division between the older folded or faulted group of rocks and the overlying unaffected younger series (Fig. 12.16).

Faults may have long histories of recurrent activity, as in the case of the San Andreas fault already mentioned. The Highland Boundary fault may have been initiated as early as the Ordovician and still causes slight earthquakes, e.g. at Comrie in Perth-shire. Such faults appear to mark major lines of crustal weakness.

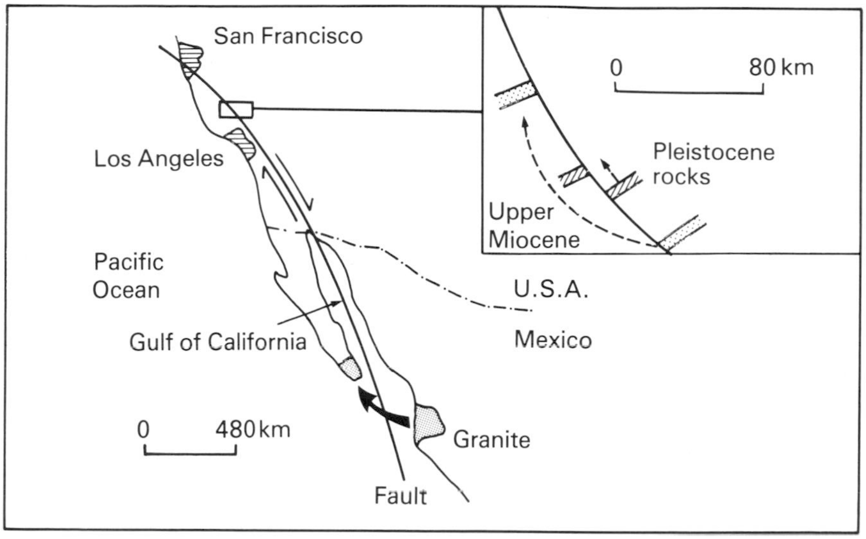

Fig. 10.14 The San Andreas fault. The enlarged inset shows some detail of the movements. Younger rocks of Pleistocene age (1 million years old) have been separated across the fault by 18 km in the course of successive movements; those of Miocene age (15 million years old) have been separated by 100 km. At the entrance to the Gulf of California a mass of granite has been moved 500 km from a similar mass of which it was once a part.

Making models of geological situations

It is often a valuable exercise to make a relief model of a particular area whose geology is known to you and to paint on it the rock outcrops. This will help you to understand the relationship between geological maps and geological sections.

(1) Simple blocks of polystyrene can be used to demonstrate some of the outcrop patterns as they occur on maps. Use felt-tipped markers to draw on the geological structure. Then carve V-shaped valleys into the top surface and redraw the outcrop pattern. Begin with tilted rocks and notice the patterns made by valleys cutting into them; then progress to symmetrical and asymmetrical folds; add an unconformity; and work up to faults of various types. Summarise your work by drawing a simple sketch-map of the situation and the section on the side of your block.

(2) More sophisticated models can be made of actual geological situations, building up the relief from an Ordnance Survey map, and adding the coloured geological outcrops.

Materials required. A baseboard which will not bend easily (e.g. 1–2 cm chipboard), cut to size; Polyfilla; polystyrene ceiling tiles; a battery-operated, hot wire cutter for the tiles; adhesive for tiles; tracing paper; modelling tools; basic colours in powder paints; varnish; turps for thinning; transfer lettering (e.g. Letraset).

Method. Determine the scale of the map to be used as a basis, and relate this to the thickness of the ceiling tiles for a vertical scale (e.g. tile thickness = 100 m on a map scale of 1:250,000; 50 m on 1:50 000; 15 m on 1:10 000). Trace out the relevant contours on a large sheet of tracing paper, and transfer them, with the trace of the coastline, to the baseboard. Having done this cut out and pin sections of the tracing paper on the ceiling tiles, and cut round the traced contours with the hot wire cutter; stick each section to the baseboard, and stick higher contour areas on the lower. Then with the map in front of you use a thick mixture of Polyfilla to round off the slopes and eliminate the stepped effect; go over each completed section with water to smooth the surface. Allow to dry. Mark on the areas of rock outcrop. For each area make up sufficient powder paint colour in a thick paste to cover the entire area and the space in the key drawn at the side of the model. When the painting is finished allow to dry. Paint the whole model with a thinned varnish. Allow to dry. Letter salient points, the key, a title. Finally add another coat of full-strength varnish.

Joints

Many cracks we see in rocks are not faults because the rocks are not displaced on either side. These partings are joints, and are due to stresses of different sorts from the earth movements we have just been discussing. They are important in affecting the detailed shape of landforms and drainage patterns, since they are lines of weakness and easy entry for weathering agents. There are several types of joint.

(1) Sedimentary rocks are criss-crossed by **shrinkage joints**, formed as the drying rock became more compact.

(2) Vertical **cooling joints** are common in lavas, where the shrinkage took place for another reason. The hexagonal columns of the Giant's Causeway and the Isle of Staffa are famous scenic wonders. The columns develop in directions perpendicular to the cool surfaces over which lavas flow, and they may be curved.

(3) Igneous intrusions may be crossed by **parallel joints** which cause great sheets of rock to come loose. The sides of the Goat Fell granite in the Isle of Arran are characterised by such 'boiler plate' formations, which are thought to have been caused by the expansion of the rock as the original pressures were released by the removal of the overlying rocks. These intrusions also have groups of joints at right angles to each other, formed during cooling, and these often affect the drainage pattern, as in the case of Bodmin Moor. Joints connected with igneous rocks are liable to be filled with pegmatite, or mineralised veins as the fluids and gases continue to circulate through the rocks at a late stage in the cooling process.

(4) Other joints are the direct result of the stresses occurring in earth movements, particularly where folding is involved, and are illustrated in Fig. 10.16.

Structural regions of the continents

The field evidence often provides us with small- or medium-scale structures which may seem to exist in isolation. As we make a wider study of the same region we find that certain patterns of structures (and often rock-types as well) are repeated, giving that region an overall character. This in turn enables us to relate that region to general tectonic conditions in the earth's crust.

We can distinguish at least five major types of structural region:
(1) Young fold mountains.
(2) Block mountains.
(3) Shield areas.
(4) Platforms and basins.
(5) Continental margins.

Young fold mountains

Young fold mountains, like the Himalayas, Alps, Rockies and Andes, are the most impressive features of continental relief. They mostly exceed 3000 m above sea level, and include the earth's highest point (Mount Everest, 8848 m). Young fold mountains are composed of long, narrow ranges and are termed 'young' because they have mostly been uplifted in the last 50 million years, and are composed of rocks formed during the Mesozoic and Cainozoic eras (i.e. over the last 225 million years).

Within the group of young fold mountains there is a variety of structural situations and rock-types but several features are common (Fig. 10.17).
(1) Each fold mountain zone has a great thickness of sedimentary rocks compared with adjacent regions (e.g. the Alps have over 10 km of Mesozoic and Cainozoic rocks, whereas there are scarcely 2 km in central Europe north of the young fold mountains).
(2) Massive granite intrusions, up to several hundred km long and orientated along the trend of the mountain ranges, are another common feature of young fold mountains.

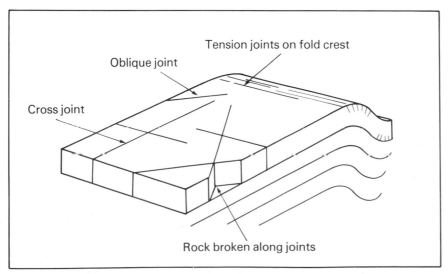

Fig. 10.15 Joints. This diagram shows where lines of weakness occur in a rock following folding: these lines of weakness often form joint openings.

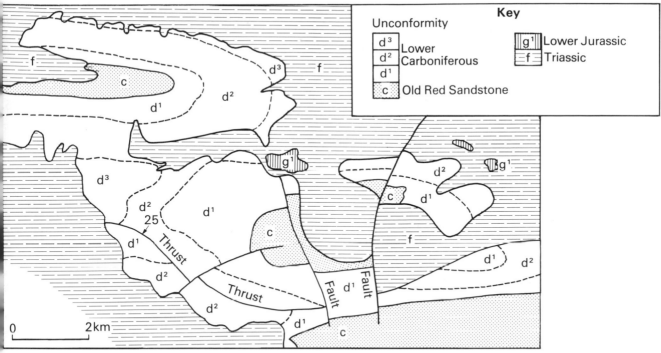

Fig. 10.16 Folds and faults. Indicate the positions of the axes of the anticlines and synclines on a sketch-map of this area. Draw in a major unconformity. Describe the nature of at least two different types of fault. Suggest dates for the formation of the folds and faults. This is a sketch-map of the geology of part of the area near Wells in Somerset, and is based on the maps of the Institute of Geological Sciences. (*Crown Copyright Reserved*)

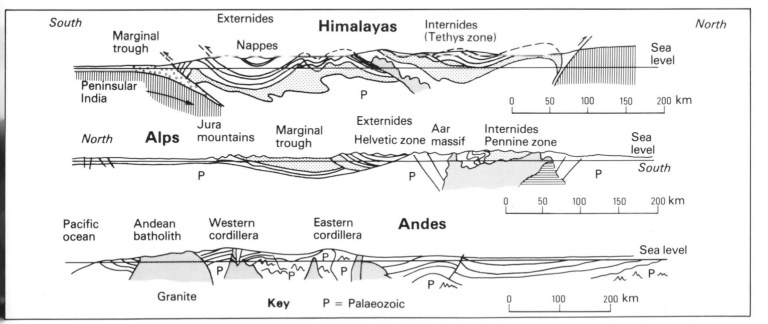

Fig. 10.17 Sections across three major fold mountain systems. The Himalayas and Alps both have strongly developed 'Internide' zones, where the rocks are highly deformed and altered by comparison with the 'Externides'. The Andes are rather different (see also Chapter 3). Can this difference be explained?

(3) Rock structures include large-scale recumbent folds, thrusts and nappes: only the most complex forms are found following conditions of extreme pressure and deformation.

(4) Two major divisions can often be recognised. The 'internides' are where earth movements are most intense, granite intrusion and metamorphism are most common, and the sedimentary rocks are mainly greywacke-shale alternations. The 'externides' are less altered and have a greater range of sedimentary rocks, although the structures are still complex (Fig. 10.18).

Block mountains

These include areas like central and northern Wales, the southern Uplands of Scotland, the Central Massif of France, the Rhine Highlands and the Appalachian mountains of the USA. They are all composed of rocks deposited mainly during the Palaeozoic era, and many contain structures which resemble the young fold mountains, unusual thicknesses of sedimentary rocks compared with adjacent areas, and rock associations similar to those of the internides and externides. In short, these areas are just older fold mountains, where the rocks are indurated by age, and where subsequent earth movements have led to their being broken into faulted blocks, parts of which have been uplifted, whilst other parts have subsided beneath more recent sediments.

The study of the rocks in these ancient and worn-down mountain ranges before the advent of plate theory led to the concept of the **geosyncline**. A geosyncline was thought of as an area of crustal subsidence in which the great thicknesses of sediments found in fold mountains could accumulate. Eventually the margins of the geosyncline began to come together, compressing the rocks between and causing them to be folded and uplifted as fold mountain ranges (Fig. 10.19). An extensive literature was built up over the years on this topic, and divisions of the geosyncline were made, including the eugeosyncline (features very similar to the internides) and the miogeosyncline (similar to the externides). It was always difficult, however, to cite a case of a modern example of a geosyncline: areas like the Mississippi delta and ocean trenches were suggested, but did not have many of the requisite features. The idea is now unnecessary, since it has been replaced by plate theory.

A simple anticline, cut open by erosion, Iran. The anticline plunges towards the top left. Notice the part played by stream erosion in destroying the rocks in this arid region. Make your own sketch-map to the situation, numbering the layers of rock in the order in which they were formed. (Aerofilms)

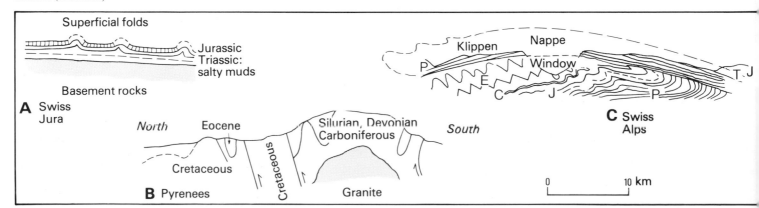

Fig. 10.18 Structures in different young fold mountains in Europe. **A** Simple folds in the Swiss Jura mountains: these affect only the surface rocks, which have slid over the underlying salty muds. **B** The Pyrenees. Uplift has been caused by the injection of igneous rock, and this has led to the folding of the marginal rocks. **C** The Alps. Nappes are slices of rock which have travelled over the tops of younger rocks and have become detached from the rest of the rocks with which they were originally formed. This process probably occurs during uplift, when these rock masses break away from the highest areas due to gravitational force. (P–Permian; T–Triassic; J–Jurassic; C–Cretaceous; E–Eocene)

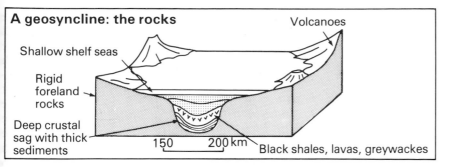

A geosyncline: the rocks

Volcanoes

Shallow shelf seas

Rigid foreland rocks

Deep crustal sag with thick sediments

150   200 km

Black shales, lavas, greywackes

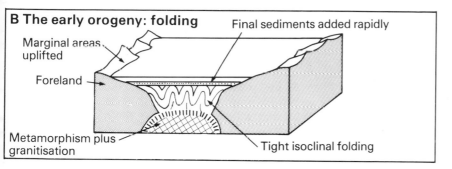

B The early orogeny: folding

Final sediments added rapidly

Marginal areas uplifted

Foreland

Metamorphism plus granitisation

Tight isoclinal folding

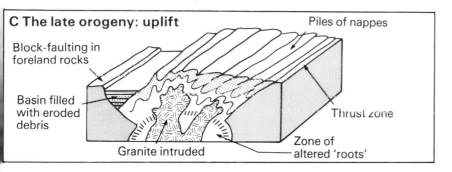

C The late orogeny: uplift

Piles of nappes

Block-faulting in foreland rocks

Basin filled with eroded debris

Thrust zone

Granite intruded

Zone of altered 'roots'

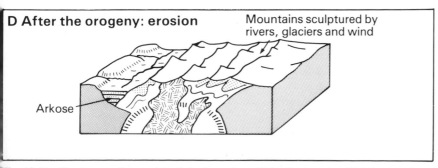

D After the orogeny: erosion

Mountains sculptured by rivers, glaciers and wind

Arkose

Fig. 10.19 The 'geosyncline model' of the formation of folded mountains. This has now been replaced by plate theory.

A The Geosynclinal Phase. A geosyncline is formed when a long, narrow section of the earth's crust becomes weak and sags downwards, forming a deep ocean trench or an area of shallower seas with long-continued subsidence. At first in the deeper waters the sediments may accumulate slowly as fine mud, containing a high proportion of undecomposed organic matter in the deep, stagnant waters. The surrounding lands tend to rise and sooner or later great quantities of sediment are poured into the geosynclinal sea. Volcanic action is common, and thick layers of spilite lava and tuff form. Earth movements affect the surrounding land masses to an increasing extent, and the intensity of erosion works up to a climax when thick piles of greywackes build up. The trench is eventually filled in with up to 10 km of sediment and lava. It is a very gradual process, and usually takes anything from 100 to 200 million years.

B The Early Orogenic Phase. Finally the geosynclinal sag ceases to subside, and the sediments which have filled it begin to be compressed into folds, thrusts and nappes. All this begins during the later stages of the geosynclinal phase when there are a series of short-lived earth movements separated by quieter periods of continued sedimentation (a series of alternating shales and coarse greywackes known as the **flysch facies**). Small-scale, local unconformities are a feature of the rocks formed at this time. The most deeply buried rocks become metamorphosed under conditions of great heat and pressure, and they are often granitised or impregnated partially by outside fluids to form migmatites. At this stage the mobile geosyncline lies between two areas of resistant, foreland rocks, and as they move together like the jaws of a vice the rocks in the middle are compressed.

C The Late Orogenic Phase is the mountain-building stage. The folded and metamorphosed sediments and lavas are uplifted into the towering chains of folded mountains with which we are familiar. The whole process is probably caused to a large extent by isostatic adjustment and is accompanied by powerful thrusting movements as the soft rocks are squeezed upwards. As soon as the mountain ranges are formed erosion begins, and marginal troughs and basins are rapidly filled with arkose-type deposits (i.e. **molasse**). The intrusion of granite from below takes place during this late stage. The whole orogenic phase also lasts a long time, partly overlapping the period in which the rocks accumulated: the deformational movements during the Alpine orogeny lasted from the late Cretaceous to the Miocene period (i.e. up to 60 million years), but rocks were still being laid down and earthquakes in these areas today testify to continued, but waning, activity. The axis of a geosyncline may migrate so that folding in one part is paralleled by subsidence and sedimentation in another.

D The Post Orogenic Phase. As the mountain ranges become more stable the igneous activity and erosion begin to slow down, but it is a long time before a mountainous area becomes 'quiet'. During this stage there is an unusually large proportion of dry land, and continental deposits including desert sand and evaporated salts are characteristic. Eventually the area is completely worn down and shallow seas transgress the exposed mountain 'roots'.

Precambrian shields

The shield areas have an outcrop over 25 per cent of the world's continents, and are formed of Precambrian rocks over 570 million years old. Much of Africa is formed of such rocks, as is the north-eastern part of Canada and central Scandinavia. Close study of the often highly-metamorphosed rocks in these areas shows that they are also arranged in a series of elongated structures similar to those in the young fold mountains and block mountains (Fig. 10.20). There are, however, some major differences in the rock-types, and the processes of sedimentation must have been different in Precambrian times: the vast deposits of iron being tapped in the shield areas of Canada, west Africa and western Australia were formed in distinctive banded deposits in environments very low in oxygen. Other evidence also suggests that Precambrian conditions were short of oxygen and hence of living forms which became such an important constituent of rocks from the Cambrian period onwards. And yet the study of shield regions demonstrates that the major features of the geological cycle were in operation.

Platforms and basins

The thinner sequences of rocks in areas between fold mountain zones are often of the same age and are composed of rocks similar to those found in the externides of young fold mountains – quartz sandstones, fossiliferous shales, and limestones. They are areas where the rocks have been disturbed only slightly, if at all, by earth movements (Fig. 10.21), and the wearing away of the softer sediments in the sequence results in landforms dominated by cuestas (escarpments) and vales. South-eastern England, the plains of northern France and the Gulf Coast plains of the USA have this type of situation.

In many cases the downwarping of areas of older rocks led to an invasion of shallow seas in which thin layers of sediment were formed, but the closeness of the underlying rigid 'basement' prevented earth movements having very much effect.

Continental margins

The continental margins include the continental shelf, slope and rise (Fig. 5.18). Recent studies have confirmed that these should be studied as a part of the continent, rather than as part of the ocean basin, and it is significant that the sediments forming in the rise and shelf are similar to those found in the internides and externides of fold mountains. It is in the continental margin environments that the rocks which will form the next groups of fold mountains are accumulating, not in another form of hypothetical geosyncline.

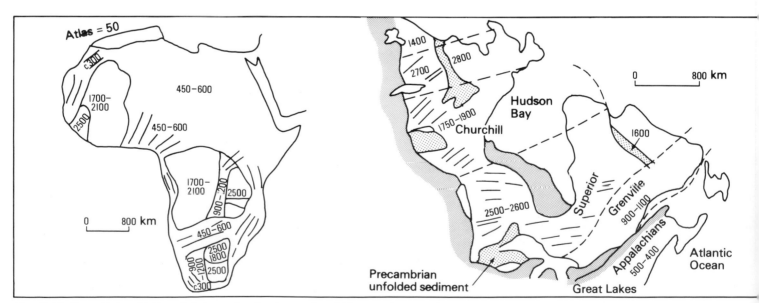

Fig. 10.20 Precambrian shields and their trends related to ancient fold mountain ranges in Africa and northern Canada. The figures are radiometric dates in millions of years for the rocks of the respective zones.

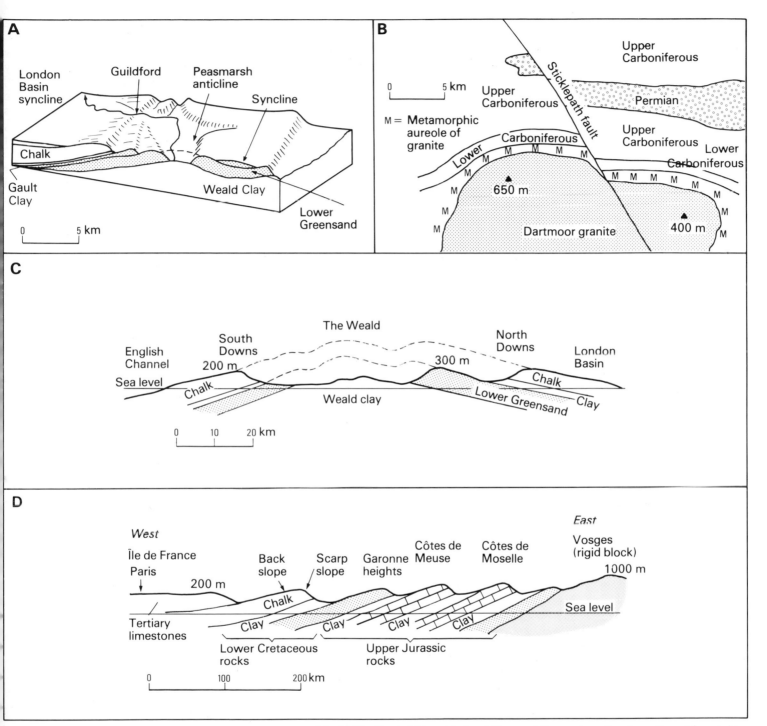

Fig. 10.21 Simple folds and faults. **A** Open anticlines and synclines in south-west Surrey: how do these relate to the relief features? **B** A tear fault in central Devon: how has this fault affected the rock outcrops? **C** The Weald of south-eastern England is a dome-like structure with smaller folds producing a 'corrugated' effect. The rocks are of Cretaceous age, and their erosion has produced a sequence of cuestas and vales. **D** Cuestas in the eastern part of the Paris basin, France. The rocks range from Jurassic to Tertiary age.

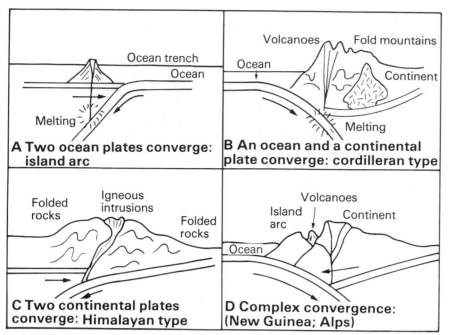

Fig. 10.22   Types of convergent plate margins which give rise to characteristic fold mountain arrangements.

Mountain ranges and plate tectonics

There is a clear relationship between many of the major structural regions of the continents: they have either been mountain ranges which have been worn down by erosion, or provide the site for the thick accumulation of sediments which may develop into mountains in the future. Mountain ranges are formed along convergent plate margins, and the present young fold mountains have been raised in a number of distinct ways (Fig. 10.22).

Two ocean plates converge

When this occurs volcanic activity is engendered, producing island arc accumulations of lava and ash on the continental sides of ocean trenches. The Japanese islands have been formed in this way (Fig. 10.23).

An ocean and a continental plate converge

This situation is typified by the Andes in South America (Chapter 3), and is known as the cordilleran type (Fig. 10.24).

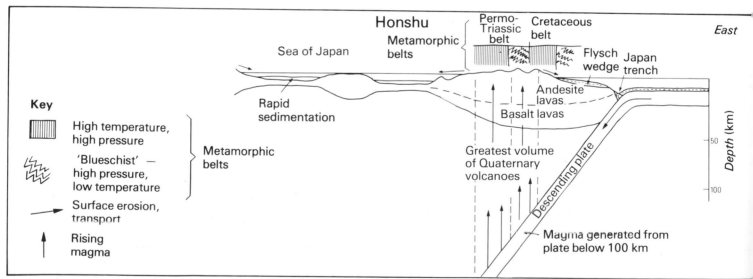

Fig. 10.23   Japan: an example of mountain formation by the building of island arcs. The two belts (Permo-Triassic and Cretaceous) suggest that Japan is formed by the collision and amalgamation of two island arcs. Compression along the line of the trench has produced metamorphic rocks characteristic of high pressure and low temperature conditions ('blueschists'). The volcanic and earthquake activity associated with Japan is related to the descending plate and the friction it causes.

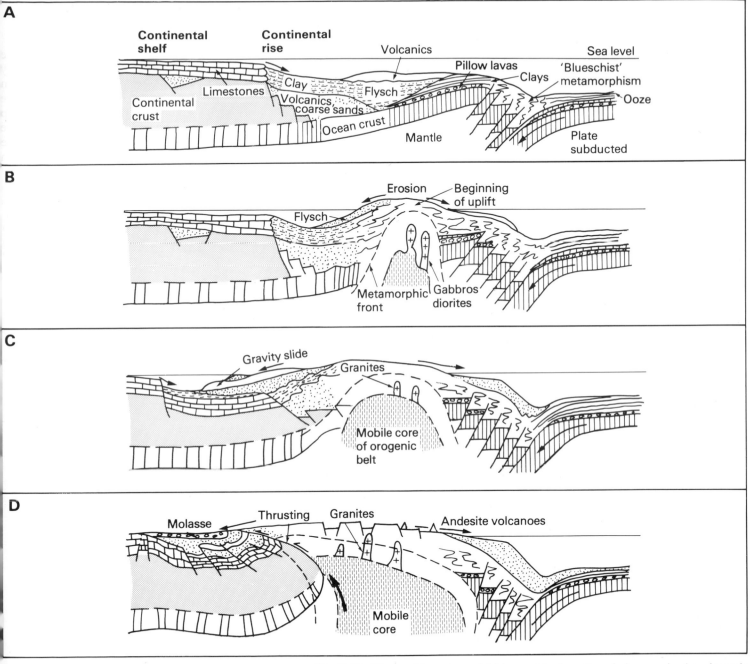

Fig. 10.24 The cordilleran type of fold mountain formation. **A** An oceanic plate descends beneath the continental rise, leading to the eruption of submarine volcanic rocks and metamorphism at the plate margins. **B** The rocks overlying the descending plate are melted and begin to rise, leading to an updoming of the area and intrusion of gabbros and diorites with high temperature, low pressure metamorphism. **C** The mobile core of the mountain-forming belt grows and high temperature, high pressure metamorphism becomes extensive. The continental shelf subsides and gravity slides occur to form nappes. **D** Mobile thrust sheets are driven towards the continent and granites are emplaced at high levels in the mobile core zone, producing further uplift, fracturing and erosion.

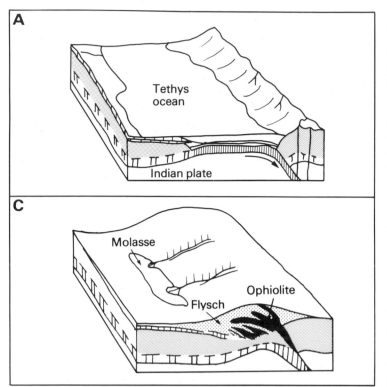

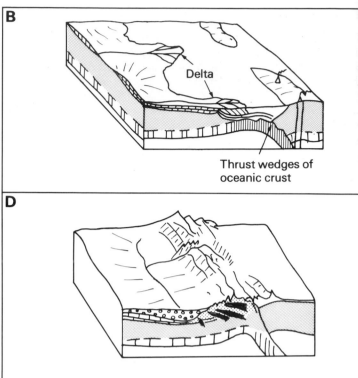

Fig. 10.25 The Himalayan type of mountain formation results from the collision of two plates carrying continents. **A** Approximately 65–70 million years ago: the Indian plate brings its continent northwards and approaches the Asian plate, causing the Tethys ocean to contract. **B** Approximately 30–60 million years ago. As the two continental masses get closer thrust wedges are produced. **C** Approximately 2–30 million years ago. The first ranges are produced by the collision, and erosion leads to the formation of molasse deposits in marginal basins. **D** Today. Erosion has already had a major effect on the range.

A LANDSAT satellite view of the Rocky mountains, taken from 914 km above the surface. It shows the snow-capped Wyoming range (right), the upper Snake river and Palisades reservoir (top centre), and some lower ranges to the west. (NASA)

Two continental plates converge

The Himalayas were formed as the southern part of the Indian subcontinent moved northwards and plunged beneath the main Asian plate, crumpling the rocks between (Fig. 10.25).

Complex situations

A variety of situations is possible. New Guinea, for instance, is formed of an island arc and continental fragment which have collided. The Alpine ranges in Europe have such complex trends (Fig. 10.26) because the collision of Africa and Europe has trapped fragments of continental rocks and island arcs between the main ranges.

The study of mountain formation has thus been given a special impetus by plate theory. Before the 1960s geologists had a fair knowledge of the continental rocks, but the geosynclinal theory was not an adequate interpretation. We can understand why that was the case, since it was based on evidence from 29 per cent of the earth's surface, and the lack of evidence from the ocean basins was a handicap in discovering patterns of worldwide significance.

Mountain building and continental accretion

The continental masses have grown over geological time, but not every type of mountain-forming process leads to such additions. When continents like Africa and Europe collide to form the Alps there is often no addition to the areas of either: ocean-floor sediment is squeezed between them and over older rocks. The cordilleran type of mountain building may cause the uplift of continental margins, but these are often driven over the top of older continental rocks in the course of uplift. The major type of accretion to continental areas is caused by island arc formation, and the welding of these arcs to continents. This is happening at present in southeast Asia.

Searching farther back into the past, it is clear that older fold belts (i.e. in block mountains and shield areas) were also at convergent plate margins when they were formed. It is thus possible to reconstruct past plate and continent-ocean histories, and this will be the major theme of Chapter 11.

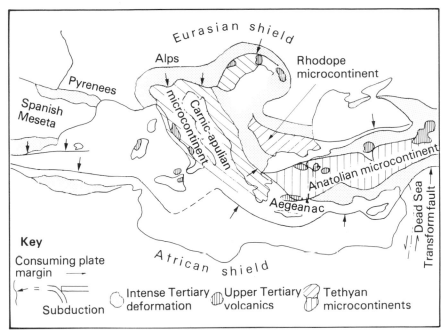

Fig. 10.26 The Alps. These have been formed by a complex series of events, which have interrupted the smooth evolution of a Himalayan type of situation (with the African and European continents colliding). Smaller sections of continent and island arcs have been caught up in the collision, leading to S-shape fold mountain trends. The Mediterranean Sea is the remnant of an extensive ocean.

The coastal range of southern Alaska. The high mountains are being eroded by glaciers and the sediment plumes extending up to 50 km out to sea reveal the way in which the rock debris is transported. (USGS)

II
Earth history: the grand view

The study of minerals, rocks, fossils and rock structures all provide us with evidence for interpreting what occurred in the remote geological past. There are two levels of scale at which we can carry out such studies.

(1) The world scale involves reconstructing the positions of continents in the past and providing a picture of the overall development of the earth's surface features through time. Interpretations of this type are based mainly on our knowledge of major structural regions in relation to plate tectonics (Chapter 10), and of ancient magnetic fields and their impact on the rocks formed at the time (Box 11.1).

(2) The local scale refers to individual quarries, cliff faces and stream beds, which have rock outcrops providing a different range of evidence and enabling us to work out the most detailed picture of the past. Such evidence includes information about rock-types and their conditions of formation, together with the order in which they were formed (Chapter 12).

Both levels of scale must be seen together, and results from one type of investigation used for testing against the other. Thus conclusions concerning the movement of continents, the formation of mountain ranges, or the opening of oceans, can be compared with the local rock successions of the same period in time. In this chapter we look at the main events of earth history, whilst in Chapter 12 some of the detailed evidence is studied in the context of the British Isles.

The earliest events

The oldest rocks known on the earth have been dated as 3850 million years old, and occur in West Greenland and north-central USA. These are in areas of shield rocks which may be regarded as the 'basement' on which the continental masses have

developed: such shield areas are found in every continent.

One of the early events, for which there is little evidence as to the precise sequence, was the formation of the earth's atmosphere and oceans. It seems possible that the original atmosphere was lost to space, since the 'noble gases' (argon, neon, zenon) occur in much smaller quantities than in the atmospheres of other planets, but this atmosphere would have been replaced by gases produced as a result of internal earth movements leading to volcanic activity, and then condensation of water vapour on cooling would lead to the formation of oceans on the surface.

The earliest events related to rock formation are likewise unclear, since areas of rocks older than 2700 million years are small and scattered, and none provide any evidence for the most ancient events in the earth's history. These may be similar to those affecting the moon and Mars (Chapter 13). The lack of certainty about the early events has led to several versions of what could have happened during the years 3850–2700 million years ago.

(1) It has been suggested that the earth heated up by release of internal radioactive energy, resulting in the separation of the denser rock materials concentrating towards the earth's centre, leaving a low-density granitic shell as an outer crust. Volcanic activity, perhaps associated with impact cratering (cf. the moon), brought denser rocks to the surface and resulted in an unstable situation which was resolved by overturning of the surface layers. This would account for many of the intermixtures of granitic and darker basic rocks occurring in these ancient portions of the shields.

(2) Another suggestion is that the process of plate tectonics began well before 3000 million years ago. A crust of basic, dense ocean plate materials began to move, leading to island arc situations, and the continents began to grow from such nuclei. Collisions and continuous igneous activity added to the

How was the earth formed?

The answer to this question is bound up with the origin of the whole Solar System – of which the earth is one member – and therefore with the science of astronomy. We still understand so little about conditions in the vast expanses of space that our ideas about the origin of the earth cannot be very definite. That is why there are so many theories about it.

For many years it has been assumed that the earth cooled from a hot, molten mass, and that contraction took place as it solidified. Recent research, especially connected with radioactive minerals, has caused scientists to change their minds. One of the most widely accepted modern theories concerning the origin of the earth and the solar system generally suggests that the sun was one of a series of stars developing out of a great contracting cloud of dust and gas. Some of the gases and dust left over formed a giant disc around the sun, but this soon broke up again into large masses. These masses grew by the accumulation on them of many **planetismals** (formed of fine silicate dust, water and ammonia) to form our present **planets**: moons and meteorites may be the remains of planetismals which failed to join up with a planet. All this happened at relatively low temperatures, and the earth was originally a cool body, solid throughout.

The next stage in the earth's development is also one which has given rise to a lot of discussion based on the few facts available. Some scientists suggest that the original dust forming the earth contracted as the small but important quantities of radioactive elements decayed, releasing energy and melting the central core. These reactions also resulted in geochemical separations as the heavier minerals sank in the molten interior to form the heavy nickel-iron core surrounded by layers of silicate minerals of decreasing density. Water was then formed on the surface of the earth by the condensation of water vapour. The remaining gases involved formed the first atmosphere composed of ammonia (NH_3), water vapour (H_2O) and methane (CH_4), but these broke up as sunlight decomposed them. Most of the hydrogen escaped into space, leaving a preponderance of nitrogen and oxygen. Volcanic action added a small proportion of carbon dioxide, which has been a most important ingredient for the development of plant life.

Many scientists feel that it was not until plant life became established on the earth that oxygen entered the atmosphere. It is thought that in the earliest days all life was restricted to the sea, where it was protected from the ultraviolet rays of the sun, which would have burnt it up. Oxygen was gradually liberated in increasing quantities by the processes of photosynthesis, until life became possible nearer to the ocean surface. The rate of oxygen accumulation increased markedly at this stage. Throughout the Precambrian the oxygen quantities available would supply only sufficient energy for the life processes (metabolism) carried out by single-celled creatures. By the beginning of the Cambrian period, or just before, a wide range of many-celled (metazoan) creatures existed, and it is thought that a critical level of oxygen content in the atmosphere was passed: one per cent of the present level is suggested.

The first land life emerged in the late Silurian period, nearly 200 million years later. At this stage there must have been sufficient oxygen to build up the protective ozone layer in the upper atmosphere, which absorbs the ultraviolet radiation as it enters the atmosphere. After this life could develop on the earth to the stage where we find it today. This facet of the earth's development illustrates the fact that life has not been evolving against a static background and in a static atmosphere, but that all these aspects of our planet's existence have been evolving together and affecting each other on the way.

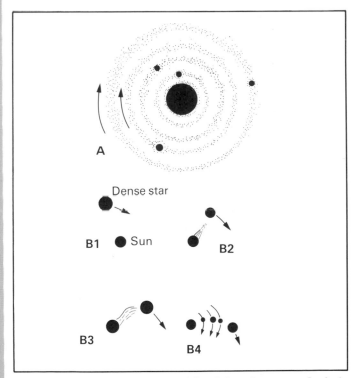

Fig. 1 Two ideas concerning the origin of the solar system. In **A** the planets are accreting out of a mass of dust and gas whirling around the proto-sun. **B** shows an older theory which envisaged a series of stages which might have occurred if a large star had passed close enough to the sun to pull away some of the material into a filament; such a filament might then have broken up into planet-like masses some of which were still held in orbits around the sun. Refer to *Scientific American*, September 1975, for greater detail on the origin of the solar system.

initial islands, and weathering of the rocks and minerals exposed to the atmosphere led to the formation of new, less dense minerals. This view accounts for the extensive occurrence of 'green-stone' rocks, very similar to those forming along the island arcs of today.

(3) Others resist the suggestion that the plate tectonics mechanism had much effect until the end of the Precambrian, referring to the observations that African rocks of this age show little effect of tectonic events in terms of folding or metamorphism resulting from crushing.

By 2700 million years ago increasing masses of continental rocks existed, moving together with respect to the magnetic poles without fragmentation or dispersal. Mobile belts, in which rocks accumulated and were deformed, appear to have been internal to these continental masses. This was possible, since the crust was less thick and rigid than at present, at a time when radioactive release of energy was more intense and geothermal gradients were greater (30°C/km as compared with 10°C/km today). Faster melting of rocks at shallower levels could be related to smaller convection cells and smaller crustal units. Any disturbance would have resulted in melting, upward movement of molten rock and local thickening of the continental crust. At times of more rapid continental movement – and the process seems to have taken place in fits and starts – there was greater activity, and where the crustal rocks reached 25 km in thickness granites could have been generated.

This is a general (and still disputed) picture of the situation during much of the Precambrian time to approximately 1000 million years ago. Small, thin continental masses were gradually built up into major concentrations of low-density surface rocks. The maps of Africa and North America (Fig. 10.20) show how the rock groups of different ages are arranged in narrow zones, suggesting groups of ancient mountain ranges being fused to each other and adding to the sizes of continental areas. Such maps also have implications for the discovery of mineral resources.

Our knowledge of the development of life on earth is extremely fragmentary for this period. Organism-like structures have been found in rocks over 3000 million years old; by 2000 million years ago there were photosynthesising algae releasing oxygen into the atmosphere; and by 1000 million years ago single-celled plants and animals were reproducing precisely the features of the parents in their young. A vast period of time thus saw the extremely slow development of living forms at a primitive stage. A most important related development was the increasing oxygen content of the atmosphere, reflected in the types of rocks which were produced: low oxygen iron compounds were commonly formed in the earlier stages, but later red iron oxides became more common. During this time the ozone layer in the atmosphere, protecting the surface from harmful solar radiation, was built up. Gradually sufficient oxygen built up to support the higher metabolic rates of multi-celled animals.

The last 1000 million years

Major differences can be drawn between the early Precambrian phase, and the last 1000 million years, in terms of rock-types, structural features and the presence of a wide variety of fossils in the rocks. The mechanism of plate tectonics has functioned fully during the latter period of time, as major continental masses have moved around over the earth's surface on the backs of plates, whilst fold mountain ranges have been produced by colliding plates. Two distinct periods of activity can be recognised.

(1) Late Precambrian to the end of the Palaeozoic (1000–225 million years ago), when the continental mass broke into several pieces, but eventually recombined to form a single major continent, known as Pangaea.

(2) During the last 225 million years the continents have again broken apart, and are now highly dispersed.

The earlier of these two periods was characterised by continental distributions very different from those we would recognise today (Fig. 11.1). Few of the continents resembled their present shapes: the British Isles area, for instance, was cut in two by an earlier ocean on the site of the present Atlantic. This ocean widened to approximately 2000 km (though some would say only 500 km) in the early Cambrian, but then closed until the collision of its margins resulted in the formation of the Caledonian mountains. Eastern USA and Canada, Greenland and Scandinavia were also involved in these events. The closing of this ocean gave rise to an extensive 'Old Red Sandstone' continent incorporating all these areas about 400 million years ago. Then a re-orientation of this land mass brought it close to the combined Africa and South America continent, leading to further collision and eventually an amalgamation into a single world continent. These later events can be related to the formation of the Hercynian mountains across central and southern Europe, and to the second phase of folding in the Appalachian mountains of North America.

Pangaea seems to have been relatively short-lived and broke up about 225 million years ago, giving rise to the present patterns of continental distributions (Fig. 11.2). Collisions led to the young fold mountain chains of the western Americas and the Alps-Himalayas-Indonesia line. The British Isles were again greatly affected by the movements, since the North Atlantic opened up slowly through the Jurassic and Cretaceous, and then rapidly in the Cainozoic as the split between Greenland and Scandinavia became the major extension northwards. During the last 200 million years the area in which the British Isles has developed, which had been situated south of the Equator for most of the first phase and had moved to the Equator by the Carboniferous period, now moved through the northern subtropical belt and desert latitudes in the Triassic to nearly 50 degrees North by the Jurassic. Subsequently there have been smaller movements southwards and northwards.

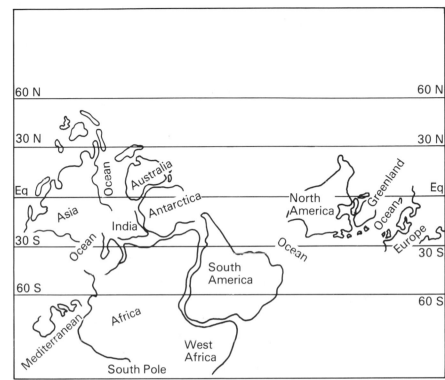

Fig. 11.1 A reconstruction of the Cambrian world, based on evidence from studies of 'fossil' magnetism in the rocks. Notice the latitudes of the British Isles and various parts of Europe. There seems to have been little land in the northern hemisphere at this time.

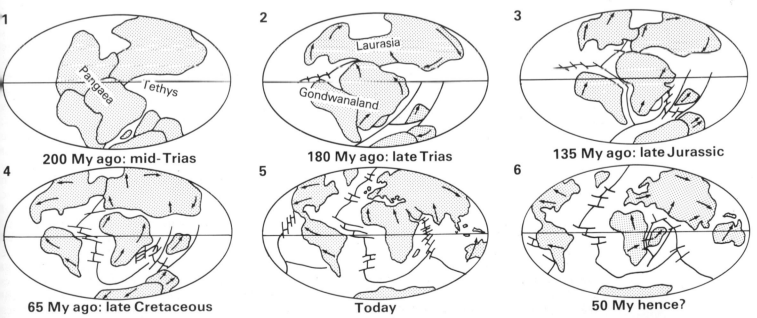

Fig. 11.2 The generalised development of the major continents on the earth's surface over the last 200 million years. These maps are based on a wide range of observations, but also include some reasoned guesswork.

These two major phases must be related to the continuing development of life on earth. By 1000 million years ago there were still only single-celled plants and animals, so far as we know from fossil evidence. By 600 million years ago there was a diverse group of multi-celled (metazoan) creatures living in the seas, and by 400 million years ago the rocks formed on the 'Old Red Sandstone Continent' included fossils of land plants and freshwater fish-like animals. It may be significant that the development of metazoan animals can be correlated with a major phase of continental break-up, isolating different groups, and also with a late Precambrian ice age affecting much of the world. Invertebrate animal fossils dominate the Lower Palaeozoic (Cambrian–Ordovician–Silurian) rocks, but by the Devonian fishes and amphibious vertebrates,

together with land plants, had become important. After the Carboniferous reptiles became the dominant vertebrate forms and maintained this position until the massive extinctions towards the end of the Cretaceous. The causes of this are little understood, but the results are clear. The mammals, which had been insignificant and small since their origins in the Triassic, took over the niches occupied by the departed reptiles, and diversified rapidly as the continents spread apart in the early Cainozoic.

Plate tectonics and continental movements can be related to the diversification of living forms on the continents, and to the distribution of distinctive groups in widely separated areas. Thus the marsupials are found in Australia and South America – a fact explained by the connections of both continents to Antarctica until a few million years ago.

How old is the earth?

Over the last 200 years scientists have realised that the earth is very old. We can all observe that if our local river is wearing down its valley it is doing it so slowly that we cannot see what is happening. Rivers become swollen and dis-coloured by mud after a storm, and this mud must have been washed into the stream by the rain. As it will never be returned to the field from which it was taken, the land surface has been lowered very slightly. Over many years the process becomes important, and it has been calculated that the surface of Great Britain is being worn down at a rate of 30 cm every 3000 or 4000 years. Even this rate is rapid compared with the deserts, where the rocks are scarcely affected over periods of time involving tens of thousands of years. Egyptian monuments nearly 6000 years old are nearly as fresh as when they were made, whilst in Britain we have difficulty in maintaining buildings a few hundred years old.

This slow destruction of rocks and relief at the earth's surface is part of the series of events which have affected the world since it was first formed. Periods of rock destruction which must have lasted for a very long time have been followed by the formulation of new rocks and by the slow uplift of new mountains accompanied by the dramatic outpouring of volcanic lavas. As many as a dozen of these cycles have been recognised in some parts of the world.

Having realised that the earth is very old we would like to have some idea of its actual age in years. Scientists have estimated geological time in a number of ways. Some worked out the total thickness of rocks formed as sediments in the sea, estimated the time taken to form a certain thickness, and multiplied this figure by the total. If, for instance, we found a layer of mud 50 cm deep covering the wreck of a Roman galley containing dated coins, we might assume that 50 cm of mud takes, say, 2000 years to form. Our calculation of the total thickness of sedimentary rocks in our local area might add up to 25 000 m, and by combining these figures we would reach a date for the formation of the first sedimentary rocks at 100 million years ago.

Another result was obtained by assuming that the oceans were originally fresh water which had condensed from

The geology of the British Isles area

For a relatively small group of islands, this area includes a remarkable variety of geological features. This fact has been recognised for some time, but the new interpretations relating to plate tectonics emphasise the point. Not only do the British Isles contain most rock types, mineral resources and representatives of almost every geological period and epoch, but they have at times been the sites for both constructive or destructive plate margins, island arcs, fold mountains and most major structural region types; they contain a small area of shield rocks in the extreme north-west; and they have passed through most major world climatic regions – glacial, arid, equatorial and temperate! This makes the geological study of the British Isles particularly fascinating: the microcosm of geology makes it possible for the enthusiast to examine the widest variety of features with the minumum of travel. A study of British geology equips a student to approach the interpretation of geological problems throughout the world. This fact also partly explains the important role played by British geologists in the development of the subject from the early nineteeth century.

The British Isles are situated today on one of the world's most extensive areas of continental terrace, surrounded by shallow seas to the west and north, the North Sea to the east and the English Channel to the south. Exploration of these submarine areas since the 1960s has led to the discovery of important reserves of oil and gas, and has extended our knowledge of the underlying geology. The formation of the modern Atlantic Ocean in the second major phase of continental separation resulted in a series of 'attempts' to develop constructive plate margins northwards from the southern opening between Africa and South America. Routes to the west of Greenland, on either side of Rockall Bank, between Ireland and Britain, and through the centre of the North Sea, appear to have been possible at one stage or another. Each left its mark in rift-valley-like troughs which became the sites for the deposition of thick marine sediments in which oil accumulated. The major trough which follows the median north-south line in the North Sea became a major trap for oil-bearing horizons, and both

clouds of gas, and working out how much salt is being supplied in river solution at the present time. Both of the answers were wrong because their authors did not make allowance for the facts that the rate of rock-formation and salt-accumulation in the oceans has varied throughout geological time, that a large quantity of salt has been precipitated out of the waters, and that both processes are going on at a faster rate than usual at present. These ages were thus too small.

The best methods of dating the rocks accurately involve using the results of a natural process which has operated at a known rate from a definite starting-point in time. Varves are formed in lakes from the debris brought in by glacial meltwater. The coarser material (silt and sand) settles quickly in summer and is followed by the finer clay in winter. Each varve represents a year's deposition. By counting such annual deposits in Sweden it has been calculated that ice retreated from southern Sweden 13 500 years ago, and the far north of Sweden 8700 years ago. This method, however, can only be used for short periods of time.

In recent years the science of nuclear physics has come to the aid of geologists. Some elements like uranium and thorium are unstable, and gradually break down into more stable elements, such as lead. There are very good grounds for believing that the rate of disintegration, which can be determined, is not affected by external conditions such as heat, cold, pressure, chemical changes, etc., and has remained the same throughout geological time. If the amounts of uranium and lead are measured the age of the rock can be calculated. The uranium–lead method is a good one, but rocks containing the minerals are not common. On the other hand, the potassium–argon method uses a more common element, but is more difficult to handle.

The use of these radioactive mineral 'clocks' enables us to give reliable dates to rocks for the first time. The oldest rock occurring at the surface of the earth is now known to be 3900 million years old, but the planet was first formed nearer to 4500 million years ago. The date of the planet's origin is unknown because the oldest rocks we can examine at the surface were certainly not the first to form: they are metamorphosed sediments and volcanic rocks, and the sediments at least must have been derived from even older rocks.

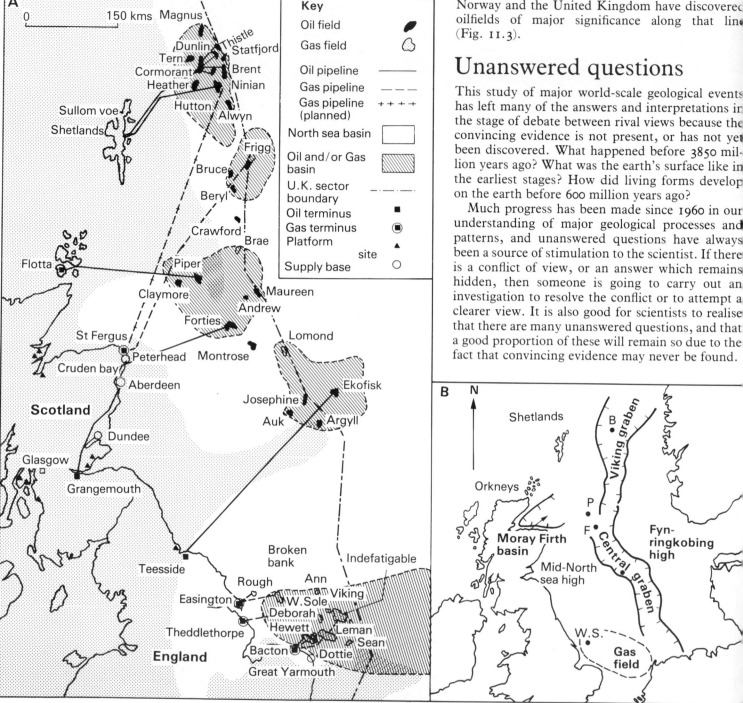

Norway and the United Kingdom have discovered oilfields of major significance along that line (Fig. 11.3).

Unanswered questions

This study of major world-scale geological events has left many of the answers and interpretations in the stage of debate between rival views because the convincing evidence is not present, or has not yet been discovered. What happened before 3850 million years ago? What was the earth's surface like in the earliest stages? How did living forms develop on the earth before 600 million years ago?

Much progress has been made since 1960 in our understanding of major geological processes and patterns, and unanswered questions have always been a source of stimulation to the scientist. If there is a conflict of view, or an answer which remains hidden, then someone is going to carry out an investigation to resolve the conflict or to attempt a clearer view. It is also good for scientists to realise that there are many unanswered questions, and that a good proportion of these will remain so due to the fact that convincing evidence may never be found.

Fig. 11.3 British oil and gas. **A** The main fields in the British sector of the North Sea (1975). Further searches are being made in the Irish Sea and English Channel. **B** Relate the occurrence of oil in particular to the central graben (rift valley) of the North Sea. (B-Brent; P-Piper; F-Forties; A-Argyll; WS-West Sole) The graben was probably formed as a part of the movements leading to the opening of the Atlantic Ocean, and became the site for the deposition of thousands of metres of marine sediments, in which the main oil and gas deposits have been trapped.

12
Earth history: the details

Just as the theory of plate tectonics led to a revolution in thinking about the major features of earth history in the 1960s, so the development of ideas in relation to the finer details of geological evidence constituted an earlier revolution in ideas at the end of the eighteenth century and during the early years of the nineteenth century. This earlier revolution influenced the work of Charles Darwin and much modern scientific thinking in relation to the extent of time.

It is now fairly easy for us to adopt the view that when we examine a quarry or cliff section we are looking at the results of geological processes which operated many millions of years ago. We may be able to conclude that the rocks under study are sedimentary in origin and were deposited over a certain length of geological time under particular conditions; that at a later date the soft sediments were hardened and fossil shells were dissolved away to leave moulds and casts; and there may be evidence for folding, uplift and erosion. But these conclusions were not obvious interpretations as recently as 200 years ago. They have been made possible by studies of the associations within layered arrangements of rocks, or strata, a division of geological investigation known as **stratigraphy**. The principles on which such studies are based include ideas from science and history.

The principles of stratigraphy
Early ideas

Man's earliest writings show that he has always been interested in his past and origins, and at various times in his history he has related the evidence he has found in the rocks to these problems. Greek philosophers as early as 500 BC had decided that what we now call fossils had once been living animals, but as with many of their discoveries, the practical implications were not clear at the time. True geological studies were held up until the mid-seventeenth century AD by three major factors.

(1) Observers used **the wrong method**. They made a few, often unconnected, observations from nature and spent many hours speculating on their possible origin, often incorporating concepts which could not possibly have been observed or measured. Thus if something was beyond explanation at a particular moment they would bring in the supernatural. This approach brought with it a series of strange ideas and misconceptions, as the following passage, written in the mid-fourteenth century, shows:

It often happens that in one place or another the earth shakes so violently that cities are thrown down and even that one mountain is hurled against another mountain. The common people do not understand why this happens and so a lot of old women who claim to be wise say that the earth rests on a great fish called Celebrant which grasps its tail in its mouth. When this fish moves or turns the earth trembles. This is a ridiculous fable and of course is not true. . . . We shall therefore explain what earthquakes really are and what remarkable consequences result from them. Earthquakes arise from the fact that in subterranean caverns and especially those within hollow mountains, earthy vapours collect and these sometimes gather in such enormous volumes that the caverns are not able to hold them. They batter the walls of the cavern in which they are and force their way out into another and still another cavern until they fill every open space in the mountain. . . . If they cannot reach the surface they give rise to great earthquakes. This unrest is brought about by the mighty power of the stars. (*Conrad von Megenberg*, d. AD 1347)

(2) Geological studies are based on a **correct knowledge** of chemical, physical and biological processes, and it was not until these sciences began

to understand the true nature of their materials that advances could be made in geology. Rock-forming minerals were only analysed chemically in the nineteenth century; the discovery of radioactivity did not take place until the very end of the nineteenth century; and the theory of evolution by natural selection was a mid-nineteenth century proposal. All have made important contributions to our understanding of the past.

(3) There was a tendency to confuse a science dealing in origins with **religious beliefs**, and even with unwarranted additions which have been made to these beliefs. The belief that the earth's history was confined to 6000 years was not a matter of biblical interpretation and simple mathematics, but was the legacy of an early Christian administrator who liked rounded numbers! And yet it handicapped and restrained the progress of geology for hundreds of years. It was only when it could be seen that there was a science of geology based on observation and measurement, subject to the limitations of the observers and their instruments, and distinct from the search for meaning and purpose in life (the province of religion), that progress was possible.

The beginnings of modern geology

Nicolaus Steno (1638–87), a Danish priest, put forward the basic tenets of stratigraphical studies in 1669. In conflict with the 'Flood Geologists' who held that the rocks were formed when a mass of sediment settled out of the declining waters of Noah's flood, he stated that each layer of rock was formed on top of earlier strata by a separate series of sedimentary processes: the lowest layers are therefore the oldest. This idea seems obvious to us, but he had to give it an imposing title: *The Law of the Superposition of Strata*. This principle is applicable in each local area, although there are areas where intense folding causes some of the rocks to be upside down. We may be able to use internal structures, such as cross-bedding or graded-bedding to determine whether the rocks are the right way up or not. On the whole the oldest rocks have the fewest fossils and the greatest degree of folding. We have been into these factors in detail in Chapters 8 and 10.

Other principles which Steno referred to in his work include:

(1) **The principle of original horizontality.** Most sediments and lava flows were originally horizontal (Chapter 10), although there are exceptions to this rule which should be kept in mind (Chapter 8).

(2) **The principle of original lateral extension.** This states that rock layers formed at the same period in time can often be traced over a wide area, although these layers may be cut into and separated by erosion.

(3) **The principle of cross-cutting relationships.** Igneous intrusions and unconformities (Chapter 8) give rise to situations where one group of rocks cuts across another. The rocks which cut across another group are always younger in age.

Steno thus provided a basis for the study of rock sequences, although his conclusions were only being applied for the first time 100 years later. This study of rock sequences was useful in local areas, but widespread correlation on the basis of rock-type alone is not satisfactory, since different types of sediment were laid down at the same time. William Smith, an English engineer, journeyed about the country in the course of examining the ground as a prelude to canal- and bridge-building. He received no formal geological education and was thus free from ideas such as Flood Geology, which were still being taught at the end of the eighteenth century. He found that, although a rock of a certain age might change its nature from sandstone to clay over a distance, it would still contain some of the same fossils (Fig. 12.1). Using this principle of **dating the rocks by the fossils they contained**, he was able to publish a very reasonable geological map of England and Wales in 1815, bearing many resemblances to the maps of today. Many of his names for the rock horizons are still in use. He anticipated Darwin's ideas on evolution, which were the basis of his own practical scheme of working, by thirty years.

Whilst the sequence of rock layers could be established and correlated with other sequences in different parts of the country, there was also the matter of interpreting the conditions which gave rise to them: what were the geographical situations at a particular period in the past? Flood Geology suggested that all the present dispositions of the rocks had been formed at one moment in time. Another interpretation, associated with the famous German teacher of the late eighteenth century, Abraham Gottlob Werner (1749–1817), argued that **all the rocks had been formed in the sea**: granites were chemical precipitates, as was shown by their crystalline nature, and rocks like those in the Giant's Causeway were enormous crystals formed by chemical precipitation in the sea. Werner's ideas became associated with Flood Geology, and the religious note kept them alive until the mid-nineteenth century, but neither were based on close observation of the evidence in the rocks. Werner never travelled out of Upper Saxony, but as soon as his own pupils began travelling farther afield, visiting ancient volcanic centres such as the Auvergne of central France and the Inner Hebrides of north-west Britain, they became convinced he was wrong.

James Hutton (1726–97) started the modern approach to the interpretation of past events as shown by the rock evidence. Not only did he conclude that **some rocks had cooled from a molten state**, and so took up sides against Werner in the debate of that time, but he went further. Even more important for the development of geology was his concept that **the forces we observe acting on the landscape today** have been in operation throughout geological time; we do not need to suggest that any other forces have operated in the past, which we cannot observe today. He expressed this in the following way:

In examining things present we have data from which to reason with regard to what has been; and, from what has actually been, we have data for concluding with regard to that which is to happen here after. Therefore, upon the supposition that the operations of nature are equable and steady, we find, in natural appearances, a means of concluding a certain portion of time to have necessarily elapsed, in the production of these events of which we see the effects. . . .

But how shall we describe a process which nobody has seen performed, and of which no written history gives any account? This is only to be investigated, first, in examining the nature of those solid bodies, the history of which we want to know; and secondly, in examining the natural operations of the globe, in order to see if there now exist such operations as, from the nature of the solid bodies, appear to have been necessary in their formation. . . . Therefore, there is no occasion for having recourse to any unnatural supposition of evil, to any destructive accident in nature, or to the agency of any preternatural cause, in explaining that which actually appears.

His was a truly scientific approach, limited to the observations he could make, and he could find no evidence to suggest that geological processes had not operated from the earliest times:

But if the succession of worlds is established in the system of nature, it is in vain to look for anything higher in the origin of the earth.

The result, therefore, of our present enquiry is that we find no vestige of a beginning – no prospect of an end.

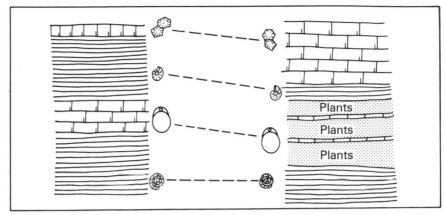

Fig. 12.1 Correlation by fossils. Two successions of rocks 400 km apart have different thicknesses of rocks, but contain similar, distinctive fossils at several levels to enable a correlation to be made.

Hutton's friend, John Playfair, popularised his views in the early years of the nineteenth century. In 1830 Charles Lyell, who had recently been persuaded that Hutton was right, published the first edition of his book, *Principles of Geology*, which was to be the basic text for geologists during the next fifty years and greatly influenced Charles Darwin's view of the world. Its subtitle: 'Being an attempt to explain the former changes of the Earth's surface by reference to causes now in operation', shows how it followed Hutton's basic theme. It has been said that Lyell's *Principles* and Darwin's *Origin of Species* created a revolution in scientific thought in the nineteenth century akin to that associated with the impacts of Galileo, Kepler and Newton in the seventeenth.

The basic concept for interpreting the rock record put forward by Hutton and elaborated by Lyell earned the nickname '**uniformitarianism**' from its opponents, and was later summarised in the dictum: 'The present is the key to the past.' The concept has maintained its place as the basis of geological studies despite continuing criticisms of its viability. Physical, chemical and biological processes have acted in the same ways over the hundreds of years covered by written historical records, and this lends support in the short term. Over periods corresponding to the geological time-scale one can point to the fact that light from a nebula, which we believe started its journey to the earth hundreds of millions of years ago, contains the same characteristic group of lines as is seen today in a hydrogen spectrum in the laboratory. This suggests that the processes may have acted in the same ways throughout time. The differences in the geography of the past – i.e. fluctuating ice-sheets, moving continents, changing positions of mountains, completely different groups of animals and plants – do not affect the case, since they can all be seen as part of the grand scheme of a planet developing under the processes which are still in operation.

James Hutton and William Smith can thus be regarded as the true fathers of geology and stratigraphy in their modern aspects. Before 1780 there had been long ages of groping in the dark for the keys which would unlock the door to the ideas needed to interpret the mass of geological facts gathered over hundreds of years. By 1830 Charles Lyell could produce a systematic account of the geological information which was to hand in terms which are relevant today. The way was open for the research which has led to our modern understanding of the geological history of our lands. Playfair wrote the following in his Introduction to the *Illustrations of the Huttonian Theory of the Earth* (1802) which began the revolution:

... we shall see abundant reason to conclude that the earth has been the theatre of many great revolutions, and that nothing on its surface has been exempted from their effects.

To trace the series of revolutions, to explain their causes, and thus to connect together all the indications of change that are found in the mineral kingdom, is the proper object of a *theory of the earth*.

But, although the attention of men may be turned to the theory of the earth by a very superficial acquaintance with the phenomena of geology, the formation of such a theory requires an accurate and extensive examination of these phenomena, and is inconsistent with any but a very advanced state of the physical sciences. There is, perhaps, in these sciences, no research more arduous than this; none certainly where the subject is so complex; where the appearances are so extremely diversified, or so widely scattered, and where the causes that have operated are so remote from the sphere of ordinary observation. Hence the attempts to form a theory of the earth are of very modern origin, and as, from the simplicity of its subject, astronomy is the oldest, so, on account of the complexness of its subject, geology is the youngest of the sciences.

The last statement may raise a smile, but it illustrates the feeling that these early geologists had that they were at the beginning of great discoveries. Geology is still a growing, young science, and advances in our knowledge have been tremendous over the past twenty years. It has been estimated that 90 per cent of the geologists who have contributed to our knowledge are still living.

The geological column

In Chapter 1 you were introduced to the table of periods of geological time, which is often known as the Geological Column (Fig. 1.8). It shows the relationships of the geological time units which were built up in the early years of the nineteenth century. At first they were based on the major rock-types and the obvious unconformity divisions between. Thus the Old Red Sandstone of the Welsh borders could be distinguished from the overlying grey limestones and coal seams included in the Carboniferous division, and the brick-red New Red Sandstones on top of the Carboniferous. The rocks which covered the New Red Sandstone were fossiliferous clays, sandstones and limestones, and these were named after fine exposures found in the Jura Mountains of France – the Jurassic system. Above these was another similar group including the distinctive white chalk, and this system of rocks was named the Cretaceous (Latin *creta* = chalk).

The youngest rocks of all were named after the uppermost of an earlier division of all rocks into Primary, Secondary and Tertiary groups. In 1833 Charles Lyell put forward a threefold division of these Tertiary rocks based on the number of fossil groups having living representatives. Thus the Eocene ('dawn of recent time') had 1–5 per cent of present-day species of animals; the Miocene ('moderately recent time') had 20–40 per cent; and the Pliocene ('most recent time') 50–100 per cent. Later the Pleistocene ('extremely recent time') replaced part of the Pliocene system, and the Oligocene ('slightly recent time') had to be introduced between the Eocene and Miocene.

More widespread studies of the two Red Sandstone formations produced one or two further modifications. The New Red Sandstone rocks were almost uniform in Britain, but on the Continent they could be divided into two major systems. Murchison and Sedgwick, two famous British geologists, were invited to Russia by the Tsar and named a thick group of fossiliferous limestones the Permian after the province in which they had seen them. In Germany the uppermost New Red Sandstone rocks could be divided according to three distinct rock types, and the whole group was named the Triassic. Before journeying to Russia Murchison and Sedgwick had visited Devon and Cornwall and found rocks older than the coal-bearing strata, but bearing little relationship to the Old Red Sandstone. Studies in the Rhine Slate Plateau of Germany confirmed that these Devonian rocks were the marine equivalents of the Old Red Sandstone land deposits.

The rocks beneath the Old Red Sandstone had posed greater difficulties, since they were very similar in lithology, were often highly folded and faulted, and contained few fossils. Murchison went to South Wales and worked down from the Old Red Sandstone base. He made two divisions, the Lower and Upper Silurian, named after the Silures tribe which once inhabited the area. Sedgwick worked in North Wales without any upward level for reference, and called the ancient rocks in that area the Cambrian system (Cambria = Wales). The discovery that Sedgwick's Upper Cambrian and Murchison's Lower Silurian contained identical fossils led to a bitter quarrel between these two friends, and it was not until after their deaths that Lapworth, who had worked out the complex geology of the Scottish Southern Uplands by using the graptolite fossils, proposed the name Ordovician (after a north Welsh tribe) for the overlapping rocks.

The Cambrian rocks, however, were the oldest in which any fossils could be found at the time, and all the most ancient schists and gneisses beneath them came to be known collectively as the Precambrian.

Thus the major divisions of geological time were established and given names with a variety of origins. It is amazing that they have stood the test of time and world-wide extension, and have been adapted to local conditions everywhere. The main modification has been in North America, where the Carboniferous is divided into the Mississippian (limestone) and Pennsylvanian (coal-bearing) systems. It has been found, however, that the original lines of division, often based on major British unconformities and rock-types, have not held true, and that fossil boundaries are more reliable. The general classification proposed by Philips in 1840 has been adopted:

The **Cainozoic** ('recent life') – Eocene to Recent.
The **Mesozoic** ('middle life') – Triassic to Cretaceous.
The **Palaeozoic** ('ancient life') – Cambrian to Permian.

These three eras are now known together as the Phanerozoic ('evident life'). The Precambrian rocks contain few fossils, and may be divided into two groups – the most ancient Azoic ('no life') and the Cryptozoic ('hidden life').

These are the major **eras** of geological time, to which **periods** like the Cambrian and Cretaceous belong. The rocks formed during these periods are

known as the members of, for instance, the Cretaceous **system**.

Fossils are the most important means whereby one rock is shown to be older or younger than another – i.e. given its age relative to other rocks – but they cannot give us the absolute age of rocks in years. This can now be done by using **radioactive minerals** contained in the rocks. The basic principle of this method was discussed in Chapter 11. The most reliable dates are those from igneous rocks, since the date recorded by such minerals is that at which the mineral was first formed (Fig. 12.2). We can now prove that although the granite composing Lundy Island in the Bristol Channel is so close to Dartmoor (280 million years old), it is of the same age as much more recent granites in Northern Ireland and along the west coast of Scotland (50 million years old). Sediments on the other hand contain minerals from a variety of rocks, and these may record several 'radioactive dates' covering a wide period of time. Certain minerals, like glauconite, are formed as the sediment is accumulating in the sea, and also have tiny quantities of radioactive elements which allow a sediment to be dated more accurately. The processes involved in metamorphism often result in the formation of new minerals containing radioactive elements, the decay of which can be used to give a 'radioactive date' to the formation of the mineral. But metamorphosed rocks often give several such dates, clearly indicating that there have been several periods of change in the history of the rock. These

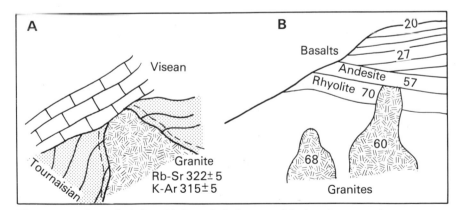

Fig. 12.2 Radiometric dating: this is particularly successful where igneous rocks are present. **A** The Vosges granite has been used to give a date for the division between the lowest Lower Carboniferous rocks (Tournaisian) and those lying on top (Viséan): the granite intrusion altered the Tournaisian rocks, but the Viséan were deposited later, after erosion had exposed the granite. The two radiometric dates given were produced from the rubidium-strontium and potassium-argon methods. **B** The Tucson mountains, in south-western USA, show how a sequence of lavas and igneous intrusions can be dated by radiometric methods. (All dates in millions of years)

are often significant in recording several phases of alteration and earth movement in an area: individual minerals within the rock give later dates than the rock as a whole, which suggests that a later phase of metamorphism affected some minerals, but not all of the rock. This is very important in studying the Precambrian rocks.

A summary of the principles of stratigraphy

(1) Our basic rule is **The Law of the Superposition of Strata**. We must first determine the order in which the rocks were formed. **Unconformities** are particularly important features of the rock succession, since they record major breaks in deposition due to uplift and erosion of the rocks formed previously. This gives us our local sequence of rocks in order of formation.

(2) **We correlate various local sequences mainly by the fossils** they contain if they are of Cambrian age or younger. This gives the rocks a series of relative ages (i.e. one rock is older, or younger, than another).

(3) If the rocks do not contain fossils we can work out the sequence by various criteria showing which is the right way up. **Radiometric dates** based on radioactive minerals give us numerical dates for rocks and are especially useful where the rocks are of Precambrian age, since there are not enough fossils to enable correlation to be carried out.

(4) When we come to **interpret the conditions** in which rocks were formed we use our knowledge of the present geological processes and compare their products with the rocks. We can tell whether a sea was advancing or retreating (Fig. 12.3); whether the rocks were involved in a full orogenic cycle; or sometimes we can work out the prevalent climatic conditions on the continents of the times.

Stratigraphy and economic geology

Just as almost every other aspect of our geological studies has had an application in terms of an economic use, so the methods of stratigraphy are finding increasing uses in the search for new sources of minerals. Rocks formed in particular environments in the past often form distinctive oil reservoirs. Thus many of the Libyan oilfields occur in Cainozoic shoreline sands, and the Canadian oilfields of Alberta in Devonian reefs. The geologist in these areas looks for associations of rocks which suggest that such conditions occurred in the past.

Correlation of rocks by fossils is another method used for tracing likely oil-bearing formations across country. In addition there are some resources which occur in formations of a particular age. Most of our oil comes from rocks formed in the last 65 million years, whereas most of our coal is obtained from rocks formed between 320 and 250 million years ago; Britain's chief brick clay horizon is the Oxford Clay of Upper Jurassic age.

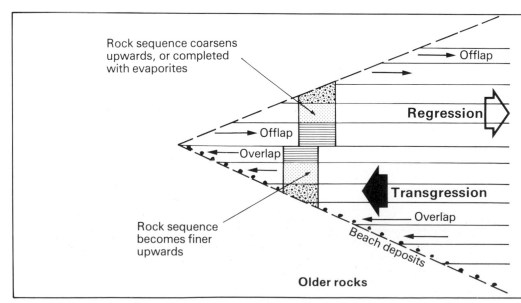

Fig. 12.3 The rock successions associated with transgressing (advancing) and regressing (retreating) seas. Such changes are commonly recorded in sequences of rocks in cliffs or quarries.

Field geology

Every geologist must be a field geologist, since all the material for his scientific observations is to be found in the rocks of the coast and countryside around him. Those who use their geological knowledge to earn a living make this the basis of all their decisions – they have to go to the place in question before drawing any conclusions. Much of the analysis of results and identification of specimens can be done only with the facilities of a laboratory or library at hand. But the basic work of the geologist must be done in the field, and there is plenty of scope for the amateur geologist to investigate the rocks in his own locality, provided that it is done in the correct way.

Before carrying out fieldwork it is important that you should be aware of safety precautions (e.g. do not venture on an exposed mountain in bad visibility or low temperatures; do not go underground unless properly equipped and led; do not attempt rock-climbing; do not go close to rock faces without head protection), and of the Fieldwork Code (e.g. always ask permission to go on land belonging to others – and all land does, even moorland; do not over-hammer exposures).

Equipment needed. A geologist's field equipment can be very simple and cheap. He will need some old but warm and rainproof clothing, a notebook and mapcase, a hand lens and implements with which he can break open the rocks to make investigation easier. It is important to examine unweathered surfaces of a rock as well as those exposed in the outcrop, and to prise out the fossils. The hardest rocks will yield only to heavy hammers weighing 1 kilogram or more, but many soft rocks will need lighter and more delicate treatment. Cold chisels are useful for extricating fossil specimens from hard rocks; a broad-bladed penknife from soft. A clinometer is used to measure the rock dips, and a tape-measure for thicknesses. Tins and boxes are essential for preserving pieces of fragile rock or fossil, many of which will need to be wrapped in paper or even cotton-wool for the journey home. All this equipment can be carried in a light rucksack.

In addition the geologist will need storage space for the specimens he has collected and labelled, and books of reference to assist him in the identification of the rocks, fossils and minerals. A list of these is given at the end of the book.

Where to look. The solid rocks are often masked by glacial deposits, soil, landslips, peat, vegetation and the works of man. Inland exposures of solid rock are largely confined to natural scars, which may be abundant in upland areas, quarries, river banks and fresh road and railway cuttings. Permission must be obtained to examine quarries and railway cuttings. Making a map in areas of complex structure with few exposures is very difficult even if the geologist has records of bores and mine workings to aid him. The line of outcrop of a hard rock can, however, often be traced by means of a relief feature such as a ridge or escarpment.

The best areas for the study of the solid rocks are the cliffed coasts, where there are excellent sections through whole series of rocks, especially where they are steeply inclined. One must always keep away from dangerous crumbling cliffs, and it is best to examine the rocks of the foreshore. Within a space of 3 km on the east coast of the Isle of Wight one can, because they are nearly vertical, study the complete succession of Cretaceous and Tertiary rocks which make up the island and span nearly 100 million years in time (Fig. 1).

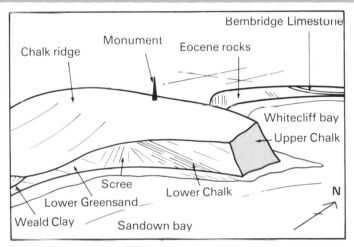

Fig. 1 The eastern Isle of Wight coast.

It is best to begin by studying an area of more recent rocks in areas like the Isle of Wight, Dorset or North Yorkshire coasts. Such rocks will be easier to study, will contain more fossils and in general will yield satisfying results more rapidly. Much has been written about these areas, and the guides published by the Geologists' Association give an excellent introduction to what you should find – or a check on what you have discovered.

If you live inland you will be able to examine rock exposures in quarries, etc., and will often be able to trace an outcrop across country in the relief. You can also place a greater emphasis on noticing the ways in which geological processes are acting today. Variations in stream-flow and load, evidence for soil creep or for glaciation in the past and the development of features like stream meanders, waterfalls and lake deltas all form profitable lines of study. It is important to make sketches of such features, adding geological notes; if you are not an artist a photograph will be just as good.

Observations to make. Observation is the essence of field-work. Spend time noticing facts about each exposure or section of cliff and record them in your notebook: such a record can take the form of a sketch or photograph as mentioned above, or an orderly list of the details. Throughout this book we have emphasised the importance of doing this and have suggested a number of questions that you can ask yourself. Here are most of them again:

(1) What is the exact position of the exposure? You will need to plot the details on a map later, so make a note or head the page of your notebook with the Ordnance Survey grid reference.

(2) What is the nature of the rocks? Write down a careful description of their colour, texture and the jointing and bedding structures affecting them. Are they sedimentary, igneous or metamorphic?

(3) If they are sediments, what is the dip and the strike of the series of rocks? Does it vary? Are there any unconformities? Are there any folds or faults affecting the rocks, and if so of what type? Is it possible to work out the succession of rocks from oldest to youngest?

(4) If the rocks are igneous, can the margin of the intrusion or lava flow be seen, or is its nearness indicated by smaller crystals? What is the rock-type, and is it a lava-flow, dyke, sill, etc.?

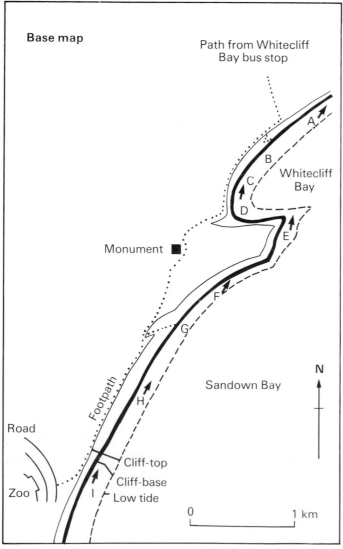

Fig. 2

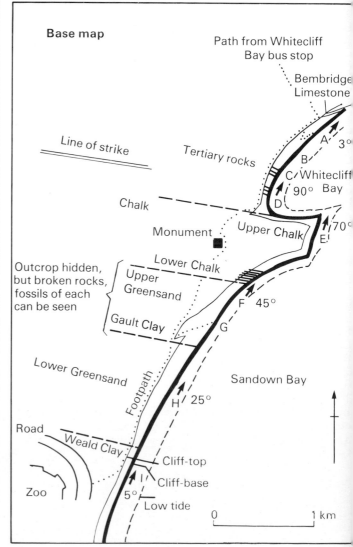

Fig. 3

(5) If they are metamorphic, is the alteration limited to any zone? See Fig. 8.29 for a field sketch of the Cleveland Dyke and see how many of the questions in (3), (4) and (5) are answered.

(6) What fossils do the rocks contain? Do they occur in special bands? The longer you hammer at sedimentary rocks the more you are likely to obtain. They will help you to date the rocks and to understand how they were formed.

(7) What do the landforms of the area reveal about the relative resistance of the rocks to erosion?

(8) What is the evidence for recent geological activity – e.g. river valley features, glacial landforms, coastal changes?

It might help here to refer you to an exercise given to a group of young geologists and based on the section at the eastern end of the Isle of Wight shown in Fig. 1. You should be able to answer some of the questions from that diagram. Each student was provided with an outline map (Fig. 2), and the following list of questions.

(a) Make a description of the rocks at each of the places A–I in your notebook.

(b) Mark the angle of dip in the rocks where the arrows show the direction on the map.

(c) Mark in the boundaries of the rock outcrops on your map and the direction of strike in one place only.

(d) Draw diagrams of any fossils you find, and especially of those you cannot take away; note the map position of each discovery.

(e) At the end of the day draw a vertical section along the line of the cliffs and work out the structures affecting the rocks. Write a short historical account of the geology of the area.

The map, Fig. 3, shows the results plotted from investigations on the ground. Your notebook might read as follows:

Point A: northern end of Whitecliff Bay
Rocks dip at 3 degrees to the north
Rocks: massively-bedded (up to 2 m thick), well-cemented, cream or buff coloured; contain many gastropod fossils (freshwater snail type), but most of shells dissolved away leaving casts.
This rock is a freshwater limestone.
It is one of the youngest in the Bay, since it lies on top of the others.
The sea has undermined the rocks here, and there are large broken boulders on the beach.
At the extreme north of the Bay the dip of the rocks brings them down to sea level, and they form a reef extending out to sea.
(Here there would be a drawing of the gastropod and a labelled sketch of the massive bedding and sea-broken blocks.)
Point B: just south of the main cliff path.
Rocks dip almost vertically.
Rocks: softer than those at A, varying from unconsolidated sands to weak clays . . . etc.

Making a simple map. You will thus make many observations and collect many specimens as you study the rocks of a particular area, but the most important result will be to make a map of your findings. You have seen what geological maps look like and have learnt to interpret them. Now you should be starting to make your own.

The base-map, on which you will record your information, will normally be available already: the Ordnance Survey 1:10 000 is the best, since there is room to plot a lot of information, and you can still see the wider relationships. Remember, however, that the thickness of your sharpened pencil line covers 3 m on the ground. You will often have to use symbols for plotting information on the map: some of those that are used commonly are listed in Fig. 4.

When the map is ready for use you can record the information gleaned from such questions as were asked above. Detailed sections can be noted at the side of the map or references made to your notebook. The most important features to record at first are the points of contact between the layers of rock, the dip measurements, and the trends of igneous intrusions, faults and folds. Having done this it should be possible to join up the various boundary outcrops, using the principles learnt in the study of geological maps. Then you should colour the different rock formations, as a means of helping you to interpret the structures in the same way as you have done on the other maps.

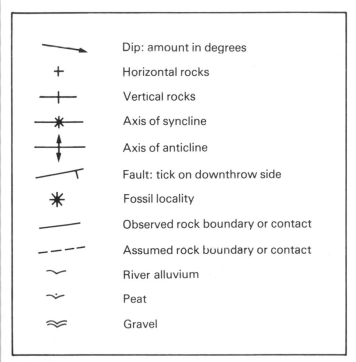

Fig. 4 Signs used on geological maps.

Applying the principles: Mesozoic and Cainozoic rocks

William Smith began his fruitful studies in the group of rocks which underlies southern and eastern England (defined roughly by a line from the Tees to the mouth of the Exe). These rocks are a series of relatively thin layers, and are gently tilted and folded: they were formed in relatively shallow seas which covered varying proportions of Britain from early Jurassic to Oligocene times. During this period of history the British Isles lay on the margins of the opening North Atlantic ocean (Chapter 11), and the activity occurring in that zone is reflected in the coming and going of shallow seas as the constructive plate margin became established and the continental terrace areas on its American and European margins began to grow on the 'trailing edges' of the parting continents. In the later stages of this rifting more violent earth activity led to volcanic eruptions in the north-western sectors of the British Isles area, together with some folding in the newly-deposited sedimentary rocks in the south east.

These rocks were formed on the older rocks of the British Isles, surrounded by the subsiding basins of the North Sea and western continental margin. The older rocks tended to be elevated from time to time, and thus the sedimentary rocks formed on what is now land are often thinner and more interrupted by breaks in deposition than those in the surrounding basins (Fig. 5.18). These basins have been studied intensively since the discovery of oil deposits in their rocks.

The study of Mesozoic rocks is carried out in the light of the overall picture outlined in Chapter 11 and of the principles of stratigraphical interpretation.

Lower Jurassic (or Liassic) rocks: correlation by ammonites

After a long phase of largely land conditions in the Permian and Triassic periods, a shallow branch of

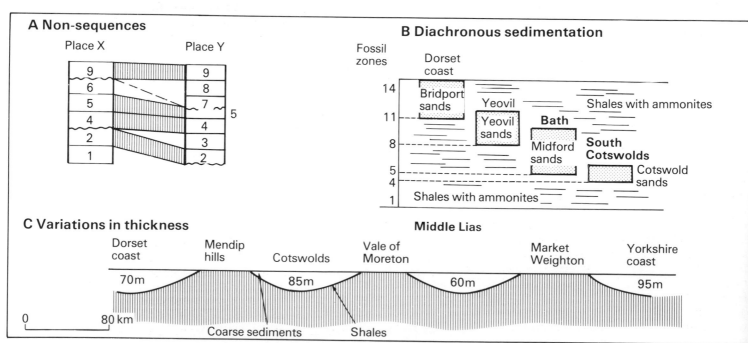

Fig. 12.4 Some features of the Upper Lias, which are important in age relationships. **A** The numbers 1–9 represent a series of fossil zones: although the layers of rock follow without visible unconformities we can tell that there was a break in deposition. How many zones do X and Y have in common? **B** The deposition of sandy sediments moved from north to south over the period of time covered by 11 fossil zones. **C** Varying depths of water and degrees of subsidence can affect the thickness of rocks which are deposited at any one time.

the ocean spread north-eastwards and deepened to inundate most of the British Isles, which had been worn down to a very lowlying landscape. Muddy waters covered the south and east and extended in shallow gulfs around the coasts of Scotland and Wales: these areas probably stood out as land, supplying fine debris to the seas. Although clays and thin bands of limestone alternate on the Dorset coast, most of the other Liassic rocks are dark shales. They contain a wealth of fossils, especially ammonites and reptile remains, but also those of other animals which like muddy sea conditions, such as oysters and belemnites. The only important variation is found in the Midlands and North, where there are layers of ironstone rich enough to be worked at such places as Scunthorpe (Lower Lias), Banbury and the Cleveland Hills (Middle Lias), and forming a low escarpment.

The Liassic rocks are very well-zoned by ammonites, and this fact enables us to reach several important conclusions concerning the conditions of sedimentation at this time. It is common, for instance, for some of the fossil zones to be missing from the succession in any one area, as Fig. 12.4 demonstrates. Two reasons have been suggested for these **non-sequences** in the rocks: either there was no deposition at the time, or the particular type of ammonite did not reach the area from which it is missing. Another important fact is that the volume of sediment deposited varied from place to place, for 2 m of shale in Dorset contains the fossils of six distinct zones, each of which normally extends through several metres of rock thickness. This is known as a **condensed sequence**. A third fact is demonstrated by the Upper Liassic rocks of southern England, where a group of sandy rocks contain the fossils of younger and younger zones as they are traced southwards (Fig. 12.4). The special conditions of deposition in which the sands were laid down gradually moved to the south, so that the layers of rock were formed across the time planes, and are said to be **diachronous**. Finally, the uneven nature of the sea floor at this time is shown by the thicknesses of Upper Liassic rocks found in different parts of the outcrop: Fig. 12.4 shows this. The dominant shales give way to coarser, shallow-water sediments at the thinnest parts of the succession.

Middle Jurassic: facies changes in space

The Middle Jurassic rocks of England illustrate the differences in rock-forming conditions which were present at the same time. We use the word **facies** to describe particular features and characteristics of a sedimentary rock: it includes the sum of its composition, internal structures and overall geometry (i.e. whether it is flat-bedded, lenticular or wedge-shaped), together with the fossils it contains. Each sedimentary facies is the product of an ancient depositional environment, and the names given to facies may indicate this – i.e. deltaic facies, reef facies. Alternatively facies are named after descriptive aspects of the rock – e.g. oolitic facies, graptolitic or shelly facies. In many ways it is better to use descriptive terms, since different depositional environments may give rise to similar features, and interpretations change: thus many rock groups described originally as estuarine facies are now reinterpreted as of deltaic origin.

Fig. 8.23 summarises the varying facies along the length of the main Middle Jurassic outcrop in England. Today's most striking relief feature in rocks of Jurassic age is provided by the thick oolitic limestones of the Cotswold Hills in Gloucestershire. Here the layers of limestone are interrupted by thin beds of marl and fine lagoonal limestones like the Stonesfield 'slate'. The best oolites are used for building stones, and the 'slate' for roofing tiles. All these rocks were formed in shallow, lime-rich seas. To the south of the shallow Mendip area, however, rocks of the same age are clays, and to the north of the shallow Vale of Moreton zone the escarpment again becomes less prominent as the proportion of clays increases. In parts of Northamptonshire the deposits are estuarine, containing fossils of freshwater shell creatures and plants, and there is another thick layer of ironstone round Corby and Kettering. This is the most valuable British source of iron ore, and is mined by open-cast methods (Fig. 8.16). The Lincoln Edge area was more like the Cotswolds, being formed of false-bedded oolitic limestone and coral reefs, but north of the Humber estuary there is the greatest contrast of all. The North Yorkshire Moors, where Middle Jurassic rocks form so many of the hilltops, were the site of a delta, very similar to that in which the Millstone Grit of Carboniferous age was formed. Coarse, pale-coloured sandstones containing thin bands of plant remains and even coal seams are interrupted at three levels by marine incursions across the deltaic flats. These rocks are very similar to the Middle Jurassic rocks of Scotland, where the east and west coasts and the islands locally have sediments of this age. Some are marine and include clays, sandstones and limestones, but others are deltaic, estuarine or fluviatile. There is even a workable coal seam at Brora on the east coast of Sutherland. Fig. 12.5 shows how the Lower and Middle Jurassic rocks of the Mendips are related to each other and to older rocks.

The Upper Jurassic and Lower Cretaceous: facies changes in time

Sedimentary facies not only change from one place to another during the same phase of geological time but are also found varying from one phase of geological time to the next in the same place. Figure 1.1 illustrates changes which can be traced in the uppermost Jurassic and the Cretaceous rocks of the Dorset coast: summarise the situation in terms of uplift in the late Jurassic and of rising sea level during the Cretaceous – or of a regressing and transgressing sea.

The **later Jurassic** rocks saw another return to widespread marine conditions, and muddy waters replaced the clear seas. On top of a thin layer of marine limestone or calcareous sand, known as the Cornbrash, the bluish Oxford Clay forms lowlying country and provides the raw materials for the largest brickmaking industries in the country round Peterbrorough and Bedford. The muddy conditions put an end to the clear limestone seas for a while, but these later re-established themselves to the south of Oxford, where a bed of Corallian Limestone forms a low escarpment. North of Oxford, however, the muddy conditions predominated, and the Ampthill Clay was deposited, adding to the width of the clay vale leading north-eastwards to the Wash. In Yorkshire the Corallian Limestones are at their thickest, and form the steep escarpment of the Hambleton Hills along the western edge of the North Yorkshire Moors. A final extension of the muddy waters led to the formation of the Kimmeridge Clay all over south and eastern England. It is very thin in Bedfordshire and round the Humber estuary, but thicker in Dorset and beneath the Fens and Vale of Pickering (Yorkshire), reflect-

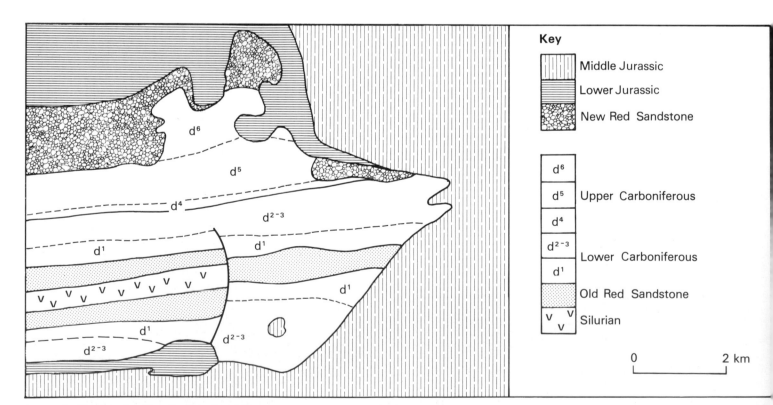

Fig. 12.5 Map exercise. How do the rocks of this area show that there was a period of folding between the deposition of the Palaeozoic formations and the Mesozoic rocks? Suggest evidence which shows that the Middle Jurassic seas extended farther than those of the Lower Jurassic, leading to an overlap situation. (Sketch-map based on the Institute of Geological Science map of the Frome area. *Crown Copyright reserved*)

ng the fact that the zones of uplift and subsidence Fig. 12.4) were still active. The seas still impinged n the coasts of Scotland, and fossils of Kimmeridge ge have been found in rocks that resemble the Old Red Sandstone on the east Sutherland coast. These emarkable rocks contain huge boulders up to 50 by 30 by 10 m in size, and it is thought that they must have been tumbled down from cliffs by omething special, like tsunami waves generated by earthquake shocks.

However, that was the end of the Upper Jurassic marine transgression, and the highest layers in the succession show us that the sea became shallower, was restricted to a much smaller area and finally excluded. The calcareous sands and oolitic limestone characteristic of the Portland Stone are only ound between the Dorset coast and Oxford Fig. 1.1). This is another of our most important building stones. The overlying Purbeck Beds are freshwater limestones containing pond-snail fossils, petrified forests and 'dirt beds' (fossil soils). Recent

borings in Kent have shown that both the Portland and Purbeck rocks are very thick there.

It is evident that at this time an area of sea had been cut off from the main shelf seas covering Europe, and had been largely filled in with sediment, leaving swampy conditions in the extreme south of Britain, whilst the north must have experienced some uplift and erosion. These conditions were continuing into the early stages of the **Cretaceous period**, for the area to the south of London was a zone of shallow lakes and lagoons crowned by the flat-surfaced deltas of rivers bringing debris from the London Platform area of land. The sand and clays, with occasional nodules of limestone and ironstone, bear a record of sun-cracks and ripple-marks on their bedding planes and contain fossils of giant reptiles and tiny water-fleas. Coarse pebbly beds in these Wealden Series rocks enable some of the ancient rivers to be traced, and the succession of sediments show a regular rhythm as the level of the Wealden lake rose and fell (Fig. 12.6).

Worbarrow Bay, Dorset. The rocks are dipping northwards: the oldest are the Upper Jurassic rocks at the extreme right; the central ridge is Upper Cretaceous chalk, and behind that are lowlands on Tertiary sands. Relate the succession of rocks to that in Fig. 1.1. (Aerofilms)

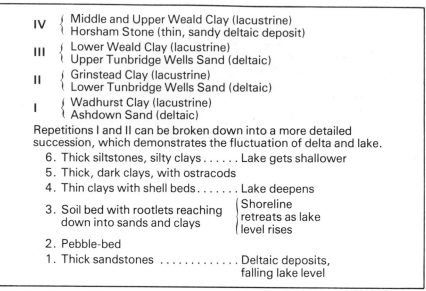

IV	{ Middle and Upper Weald Clay (lacustrine) { Horsham Stone (thin, sandy deltaic deposit)
III	{ Lower Weald Clay (lacustrine) { Upper Tunbridge Wells Sand (deltaic)
II	{ Grinstead Clay (lacustrine) { Lower Tunbridge Wells Sand (deltaic)
I	{ Wadhurst Clay (lacustrine) { Ashdown Sand (deltaic)

Repetitions I and II can be broken down into a more detailed succession, which demonstrates the fluctuation of delta and lake.

6. Thick siltstones, silty clays . . . . . . Lake gets shallower
5. Thick, dark clays, with ostracods
4. Thin clays with shell beds . . . . Lake deepens
3. Soil bed with rootlets reaching down into sands and clays { Shoreline retreats as lake level rises
2. Pebble-bed
1. Thick sandstones . . . . Deltaic deposits, falling lake level

Fig. 12.6   Repeated sediments in the Wealden Lake. Notice how the changing deposits reflect the deepening of the lake waters, or the advance of the delta.

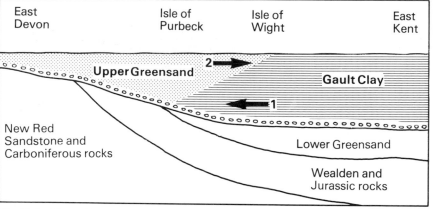

Fig. 12.7   Upper Greensand and Gault Clay. These two types of rock were formed at the same time in a transgressing sea as it advanced (1) over the tilted and eroded older rocks. This phase was followed by a period of deposition which filled the seas with coarser debris eroded from land to the west and caused the sea to regress (2).

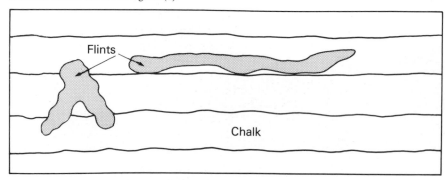

Fig. 12.8   Flints in the Chalk at Flamborough Head, Yorkshire. What features of the flints suggest that they were formed after the Chalk?

These rocks, which are now exposed in the central Weald of Kent, Surrey and Sussex, are covered in this area by the Lower Greensand, which was formed as shallow seas once again swept across the whole area and buried the deltaic and lacustrine deposits. The Greensand is formed of soft, unconsolidated brown or orange sands, sometimes with hard, cherty bands, and locally has a calcareous cement. The fact that it is seldom green is due to the chemical alteration of the green iron silicate glauconite, to the brown iron oxide as it is weathered. The presence of glauconite, however, tells us that these rocks were marine deposits. The varying importance of the different types of rock accounts for the fluctuating height of the escarpment which is normally associated with the outcrop of the Lower Greensand. It forms Leith Hill (320 m O.D., the highest point in south-eastern England), but at times disappears altogether as a feature in the landscape. The overlying Gault Clay and Upper Greensand show that the sea extended to the west of England and then was filled in by coarse sandy deposits (Fig. 12.7).

At the same time north-eastern England, on the far side of the London Platform, was subjected to quite different conditions as an arm of the sea covering Germany and Russia swept in. Thick clays containing numerous ammonites just like those found in central Europe, but different from those in the Weald and northern France, were formed. They are topped by the unusual deposit of Red Chalk.

The Chalk: unique sedimentary facies

The final episode in the Jurassic–Cretaceous phase of quiet, shallow-sea conditions was the most remarkable. It saw perhaps the most complete coverage of these islands by the sea that they have ever known. Some geologists believe that the whole area, including the mountainous north and west, was inundated by the widespread seas, in which the unique deposit of **Chalk** was formed to thicknesses reaching almost 500 m. Chalk is an unusually pure calcium carbonate rock, formed largely of microscopic plant and animal debris together with scattered larger fossil shells and their broken debris. Very little sand or mud could have been supplied to this sea, and although the lower layers contain up to half their volume of clay and have a greyish colour most chalk is dazzling white in colour with less than 1 per cent of non-calcareous material. The upper layers contain many siliceous nodules and flat masses of flint, which occur along the bedding planes but sometimes also across them (Fig. 12.8). This arrangement suggests that they were formed after the Chalk, probably as it was drying out and water was percolating through the rock. The source of so much silica, however, is a great problem. Some geologists have suggested it originated from the solution of sponge skeletons, and others have put forward the idea that the silica came from the volcanic activity associated with the early stages of the opening up of the North Atlantic ocean.

The Chalk seas were therefore the scene of slow limestone deposition, and it has been estimated that the rate of accumulation might have been something like 1 cm every 1000 years if it was anything like the similar deep oceanic oozes of today. This resemblance in characteristics led geologists to classify Chalk as a deep-sea deposit, but harder, nodular bands like the Chalk and Melbourne Rocks and Totternhoe Stone are definitely shallow-water sediments, and many of the common sea-urchin and mollusc fossils are never found in waters deeper than 200 m.

Life in Jurassic and Cretaceous times

These two periods have been called 'The Age of Reptiles', for it was at this time that that group of animals dominated the scene. On land the dinosaurs multiplied into many different varieties, including small, fleet-footed types and great lumbering monsters. The largest *Brontosaurus* was 30 m long and probably weighed 35 tonnes: it had to live in swampy conditions which helped to buoy up its weighty body, which its legs could not otherwise have supported, and to reduce its diet of plant leaves to an edible pulp. It was a herbivore (plant-eater), but some of the normally more active carnivores (meat-eaters) like *Tyrannosaurus* grew to 16 m long. Some reptile groups lived in the seas, and the ichthyosaurs and plesiosaurs were up to 10 m and 20 m long respectively. Pterosaurs developed web-like wings which were used for gliding, soaring and eventually for proper flying. The warm-blooded, feathered birds also arose during the Jurassic period and probably had a reptile ancestor. At the end of the Cretaceous, however, most of the important reptile groups became extinct. It is difficult to see the reason for this, but it may have been connected with the physical conditions such as the spread of the Chalk seas – though these did not cover the whole world – or with the changing vegetation.

The land plants had changed little during the Mesozoic era. The coniferous plants, which had first been found in the Coal Measure forests of Carboniferous age, had developed with their close relations the ginkgos and cycads, and with many ferns. Towards the end of the Cretaceous the first fossils of flowering plants are found in the rocks, and this group soon became the most important of all. Fossils of pollinating insects are also found from this time onwards,

but it is uncertain which appeared first. Perhaps the large, herbivorous reptiles had become too specialised in their diet and could not readapt themselves to these new plants: the carnivores would then be deprived of their staple diet.

The Jurassic and Cretaceous systems were also particularly important for the development of the invertebrate animals – insignificant no doubt when they were alive, but of the greatest importance to geologists today. Just as this is the 'Age of Reptiles' in one sense, so it is 'The Age of Ammonites' in another. These animals living in coiled shells spread northwards from the Tethys Ocean, with one family succeeding another: each had new features which had evolved in the deep geosynclinal trough across southern Europe. They provide us with some of the best-established correlations in the whole succession of rocks formed in past ages, although they tended to avoid seas rich in lime. The Lower Jurassic rocks and the Oxford and Kimmeridge Clays contain plenty of fossil ammonites, but there are few in the oolitic limestones and in the Chalk. The other mollusc groups were also extremely important, and the sea urchins had reached a high level of development by the time the Chalk seas arrived: they are used as zone fossils for certain Chalk horizons (Fig. 9.21). Whilst the ammonites are used for zoning the Jurassic and Lower Cretaceous, a variety of fossils – sea urchins, brachiopods, free-swimming crinoids and belemnites – are used in the Chalk.

Some of the fossils that had been important in the Palaeozoic were only represented in the Mesozoic by their most advanced species, such as the terebratulid and rhynchonellid brachiopods, and the *Pentacrinus* crinoids with their star-shaped ossicles. The corals were completely changed, and nearly all the fossils of this age belong to the scleractinian group.

At the end of the Cretaceous many groups of animals became extinct besides reptiles. These included ammonites and belemnites.

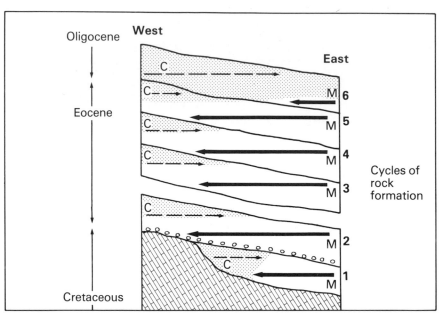

Fig. 12.9 Lower Tertiary sediments. The rocks can be divided into six cycles of sedimentation, all following a similar pattern, which began with seas advancing from the east (thick arrows) bringing marine conditions (M) and resulting in marine deposits containing fossils. Sediments from the west then gained the upper hand, filling the seas and providing continental conditions (C). The London Basin preserves a record only of the first two cycles, but all six are found in the Hampshire basin and the Isle of Wight. Compare this diagram with Fig. 12.7. How can we tell that a rock had a marine, or continental, origin? What would be the difference between the rock successions at the eastern and western ends of the area affected?

The formation of the Chalk ended the Mesozoic sedimentation, and it was followed by a break and considerable uplift during which a lot of Chalk was removed by erosion. This was the herald of the Alpine earth movements which were beginning to raise up mountains far to the south, but their main effects were not to be felt for some time in Britain.

The Lower Tertiary: marine and continental conditions

There is a complete break in Britain between the Tertiary rocks and the underlying Chalk, although the unconformity which was produced is nowhere very striking. The northern and western parts of Britain were also affected by these movements, which led to the initiation of major river systems draining towards the east. South-eastern England was not uplifted, but two areas were warped down and were drowned by part of a sea that extended across to the Paris Basin and northern Belgium. These two areas, the Hampshire and London Basins, were separated by an island due to the updom-

ing of the Weald, and they have somewhat different Tertiary sequences. Both record the fact that these basins occasionally subsided, making room for the sea to advance over the top of sediments that had been filling in the previous depth of water. The rocks of the London Basin only testify to two such cycles, whereas the Hampshire—Isle of Wight area has five, extending over a longer period of time. Whereas the London Basin rocks are all **Eocene** in age, those round the Solent estuary are Eocene and **Oligocene**. Each cycle began with the formation of pebble beds, marine clays or sands, containing fossils of fish and marine molluscs, and ended with the extension from the west of coarse deposits with evidence of shallower water and even continental conditions of sedimentation. Fig. 12.9 contains the details. The last sediments preserved are early Oligocene, and it seems probable that by the middle of the Oligocene period all formation of new sediments had ceased in south-eastern England.

Close study of these rocks in areas like the Isle of Wight shows up some of the differences between rocks which were formed on the land and those formed in the sea.

(1) **Rocks formed on the land.** These include river-laid sands and pebbles; freshwater lake clays and limestones which often include the distinctive, thin-shelled snail shells; plant and leaf beds which do not show too much disturbance or other indications of transport. Except for the lacustrine clays and the plant beds, land deposits seldom contain fossils. Bones of land mammals are extremely rare.

(2) **Rocks formed beneath the sea.** These are characterised by the fossils of marine animals: we know, for instance, that certain types of gastropod and lamellibranch shells are only found in sea water today, and many of these had close relatives in the early Tertiary. Rock types of marine origin vary from clays to sands, and those formed close to the shore may contain drifted and broken plant remains.

Discuss the significance of these characteristics, and add to the list as you study rocks formed at other periods of geological time.

Volcanic activity in the north-west

Other events were taking place at the same time in the extreme north-west of Britain. As the south-east was subject to occasional uplift and subsidence, so the north-west of Scotland was part of a great belt of **volcanic activity** extending from Iceland into Ireland as the North Atlantic split open (Fig. 12.10). Lava flows were poured out, building up tremendous volumes of basalt rock in such places as

Northern Ireland (e.g. Giant's Causeway) and Mull (over 2000 m thick). Such thicknesses required many eruptions, as the individual flows were never more than 30 m thick. Between the eruptions the lava surface was weathered, and soils were formed. In some of these soils sub-tropical plants of Eocene age have been found as fossils on Mull, and Oligocene plants occur in clays formed on top of the Antrim basalts. Volcanoes dominated what are now the islands of Skye, Mull and Arran, but today they are only represented by their deeply eroded stumps. These volcanoes often seem to have collapsed at the end of their active life, carrying down fragments of layers of rock that have otherwise disappeared from the area. Such cauldron subsidence left a complicated pattern of volcanic and intrusive igneous rocks in these areas. Intrusive masses of granite and gabbro are also found in the west of Scotland and Northern Ireland. Radiometric dates for the granites are all around 50 million years – i.e. Eocene in age. The island of Arran was widened by over 2 km in the course of dyke intrusion and an almost parallel group of dykes spread across southern Scotland and into northern England: the Cleveland Dyke (Fig. 8.29) is the best known of these.

Uplift in the Middle Tertiary

There are no rocks in the British Isles which can be assigned with certainty to the Miocene period. This was the time when the effects of the great Alpine earth storm reached these islands. Nothing happened in Britain to parallel the upheavals which produced the Alps and Pyrenees, but an indelible impression was left behind. To the south of the Thames there were thick layers of soft Jurassic, Cretaceous and early Tertiary sediments. These were folded into short, east–west folds, plunging at each end and only a few kilometres long. Farther north these younger sediments were not so thick, and the resistant, ancient rocks on which they rested yielded to the pressures in a series of faults. The whole area was tilted up to the north and west once again, giving rise to the distribution of relief we find today, and helping to expose the oldest rocks in these most heavily eroded areas.

The Upper Tertiary and Pleistocene: mainly land

The most recent rocks in Britain, apart from the thin mantle of drift deposits left by ice, wind and river action, are the deposits of East Anglia, for

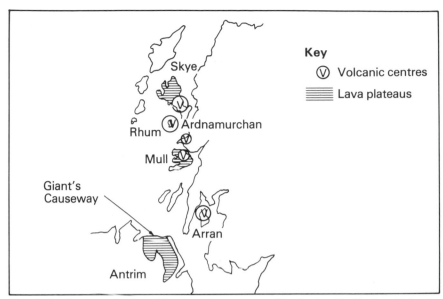

Fig. 12.10   Tertiary volcanic activity along the west coast of Scotland and in Northern Ireland.

long known as the Crags: this was an area which was warped down towards the North Sea during the Pliocene and Pleistocene whilst most of Britain was land. The rest of the country was still undergoing erosion, which left no tangible results in the form of rocks and fossils. The oldest of these Crag deposits is the Coralline Crag, formed of colonial polyzoan fossils and lamellibranchs embedded in limy sands. The faunas in the later Crag deposits, now included in the Pleistocene, indicate that the climate was getting cooler. We are probably still in the middle of the Ice Age which began in the Pleistocene period, for the glaciers have advanced and retreated several times, and at the present moment we are experiencing the later stages of ice retreat. We cannot tell whether the ice will eventually advance again, but, if it does, it will probably not do so for many thousands of years.

The fossils in the Crag deposits of East Anglia record the drop in temperature at the end of the Pliocene and the beginning of the Pleistocene period. They have thickened shells, and include varieties which are found today off the coast of northern Norway. The ice advanced from the highlands of this country and Scandinavia to cover much of northern Europe. At its greatest extent it reached the Bristol Channel–Thames line (Fig. 12.11). The fluctuations of the position of the ice-sheet front can be traced by studying the layers of boulder clay and the morainic deposits formed at the time. The succession of pollens in

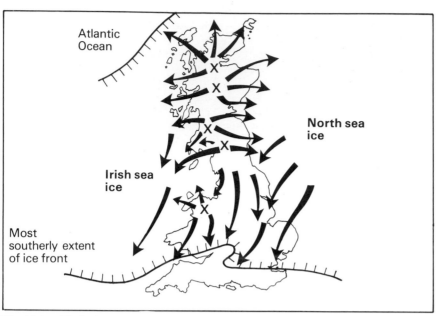

Atlantic
Ocean

North sea
ice

Irish sea
ice

Most
southerly extent
of ice front

Fig. 12.11   The Pleistocene ice sheet. This map shows the southern limit of ice advance, and the main centres of ice dispersal (X). Can you name these centres?

peat deposits, and the variation of sea level as recorded in the terraces of river valleys like the Thames, tell us how the climate warmed up or got cooler. At least three periods of glacial advance affected Britain, and it seems that the last of these left a complicated pattern of retreat deposits with occasional short-lived readvances. The warmer interglacial periods between saw higher sea levels, and one at least brought a warmer climate to Britain than that which we enjoy today. Some of the main events are summarised in Fig. 12.12.

The causes of Ice Ages

Why did the ice advance and cover much of North America and Europe? Why did it fluctuate with advances and retreats during that period? We are beginning to find some of the answers to these problems.

One of the best theories suggests that the changes began as great ranges of fold mountains were

A SUMMARY OF PLEISTOCENE EVENTS				
Event	*Ice Movement*	*Changes in Sea Level*	*Deposits*	*Landforms, Drainage*
8. Postglacial time	Retreat	Rise	Deep Thames channel filled in; Fens silted up	Low Raised Beach of Scotland (8 m)
7. 3rd Glaciation: Devensian	Advance down Irish Sea and to east coasts	Fall to level below present (*c.* — 50 m)	Marginal lakes — L. Lapworth in Shropshire, L. Pickering in Yorkshire	Retreat of ice at end broken by Scottish Readvance (High Raised Beach, 30 m) and by Highland Readvance Deep Thames channel carved
6. Last Interglacial: Ipswichian	Retreat	Rise		Avon, Severn Valley terraces; Taplow (16 m) and Flood Plain terraces of Thames
5. 2nd Glaciation: Gippingian	Advance	Fall	Second series of drift deposits over area shown in Fig. 12.11. Ice fanned out from English Midlands	
4. Great Interglacial: Hoxnian	Retreat	Rise		Boyn Hill (30 m) terrace of Thames Goodwood Raised Beach (Sussex)
3. Earliest Glaciation in Britain: Lowestoftian	Advance	Fall	North Sea Drift and Lowestoft till in East Anglia (ie. two phases of advance). Oldest drifts	Thames diverted southwards through Finchley depression, and then into present valley
2. Climate getting colder	(Advances and retreats farther north)	Fall to *c.* 100 m	Arctic Freshwater Bed Cromer Forest Bed Series: peats and deltaic series — slightly warmer Crags with increasing numbers of Arctic molluscs	
1. Marine transgression: Calabrian		Rise to 200 m	Red Crag	200 m Raised Beach on Downs, Chilterns, and around Wales

Fig. 12.12   Use the facts summarised on this chart to make your own diagrams of (1) the positions of sea level at the various stages, and (2) a cross-section of the Thames valley with its terraces.

uplifted during the Tertiary era to isolate the whole Arctic area from winds bringing warmth from the Equatorial regions. Previously the world's climate had shown smaller variations in temperature, and Britain, for instance, had a sub-tropical climate at the beginning of the Tertiary: fossils of rhinoceros and crocodile testify to this.

The cooling of the Arctic area followed, and this affected the ocean waters of that zone. Masses of cold water began to move southwards and lowered the temperatures of the high latitude oceans. Since the ocean temperatures control the warmth of the air blowing on to the continents, temperatures began to fall over the land masses affected by these winds.

At the end of the Tertiary era one of the last major mountain uplifts closed the gap between North and South America. The warm surface waters of the Equatorial oceans were diverted northwards as the warm current we now know as the Gulf Stream (Fig. 5.25). This led to greater evaporation from the oceans, and increased precipitation over the cooling lands. When this fell as snow rather than as rain the ice-sheets began to form. At first the sun's rays were reflected by the ice and were not able to heat up the air above it: strong cold winds blew outwards from the ice masses and the ice advanced towards the seas. As the ice met the sea the contrast in temperatures increased the storminess and the ice-sheets increased in volume, whilst the sea level was lowered considerably. This further intensified the bitter cold, but in so doing the processes of evaporation and precipitation were slowed down as the oceans began to freeze over. The ice-sheets, deprived of their snow supplies, began to flatten out, but still maintained the same areal coverage for some time. At last, however, the melting began to gain and vast quantities of icy water were poured into the oceans. The warming process took thousands of years, but eventually the

Life in the Cainozoic era

This was the beginning of '**The Age of Mammals**'. After over a 100 million years of insignificant existence on the earth an unprecedented evolutionary radiation took place, giving rise to an immense variety of species from the earliest Eocene. At first primitive representatives of the major mammal groups often grew to enormous sizes, whilst others, like the horse, still remained small, fox-like creatures. We can trace the development of many of the groups we know today (e.g. Fig. 9.27) in some detail, and relate them to important changes in the habits of the animals concerned. By the end of the Tertiary the horses, rhinos, wild cattle, sheep and pigs, elephants, cats, dogs and bears were all looking very similar to their modern successors. The primitive platypus, kangaroo and opossum had been isolated in the island continents of Australia and South America.

When one examines the fossils of British Tertiary rocks, one is impressed by the abundance of three groups in particular. Two of them are molluscs, the lamellibranchs and gastropods: dozens of varieties can be obtained from most exposures of Tertiary rock. Neither of these is very helpful to the geologist attempting to zone the rocks and correlate them with others in a distant era. The third group is more important to such a person, and is very valuable to the oil geologist. It is the Foraminifera, a group of protozoans – fossils of animals which evolved rapidly and became widespread in distribution as part of the plankton of the time. Most of them are microfossils, but some are up to 2 cm across.

The most interesting fossil evidence in the Pleistocene rocks relates to **the advent of human beings**. The earliest man-like creatures seem to have lived in the warmer lands of Africa, but spread to most parts of the world in the interglacial periods. At first there were several species of primitive man, but during the final advance of the ice all but Homo sapiens became extinct. The fossil evidence includes skeletons (usually very fragmentary) and flint tools, the latter showing how man's mind was developing so that he was able to become more and more independent of the natural environment and his own bodily limitations.

The fossil skeleton of man shows very few differences from that of other animals belonging to the primate group, such as the ape or monkey. It seems that man and the other primates must have had a common ancestor. The main changes are twofold – to allow man to stand erect, and the enlargement of his brain, especially in the vital frontal portions. The latter gave him an outstanding advantage by enabling him to consider the future in the light of past experience, and to think out rational answers to problems. It is not clear when the decisive characteristics which divide man from the other animals were acquired, but by the later part of the Old Stone Age (approximately 35 000 years ago) he was producing most wonderful works of art with a clear religious significance in the caves of southern France.

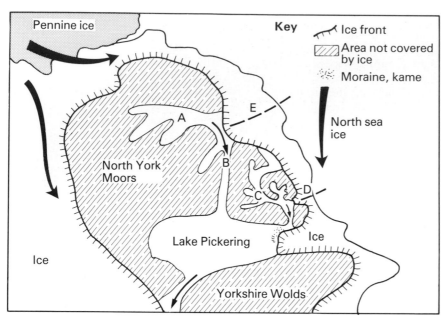

Fig. 12.13 Ice and North Yorkshire. Notice the two directions of ice source. A, B and C are three meltwater lakes and overflow channels; E is the course of the River Esk, which was resumed when the ice melted; D is the former course of the River Derwent, a river which was diverted south permanently by the ice.

oceans got warmer until the Gulf Stream began to take warmth northwards again at the end of the interglacial period.

This explanation accounts for many of the problems, but is not completely satisfactory. Further work will no doubt help us to understand the causes more fully.

The effects of the Ice Age on British scenery

The British Isles were very much involved in this advance and retreat of ice, and many of our landforms have resulted from it. There seem to have been three major advances and retreats and minor fluctuations within them.

Most snow fell on the regions which at present have the highest rainfall. These include the mountains of north Wales, the Lake District and western Scotland. They must have had the thickest ice, and the most rapidly moving glaciers, and they certainly show the greatest effects of glacial erosion. All these regions have deep, U-shaped, trough-like valleys, such as the Nant Francon in north Wales and Glen Sannox on the Isle of Arran. The valley-heads are dominated by steep-walled corries, many of which are separated by arête ridges. The Snowdon 'knife-

edge', and Striding Edge east of Helvellyn in the Lake District, are two well-known examples of arêtes. The corries along the Nant Francon Valley are characteristically on the north-east-facing sides of the ridges. Hanging valleys, striated rocks and valley-floor terminal moraines are other common features of these areas, but the lakes (lochs and llyns in Scotland and Wales) along the valley floors add most to the distinctive features associated with glacial erosion. Llyn Gwynant in Snowdonia, Thirlmere and Ullswater in the Lake District, and many Scottish lochs, testify to the uneven nature of the valley floors. Some of the deepest parts of these lakes are well below the present sea level: Loch Morar in north-west Scotland has its surface 10 m above sea level, but is over 300 m deep.

All of Britain north of the Bristol Channel–Thames estuary line was covered by ice at its farthest extent (Fig. 12.11), though areas like the North York Moors show typical periglacial features, and escaped burial by the later ice, which did not extend so far south. The South Downs on the other hand may have been capped by ice at one stage, although they were beyond the ice-sheet limit. The highlands of the eastern part of Scotland were affected to a much smaller extent than those in the west because of their smaller precipitation, and although great quantities of glacial debris were deposited in the lowland areas the ice produced few outstanding features. The deposits are over 30 m thick in places where they form the soft cliffs of the Yorkshire coast south of Flamborough Head, which has been pushed back 9 km since Roman times. Most of the soils in the lowland areas are formed on the glacial drift. Drumlins are especially common in the Vale of Eden in Cumbria, and in many of the Yorkshire Pennine dales. York stands on a notable terminal moraine left by one of the later advances of the ice.

As the ice retreated northwards the meltwaters were often trapped behind rock barriers, and when they overflowed they carved narrow, steep-sided valleys through ranges of hills. The River Severn, which once flowed east to join the Trent system, was diverted via the Ironbridge gorge into its present southerly course, and the Yorkshire Derwent, draining the eastern Moors, was turned inland to join the Humber drainage because the sea ice blocked its former outlet at Scarborough. The North York Moors are an excellent area for the study of these overflow channels: Newtondale is the most famous and striking since it is a deep channel used by a railway line, but not by rivers in its central course. Fig. 12.13 summarises some of these features.

Erratic blocks are common on the moorlands of the northern Pennines and in the boulder clays along the Yorkshire and East Anglian coasts, where ice from three directions met: pebbles of Shap granite from the Lake District, of rocks from the Cheviot area on the Border, and rhomb-porphyry from Norway can be found. Some erratics have even been discovered on local raised beaches in north Devon, an area which was beyond the ice sheet's furthest extent. It is possible that these erratic blocks, which include igneous fragments from Scotland, were contained in icebergs which drifted against the Devon coast and melted.

Although Britain is no longer covered by ice, the recent nature of that 'invasion' has left its distinctive marks freshly etched on our landscape. Glacial landforms are more noticeable in some areas than others, but every part of the country has been affected in some way. Even the southernmost parts show evidence of the tundra conditions and local ice-caps which prevailed only 10 000 years ago, and river terraces along river valleys like the Thames record the progress of sea level fluctuations.

Applying the principles of stratigraphy to older rocks

The rocks on which the Mesozoic and Cainozoic sediments and volcanics were deposited are generally harder, more broken by folding and faulting, more affected by igneous activity, and more likely to be metamorphosed on a regional scale, as well as being older. They thus constituted the rocks which the early geologists investigated after they had established many of the basic principles of stratigraphy, and demonstrated their application in the younger rock areas.

The older rocks, of Precambrian and Palaeozoic age, form the western and northern areas of the British Isles, including the south-west peninsula, Wales, Ireland, northern England and virtually the whole of Scotland. They have provided a major challenge for geologists since they were first investigated in detail during the mid-nineteenth century, and have been the source of a large proportion of British mineral resources.

As well as demonstrating that the general principles of stratigraphy can be applied in situations involving the oldest known rocks, the rocks of Precambrian and Palaeozoic age enable us to study

rocks formed in environments strange to Britain today. They include the highly metamorphosed rocks of early Precambrian time, which were probably formed by alteration deep in the earth's crust and have been exposed at the surface only after long aeons of erosion; the rocks and structures formed by plate collisions; and rocks formed in tropical desert and swamp environments as well as in deep oceans.

The metamorphosed Precambrian rocks

Although these rocks have such a small area of outcrop in the extreme north-west of Scotland (Fig. 12.14) and the Hebrides, they have been studied intensively and a fairly clear picture is emerging of the earliest recorded geological history of Britain. All these rocks have been intensely deformed and highly metamorphosed on a regional scale: the most common rock-type is banded gneiss.

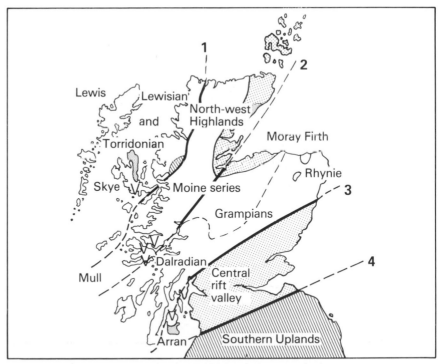

Fig. 12.14 The geology of Scotland. A map showing the outcrops of the main groups of rocks in Scotland. The unshaded areas have Precambrian rocks at the surface; oblique shading signifies Lower Palaeozoic rocks; the stippled areas have Old Red Sandstone and Carboniferous rocks; the small areas of heavy cross-shading have Mesozoic rocks; V–centres of Eocene volcanic activity; 1–Moine Thrust; 2–Great Glen fault; 3–Highland Boundary fault; 4–Southern Upland Boundary fault.

Northwest Scotland. The low, hummocky surface of Lewisian gneiss in the foreground is broken by hills of Torridonian sandstone rising on the skyline. (Crown Copyright)

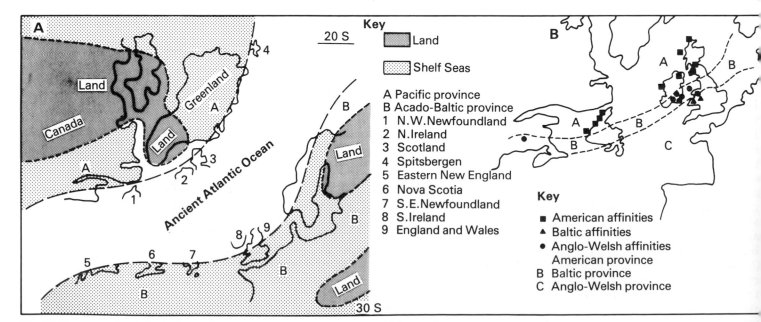

Fig. 12.15   The ancient Atlantic Ocean of Lower Palaeozoic times. The ocean which opened up during the late Precambrian had similar margins to the present Atlantic Ocean, but notice the differences (A). B shows how this pattern was arrived at – by separating distinctive groups of brachiopod (and other groups) of fossils.

They have all been included in one group known as the **Lewisian**, but studies using radiometric dating have revealed that they contain two major phases of activity. The older of these phases occurred between 2900 and 2200 million years ago, and evidence for it is found mainly in the Outer Hebrides: as well as being formed of the dominant grey banded gneisses, the rocks have structural trends from north-east to south-west and are cut by a series of basic igneous dykes intruded approximately 2200 million years ago. These dykes are three to four per kilometre, and are found over 300 km of outcrop. The second phase lasted from 2200 to 1500 million years ago, and the rocks were formed originally on top of the older group: they have associated granites and pegmatites, together with metasediments, but some tracts appear to have escaped the high-intensity metamorphism.

Although this appears a relatively simple picture, demonstrating the application of stratigraphical principles, there are still many problems of interpretation and unanswered questions. These include the relation of the rocks to earlier events; many of the internal relationships of such highly altered rocks; and the nature of their extension beneath the rest of Britain, since only in southern Ireland and northern France have rocks of equivalent age been found.

The late Precambrian and early Palaeozoic rocks: the history of an ocean

Rocks formed between approximately 1000 million years ago and 400 million years ago (i.e. late Precambrian, Cambrian, Ordovician and Silurian) in the British Isles and in neighbouring areas around the North Atlantic record the opening and closing of a major ocean. This interpretation is the result of applying the principles derived from plate theory – i.e. that ancient fold mountain belts were formed along plate collision lines at destructive margins, and that the rocks which compose the fold mountain belts were laid down in the opening ocean basins (Chapter 10). Such an interpretation resolves many of the problems facing earlier geologists wedded to the idea that continents did not move: they had to postulate an ancient continent north of Britain which has now disappeared, as well as the geosyncline idea of crustal subsidence.

The northern margins of this sea (including Greenland, northern Scotland, northern New-

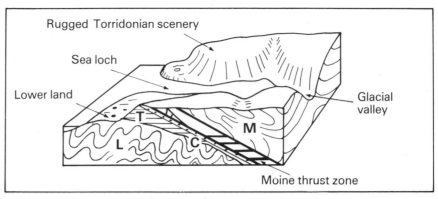

Fig. 12.16 North-west Scotland. Try and recognise two unconformities shown on this block diagram. L–Lewisian; T–Torridonian; C–Cambrian; M–Moine Series.

foundland and the interior parts of the Appalachians) record a simpler series of events than the southern. This is partly because there is a continuous outcrop in Scotland of the rocks formed during these events, and partly because there is a direct and clear relationship to an older land mass to the north, from which much of the debris forming the sedimentary rocks was derived. The southern margins, stretching from Scandinavia through Britain into southern Newfoundland and central Appalachia (Fig. 12.15), are backed by later groups of block mountains and often covered by younger rocks. This makes it more difficult to work out a simple pattern for the development of this margin.

In the north the evidence for the opening and closing of the ancient Atlantic ocean is clear, despite the age of the rocks and the complexity of their folding and alteration (Fig. 12.16). The oldest groups occur in the far north of Scotland, and have both metamorphosed and unmetamorphosed equivalents. The **Torridonian Group** rests with a marked unconformity on the Lewisian gneisses, and includes sandstones and conglomerates deposited by streams draining from a continental mass to the north-west: the stream apparently emerged from gorges carved across the Outer Hebrides area to form fans of debris across a coastal plain and continental terrace. Some deposits are thicker and may have been of continental rise type. Farther east, across the later structure of the Moine Thrust, there are poorly sorted sands and muds up to 7000 m thick, which have been altered by deep burial and folding beginning about 750 million years ago. These are known as the **Moine Series** and are mainly schistose in nature.

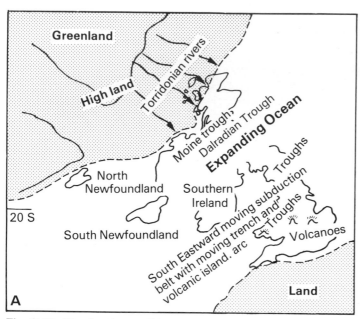
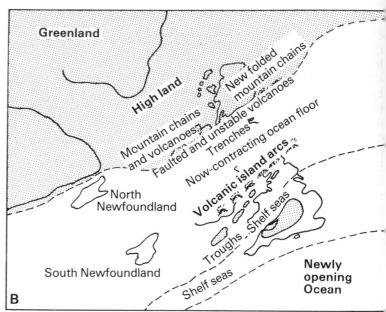

Fig. 12.17   The changing conditions in the Lower Palaeozoic version of the Atlantic Ocean. A The ocean is widening (rather like the Atlantic today) with the deposition of continental shelf and rise deposits along the edges. B Later in the Lower Palaeozoic the ocean is closing with outbursts of volcanic activity. Attempt to give a closer date to these maps by reference to the text of this chapter.

Life in the Lower Palaeozoic seas

The Cambrian period saw a remarkable growth of life on the earth. All the animals and plants of which we have a record lived in the sea, and they were dominated by primitive representatives of the invertebrate groups. There was a surprising variety of animals, and all the major groups (i.e. phyla) have fossil records extending back to these rocks.

The most important Cambrian zone fossils are the trilobites, which included some of the largest varieties, and the primitive, inarticulated brachiopods. Many other groups, such as the echinoderms and molluscs, are also represented by fossils of their most primitive known species. Sponges, coelenterates, algae and even worms have been found in Cambrian rocks.

The Ordovician saw many changes and the first appearance of some important fossil groups. Trilobites are found in even greater numbers and have a great variety of shapes adapted to a wider set of conditions. Brachiopods, too, developed many new forms with properly articulated shells. The graptolites, seemingly small and insignificant, were the most important of the newcomers and have become the main zone fossils for the rocks of this and the following Silurian period: their shapes change for every few metres of rock. In some of the lime-rich shelf seas the first of the true corals, as well as echinoids, bryozoans, molluscs and crinoids flourished. Long, straight cephalopods, up to 4 m long, and the eurypterid 'sea scorpions' were the largest animals.

The Caledonian mountain-building caused great changes in the geography of Britain and hence in the habitats available to plants and animals. Many changes were also caused by the increase of the vertebrates and the possibilities of life on dry land. The Silurian rocks saw the last of the graptolites, of the trilobites as a group of major significance, and of many of the more primitive members of major fossil groups.

The Silurian did not produce so many changes but continued the development of a rich and varied life in the seas. The Wenlock reefs illustrate the diversity present in the British shelf seas. In addition the upper Silurian Ludlow Bone Bed contains the earliest British remains of primitive vertebrates – the fish-like ostracoderms. These may have developed in freshwater conditions and therefore would have left a poor fossil record of the early stages of their evolution. Earlier fragments of such fossils have been found in Ordovician rocks in Wyoming (USA).

A further group of rocks, the **Dalradian Group**, extends across the southern Scottish Highlands and into Northern Ireland (Fig. 12.17). Some of the lower horizons of this group may be of equivalent age to the Moine Series, but Cambrian fossils have been found in the upper section. This group contains rocks which have been interpreted as tillites deposited in a major ice age at the very beginning of the Cambrian period. Similar rocks have been found in areas all around the world of that time, including low latitudes, so that this must have been a period of great climatic changes. It is thought that this may have resulted in major changes of living forms, stimulating the widespread development of hard shells – which are found first on Cambrian forms. Another feature of this group is the manner in which 'way-up' sedimentary structures (Fig. 10.1) have been used to interpret the tectonic structures – a series of nappes and slides – in this complex and altered group of former turbidites and pillow lavas.

All three groups were involved in various phases of folding from the Middle Cambrian to the Middle Ordovician as the ocean began to close. These led to uplift and erosion, resulting in Ordovician and Silurian deposits to the south of the rising land mass and mountain ranges – i.e. in the central and southern parts of Scotland. It has been suggested that Scotland at that time resembled Burma today, with parallel ridges trending north-east to south-west and basins of deposition between onshore and offshore. A northward-moving plate was subducted beneath the Grampians after the Middle Cambrian resulting in uplift, folding, faulting, granite intrusion and volcanic activity (Fig. 12.18).

The events on the southern margin were more complex still. Most of the Precambrian rocks of this area, which includes Wales, the English Midlands and Lake District, occur in scattered outcrops. They have some features in common, such as volcanic rocks and a range of sediments, but it has been difficult to establish relationships between them. Radiometric dates for most of these areas are very late Precambrian (i.e. 600–700 million years old), providing evidence of folding and uplift of continental shelf and rise deposits at that stage. There may have been an ocean plate subducted beneath Anglesey, giving rise to an area which became subject to erosion. By the Cambrian period such a land mass existed, providing sediment to basins on the north and south. Shelf sediments accumulated in the English Midlands area, including 'shelly' fossils like brachiopods and trilobites, and some corals in the late Ordovician and Silurian

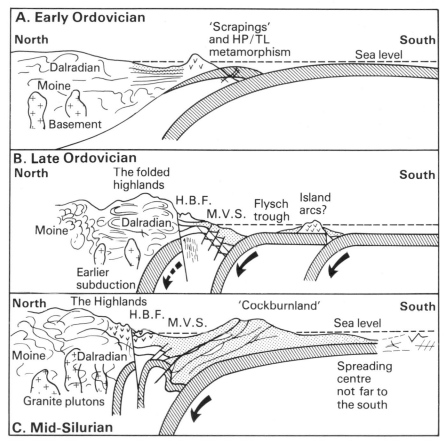

Fig. 12.18   Three stages in the evolution of Scotland in the Lower Palaeozoic, shown by north–south sections.

Silurian rocks in Shropshire: the escarpments of the Wenlock Limestone (left) and the Aymestry Limestone (right), which shale vales. (Aerofilms)

strata, whilst the turbidites of central Wales contain mainly graptolites. The Snowdonia area and the central Lake District became sites for the accumulation of large volumes of island arc volcanic rocks in the late Ordovician (Fig. 12.17 B). The Silurian period mainly saw the filling of the former ocean basin remnants with the erosional products of rising mountains formed as the oceans closed.

The Old Red Sandstone continent and Carboniferous seas

The formation of the **Caledonian mountains** across Britain by the closing of the early Atlantic ocean also gave rise to a major continental area incorporating what is now North America and northern

Fig. 12.19   The Old Red Sandstone land mass of Devonian times. Notice which continental masses are involved, and the position of the British Isles on the edge of a southern ocean (the northern part of the ancient Atlantic had been closed by the formation of the Caledonian mountains).

Europe (Fig. 12.19). The high relief of much of this continent produced distinctive deposits in the Devonian period known as the **Old Red Sandstone**, but this general term includes a variety of mainly land-deposited sediments and volcanics. Thus in Britain the **Welsh rocks** are a mixture of red and green sandstones, with shales and conglomerates. The typical, repeated sediments are those of a fluviatile plain (Fig. 12.20 A) with channel lag, point bar and overbank flood deposits (see Fig. 4.19). The climate must have been sufficiently rainy to give rise to at least seasonal streams, shifting their courses across the coastal plain.

In **Scotland**, the **central rift valley** was the site of coarser wedges of sediment filling the depression from the erosion of the mountains to north and south. Great alluvial fans built up along the margins, similar to those occurring in semi-arid mountain regions today. The deposits here locally reach over 6000 m in thickness and are interbedded with lavas, mainly andesites and basalts plus tuffs and agglomerates. The fossils in the Welsh and central Scottish rocks are similar: early plant and vertebrate remains. Other areas, marginal to the central rift, also experienced igneous activity.

At Ben Nevis and Glencoe these lavas have dropped down in cauldron subsidence and are surrounded by ring dykes: the process brought them into close proximity with large granite masses, which were being intruded at the time. Thus Britain's highest mountain summit is actually made of a downfaulted mass of lava. Some of these granites are associated with swarms of south-easterly trending dykes. The Cheviot, farther south, was an active volcano (Fig. 12.21).

In the **far north of Scotland** another basin of Old Red Sandstone deposition has been preserved round the Moray Firth, in Caithness and in the Orkney and Shetland Islands. Thick deposits of grey, flaggy sandstone form the striking cliffs of Caithness and these thin layers of rock contain groups that are repeated again and again (Fig. 12.20 B, C). They must have been deposited on a sinking plain, and conditions determined whether limestone, bituminous shale or sandstone was formed. There are scattered outliers of these rocks on the Grampians to the south of the Firth. At the same time granites welled up, raising the surface relief further and maintaining rates of erosion: over fifty such masses were intruded between 410 and 380 million years ago from the north of Scotland to the Lake District.

The lowlying southern margins of this continent gave way to marine conditions. There is still debate as to whether this zone including southernmost England and central Europe was one of relatively shallow seas, or a major ocean, but certainly the sediments formed, and the fossils they contain, reflect marine environments which extended northwards during the Carboniferous to cover much of the British Isles.

During the **Devonian** period only the southernmost part of the British Isles was covered by the sea (Fig. 12.19). Only the southern tip of Ireland and a strip extending eastwards from the south-west peninsula were in this zone.

The marine rocks were formed in a deepening sea, which was part of an extensive trough of subsidence extending across central Europe at this time: the rocks of Brittany, the Ardennes, the Rhine Slate Plateau and the Harz mountains were also formed in the same general environment. The sequence of rocks in Devon is interesting because it is typical of so many areas which have later been involved in the formation of folded mountains.

The oldest Devonian rocks are known as the Dartmouth Slates and are reddish slates and sands containing some plant and fish debris like the Old Red Sandstones farther north. These are covered by shallow water marine sands and shales, and the Middle Devonian rocks include local lenses of volcanic rocks and massive limestones which contain reef-building fossils. The Upper Devonian rocks of south Devon and most of the Devonian age rocks in Cornwall are mainly fine shales which contain the fossils of creatures which swam in deeper waters (e.g. goniatites), but few bottom living forms. This record suggests that the seas were gradually getting deeper. In north Devon these rocks are much thicker and are interleaved with red sands and silts derived from the continental area to the north: the coast was near at hand (Fig. 12.22).

The **Carboniferous** rocks of Devon have been called the Culm but are equivalent in age to both the Lower and Upper Carboniferous elsewhere. The Lower Carboniferous horizons are thin, dark, lime-rich rocks and cherts including radiolaria. These rocks may have been formed in deep waters and include volcanic layers. The Upper Carboniferous rocks record a dramatic change. They are a monotonous succession of extremely thick interbedded shales and sandstones. The sandstones are often of a greywacke composition, may have graded bedding, and the bottom surfaces show structures associated with turbidites, like load casts and flute marks. This association is very much like the flysch

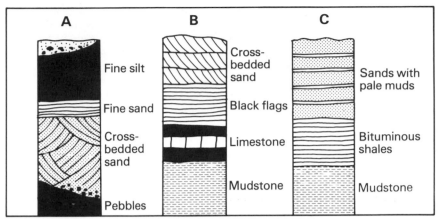

Fig. 12.20 Rhythmic sedimentation in the Old Red Sandstone rocks. Relate these sequences to different modes of formation, as suggested in the text.

Old Red Sandstone cliffs at Duncansby Head, Caithness. The cliffs are 180 m high. (Crown Copyright)

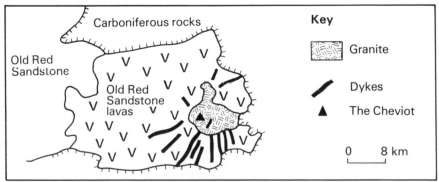

Fig. 12.21 The Cheviots. Can you work out the order of events in the formation of this complex of igneous rocks? If not, look back at the section 'Igneous rocks and time' in Chapter 8.

of the Alps. Fig. 12.23 summarises the progression.

An interpretation of this succession of rocks suggests that the Devonian and Lower Carboniferous saw the extension and deepening of the area of marine deposition, but that at the beginning of the Upper Carboniferous an uplift led to more rapid sedimentation of some thousands of metres of flysch facies. At times deltas extended southwards into the area from Wales, and this sequence of events ended in the Upper Carboniferous when the whole area was uplifted to form mountain ranges.

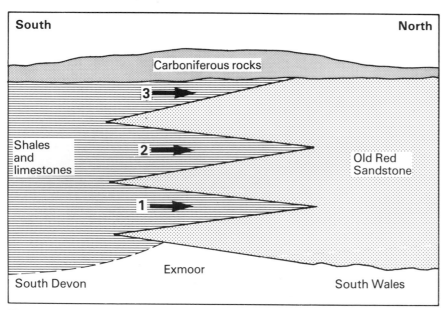

Fig. 12.22   The Devonian shoreline. The rocks of Exmoor show how the sea advanced northwards three times to deposit typical marine rocks, and how the three sandstones were formed at intervals on the land.

All younger rocks rest on top with marked unconformity		Folding and uplift
Upper Carboniferous	Flysch facies: alternate shales and greywackes	Seas filled in by rapid sedimentation after local uplift
Lower Carboniferous	Deep-sea oozes	Seas deepening with slow rates of sedimentation
	Deep-sea shales	
Devonian	Shallow sea muds, sands, limestones	
	Old Red Sandstone facies	Land conditions, or just offshore

Fig. 12.23   Upper Palaeozoic rocks in south-west England: an interpretation of their origin.

The Lower Carboniferous: a variety of facies

By the end of the Devonian period the Caledonian mountains had been reduced by erosion to a large extent, and the southern seas advanced northwards. The rocks of Lower Carboniferous age record this advance and then the Upper Carboniferous witnessed the filling in of the seas and the raising of new mountain chains. The earlier part of the Carboniferous period, then, was a time of marine transgression across the worn-down remnants of the Caledonian mountains and the Old Red Sandstone deposits. Conditions on the sea floor varied and so the types of rock formed at the time record a number of distinctive changes from place to place (Fig. 12.24). Some areas had a gently subsiding sea floor; in others there was more sudden subsidence and more rapid sedimentation; and in central Scotland, at the farthest extent of the transgression, most of the sediments were fluviatile or deltaic. The Lower Carboniferous facies are amongst the most varied of any geological period.

The best known rock-type of the Lower Carboniferous is the massive, well-bedded grey bioclastic limestone, often referred to as the 'Standard' or 'Mountain' Limestone. In the Bristol area, and in south Wales, it forms an uninterrupted succession up to 1000 m thick and contains plentiful fossils; some bands are almost entirely composed of crinoid debris, and corals and brachiopods are common. The Mendip Hills, with their caverns, dripstones, potholes and bare, dry, rocky surface, are formed of this Carboniferous Limestone, and the Avon Gorge at Bristol is cut through it.

This area has a complete record of deposition carrying on from the underlying Devonian rocks, whereas other parts of Britain usually have a definite gap between the two groups of rocks represented by an unconformity.

Central and northern England and Scotland had an amazing variety of conditions for such a relatively small area. Much of the English Midlands remained land for the whole period (an area without Lower Carboniferous rocks, known as St George's Land); the southern and central Pennine area was a deep, down-faulted basin, as was the Northumberland–Durham area: between these were shallow seas; and between central Scotland and north Northumberland there was an area of lowlying plains. Many distinctive rock facies were formed at this time.

The areas which were deep basins now have thick deposits of shale, containing thin-shelled lamellibranchs and goniatites, alternating with thin layers of black limestone. The shallower seas of those

imes have a variety of reef limestones: some are
ormed of great masses of colonial corals piled on
op of each other, whilst others are built up of algal
nd bryozoan skeletons. Along the edges of the
leeper basins a series of reefs tended to grow, and
he hard, structureless limestone stands out in the
present landscape as rounded knolls. The northern
part of the Pennines experiences a lesser degree of
ubsidence and a type of cyclic deposit known as
he **Yoredale facies** accumulated. Each cycle begins
vith limestone (clear, shallow off-shore condi-
ions), followed by shale (inshore, muddy condi-
ions giving place to estuarine and lagoonal), then
by sandstone (fluviatile or deltaic) and finally by
eat-earth and coal (swamps on delta top). Subsid-
nce was intermittent and apparently sudden,
because the next limestone succeeds the coal with
ittle or no shale intervening. Marine incursions
vere therefore repeated, but did not persist for
ong. The average thickness of the cycle is about
,o m of rock, and one such succession outcrops
eneath Hardrow Force in Wensleydale. Fig. 12.25

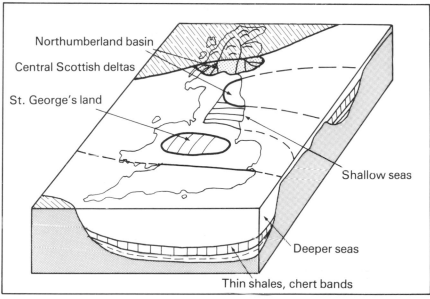

Fig. 12.24   Lower Carboniferous: mainly marine conditions. The Upper Palaeozoic seas had
their greatest extent across the British Isles at this time.

Life in the Upper Palaeozoic

The Devonian and Carboniferous periods saw far-reaching
changes in the pattern of life on earth. The vertebrate group
became prominent for the first time, and life moved out on
to the land with increasing success.

The Devonian period saw the real beginning of both these
processes. British rocks of this age contain few fossils, but
very important ones. Fish of all types – jawless ostracoderms,
placoderms with primitive jaws, sharks and early bony fishes
– inhabited the seas and freshwater lakes. The first amphibious creatures emerged, and the earliest insects joined them
on the dry land. Both were preceded, however, by the
spread of land plants. One of the Old Red Sandstone
outcrops at Rhynie in north-east Scotland is a peat deposit
of the time, which was impregnated by silica-rich waters
from a nearby hot spring, and has preserved some of the
most primitive types of land plant (Fig. 9.28). By the end
of the Devonian there were dense forests. In the seas coral
reefs and spiriferid brachiopods were common, and the
goniatites – ancestors of the ammonites – travelled widely.

The change to extensive, shallow, lime-rich waters in the
Lower Carboniferous led to a flourishing group of corals,
productid brachiopods, crinoids and bryozoans. At times
they were so abundant that they formed a large proportion
of the bioclastic limestones which have resulted. The main
difficulty here is that none of them are really adequate as
zone fossils, and the Lower Carboniferous has to be zoned
by a combination of them all – i.e. a fossil assemblage named
after a distinctive coral or brachiopod it contains. The
swamp forests of the Upper Carboniferous demonstrate how
complete had been the colonisation of wet-soil conditions by
plants, many of which grew to gigantic sizes (over 50 m
high). Large insects, including dragonfly-like varieties with
a wing-span of 75 cm, as well as spiders, scorpions and
millipedes also inhabited these areas, and the lakes contained
numerous lamellibranchs. Fishes became more and more
common, the amphibians reached the height of their development and the first fossil reptiles are also found in these
rocks.

The Upper Carboniferous Coal Measure rocks were
another headache for the zoning palaeontologist until
recently, when it has been found – rather unexpectedly –
that the lamellibranchs which lived in the deltaic swamp
pools, as well as plant remains and microscopic spores, can
be used with some degree of reliability, especially in
conjunction with each other. In addition the occasional
marine bands, which are more frequent in the lower
horizons, are found over wide areas, linking up several
isolated basins of deposition, and they can be used as 'index
horizons'.

shows a succession which has an outcrop in a valley tributary to Swaledale. It is thought that the repetition of the sediments occurred as distributaries swung across a delta surface in a shallow sea.

The Yoredale facies is also found, often in modified form, in Northumberland and the Midland Valley of Scotland, where the Lower Carboniferous also included a different type of rhythmic deposit known as the Cementstones, which are locally over 300 m thick. These consist of repeated alternations of shale and argillaceous dolomite, which may have been formed annually as wet and dry seasons followed one another. Near Edinburgh many oil-shales were formed in well-vegetated lagoons. Volcanoes were also erupting mainly olivine basalts at this time in Scotland, and many of the lava surfaces between flows have been weathered to a reddish soil. These lavas form the hills around Glasgow and many of the associated necks stick up as isolated hillocks all over the Midland Valley including Edinburgh Castle Rock and the adjacent Arthur's Seat.

The Upper Carboniferous deltas. The Midland Valley of Scotland was the farthest north the Lower Carboniferous sea reached. There the Upper Carboniferous began with a group rather like the Coal Measures followed by a return of the Yoredale facies and then by a sequence of sandstones and fireclays. The latter are used to make refractory bricks for furnace linings. Deltas advanced southwards from the Scottish Uplands and outward from St George's Land to cover much of central and northern England as the sea withdrew. At first many of the rocks in these areas were current bedded gritty sandstones, often containing plant debris, and interbedded with thin marine shales in which are found fossil goniatites, indicating temporary readvances of the sea. Many of the minerals in these sandstones have been traced back to distinctive rocks in Scotland and even in Norway. Beyond these deltas marine limestones were still being formed (e.g. in Cumbria). Volcanic activity continued here and there in Scotland, particularly in Fife, where it is now known to have occurred in Coal Measure times.

This group of rocks, known as the **Millstone Grit** in the central Pennines where they are over 2000 m thick, forms a transition between the sea of the Lower Carboniferous and the widespread coal swamps of the Upper Carboniferous.

The deltas had rapidly spread out to cover most of the country with lowlying, swampy conditions. The British Isles area was in equatorial latitudes at this stage, so plant growth was luxuriant. As mountains began to rise across central Europe more and more sediment was poured into the narrowing marine trough, but the central and northern parts of the country were covered by these extensive deltas (Fig. 12.26), beneath which there was continuous subsidence as the weight of sediment increased. In places over 3000 m of rocks were deposited to form the **Coal Measures** in the manner outlined in Chapter 8.

Britain was particularly fortunate to have so much coal laid down over its surface. Coal has been the basis of our economic and political power in the recent past – though other fuels are now taking its place. The conditions of deposition meant that each local area had a unique group of coal seams with rocks between. Those in the south, including the south Wales and Kent coalfields, have two sets of coal seams separated by a thick layer of sandstone which forms the moors between the deep Welsh mining valleys. In the English Midlands there are several small coalfields, once probably all part of

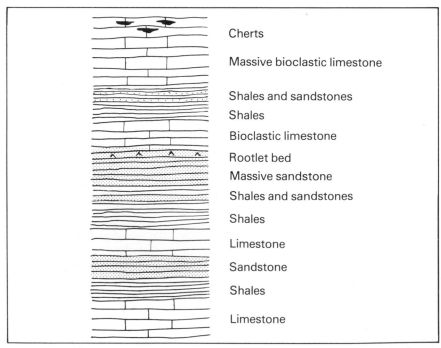

Fig. 12.25 The Yoredale series. A succession of rocks charted along a stream which is tributary to the River Swale in Yorkshire. Repetition of certain rock-types takes place in a consistent sequence, but each band is not always present. Construct a generalised unit for the Yoredales to include all the possibilities, and attempt to relate the formation of this series to the swinging mouth of a delta.

Cherts

Massive bioclastic limestone

Shales and sandstones

Shales

Bioclastic limestone

Rootlet bed

Massive sandstone

Shales and sandstones

Shales

Limestone

Sandstone

Shales

Limestone

the same basin of deposition but now divided up by faults. All their valuable seams are in the lower horizons, and the upper layers of Coal Measure rock are mostly red marls formed in an arid environment. An important addition to the Midland Coal Measures is a seam of 'blackband' iron ore, which was the original basis of the metal-working industries in the area. Farther north, flanking the Pennines, are our richest coal seams. It is interesting to compare the East Pennine and Lancashire coalfields on either side of England's 'backbone': whereas the former has many thick seams which are mined easily, they often become split into several seams to the west, and the Lancashire coalfield is also broken by faults so that it is declining in importance. It seems, however, that the deltas here drained into a sea to the west, and that the continued sedimentation was gradually submerging St George's Land to the south. The Northumberland and Scottish coalfields all have their productive seams in the lower part of the succession and the upper layers are once again characterised by 'barren', coal-less, red-coloured rocks. These were produced by erosion and deposition on the land, conditions which advanced slowly southwards.

The Hercynian mountains and New Red Sandstone deserts

The entire period of deposition, from the Devonian to the Carboniferous, had been interrupted by earth movements, giving rise to breaks in deposition, or a later start in some areas, variable thicknesses and a variety of environments reflected in varied rock facies. Towards the end of the Carboniferous the seas were filled as the southern continents moved against the northern to produce the single continent of Pangaea. Southern Britain was affected most as the greater thicknesses of sediment were folded intensely, metamorphosed and then intruded by granites. Chains of mountains, running from east to west, dominated the scenery of this zone, and it is thought that they were often topped by volcanoes, such as the one that has been completely worn away from the top of Dartmoor. The granite intrusion also led to mineralisation of the surrounding rocks, where veins of tin, tungsten, zinc, lead and copper were emplaced (Fig. 7.5), and hot fluids caused the margins of the granite near St Austell to be altered to china clay. The Dartmoor granite has been dated as 300–280 million years old by radioactive techniques.

Farther north, however, the effects were not so marked, and the folding in areas like the Malverns and the Pennines has a north–south trend. The folds are more open, and block-faulting was common. There was some local granite intrusion and mineralisation in these areas. As was the case with the Alpine orogeny the earth movements associated with the Armorican began earlier than the main time of uplift: unconformities are found in Lower Carboniferous rocks, and the effects lasted into the Permian period. The north–south folding is thought to be slightly earlier in time than the east-west trends.

Just as the formation of the Caledonian mountains gave rise to eroding mountains in some areas and deposition of the debris on their margins, so the Hercynian mountains across central Europe provided debris for another group of distinctive deposits – the **New Red Sandstone**. The British area had moved northwards during the Carboniferous, and was now in the latitudes of the subtropical deserts. Many features of the rocks reflect this position. At the same time, the great continental mass of Pangaea began to break apart as new constructive plate margins became active. These movements caused water to spill from the oceans across the continental masses from time to time, often evaporating in the desert conditions.

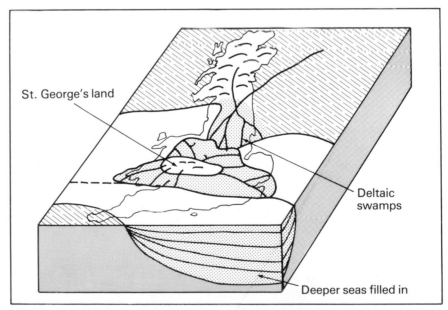

Fig. 12.26 Coal Measure times. The deltaic swamps filled the seas and submerged St George's Land.

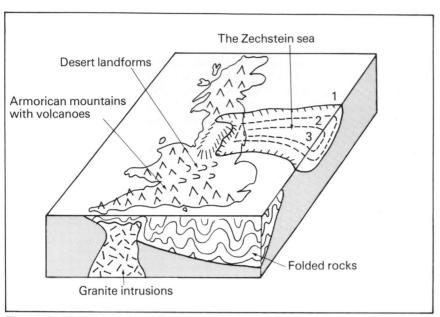

Fig. 12.27 New Red Sandstone: Permian conditions. Notice how the Zechstein sea covered a smaller and smaller area as it dried up. 1–Magnesian Limestone outcrop; 2–rock salt outer limits; 3–potash salts outer limits.

A typical series of deposits resulting from the evaporation of seawater would be:	The Whitby Borehole was sunk through the following Permian rocks:
5. Polyhalite (potassium-magnesium-calcium sulphate) 4. Rock salt 3. Anhydrite, gypsum 2. Dolomite, limestone 1. Marl	**Upper Permian Marl** Top Anhydrite (1-1·5 m) **Salty clay (3 m)** 　(3) Upper Evaporites: (65 m) 　　Upper rock salt 　　Potash rock (Sylvine) 　　Lower rock salt 　　Anhydrite 　　Limestone **Carnallitic Marl (20 m)** 　(2) Middle Evaporites: (130 m) 　　Upper rock salt 　　Potash rock 　　Lower rock salt 　　Rock salt and anhydrite 　　Anhydrite **Upper Magnesian Limestone (65 m)** 　(1) Lower Evaporites: (350 m) 　　Upper rock salt with 　　　anhydrite 　　Upper anhydrite 　　Lower rock salt with 　　　anhydrite 　　Lower anhydrite **Lower Magnesian Limestone (125 m)**

Fig. 12.28 Make your own diagram based on the facts given here. Choose a separate colour for each of the stages 1–5 on the left, and draw the rock thicknesses of the borehole succession to scale. Then write down what you have learnt about the drying up of the Zechstein Sea.

The **Permian rocks** of England are formed of coarser conglomerates, breccias and sands produced by the early weathering of the newly-formed mountain chains. They were deposited in basin-shaped areas cut off from rain-bearing winds in Devon, the Vale of Eden in north-west England and in the north-east. The first deposits contain fragments of Carboniferous rocks, but later erosion stripped off the rocks down to the Precambrian. We can find similar deposits forming today in lowlying arid areas close to mountains in such places as western USA and southern Siberia. Near the mountain slopes there are thick spreads of coarse gravel, then there are dune belts, and in the centre of these basins of inland drainage there are playa lakes, floored by salt layers (Fig. 4.20).

A great arm of the sea, known in Germany as the Zechstein Sea, extended across northern Europe, and into north-eastern England. Then it began to dry up, and the increasing salinity led first to the extinction of the fishes living in the sea, and then to the deposition of the salts held in solution. A layer of Magnesian Limestone, up to 250 m thick, extends from the Durham coast through to Nottinghamshire, forming an escarpment. As the sea dried up it got smaller (Fig. 12.27) and other salts were precipitated including gypsum, anhydrite, rock salt and potassium salts. The succession found in a borehole beneath Whitby in North Yorkshire is shown in Fig. 12.28. Many of the salt deposits of Europe, including the most famous at Stassfurt in East Germany (Fig. 8.17), were formed at this time. Scottish Permian rocks are similar to the sandstones, and are largely confined to the southwest. These rocks often cap important gasfields in older Permian sandstones beneath the North Sea.

The **Triassic rocks** of Britain continue the picture of land being worn down in desert conditions. The threefold division of these rocks found on the continent of Europe, and implied in the name of the period, is not evident in this country, because the sea which deposited the Muschelkalk Limestone failed to reach our shores. Rocks of the older Bunter Series, and the more recent Keuper Series, are very important, and cover more of England's surface than any other single age group.

The Bunter sandstones are typical desert sands, with rounded, millet-seed grains, dune-bedding structures and faceted ventifact pebbles. A study of the dune bedding has led to the conclusion that the winds came mainly from the east at this time, and this fact, together with the conclusions from studies of rock magnetism, suggests that Britain was farther south in the belt of the Trade winds. This evidence is in line with that suggesting a tropical climate for the formation of the Coal Measures, and the presence of glacial deposits of Permo-Carboniferous age in South Africa. Some places, such as the southeast Devon coast, have thick pebble beds resulting from increased erosion: these Budleigh Salterton Pebble Beds are formed of debris from northern France, and it is supposed that a river brought the spread of gravels northwards.

The overlying Keuper deposits reflect the fact that the landscape had been worn right down by the later part of the Triassic period. Fine sands are overlain by the red, sticky mudstones which provide the basis for the heavy soils of the Midlands, Cheshire Plain and Trent Valley. The 650 m of so-called 'marls' are repeated alternations of clay, dolomite and evaporite horizons due to slight fluctuations in the humidity of the climate. Fossil scorpions and drought-loving vegetation have been extracted from these layers. The landscape of the period must have been lowlying but hummocky, and the hollows were filled by evaporating lakes. The rock salt horizons in Cheshire are mined for use in the Merseyside chemical industry, and subsidence of the surface is common where the deposit has been extracted as brine.

The Palaeozoic–Mesozoic transition

The British desert deposits of Permian and Triassic age provide little evidence for the development of life during these periods, but elsewhere in the world great changes took place between these periods, especially in the seas, and the major boundary between the Palaeozoic and Mesozoic is placed here.

At the end of the Permian many of the animals typical of the Palaeozoic seas became extinct. The last of the trilobites, the rugose corals, the goniatites and many primitive varieties of other groups ceased to exist. The brachiopods and crinoids in particular were greatly affected, and the way was opened for the more intensive development of the better-equipped invertebrate groups, the molluscs and echinoids, which largely replaced many of the Palaeozoic animals. The Triassic began a new era in the story of life on earth, but there is no record of marine fossils of this age in Britain.

On land there was less change. Plants of the Upper Palaeozoic type gave way slowly to more advanced forms, and the development of the vertebrate animals continued. During these two periods the advantages which the reptiles possessed gave them the mastery of land conditions and of the less adaptable amphibians. By the Triassic the reptiles had not only conquered the land environment but some had gone back to living wholly in the water. Others developed features which resemble the early mammals, such as a single jawbone, different types of teeth specialised for tearing meat and grinding plant matter in the same palate, and a separate nasal passage: it is thought that the first mammals developed at this stage.

Carbonifer's	Permian	Triassic	Jurassic	Animal groups
				Corals
				Tabulate
				Rugose
				Scleractinian
				Trilobites
				Bivalves
				Oysters
				Myas
				Cephalopods
				Goniatites
				Ceratites
				Ammonites
				Brachiopods
				Orthids
				Productids
				Spiriferids
				Ancient crinoids
				Modern crinoids
				Irregular echinoids

Fig. 1 The Palaeozoic–Mesozoic division. A number of animal groups became extinct at the end of the Palaeozoic and were replaced by new groups in the Mesozoic, as shown by this diagram. There are other groups, however, which were scarcely affected, e.g. vertebrates and plants.

The Rhaetic transition

A thin but persistent group of rocks separates the continental Triassic rocks from the marine Jurassic. They are seldom more than 20 m thick but occur throughout the outcrop from Dorset to Yorkshire and record the transition from one environment to the other.

The first advance of the seas merely flooded the lagoons and brackish lakes that existed at the end of the Triassic, and many of the reptiles, amphibians and fish living in them were killed off to form a very thin 2–5 cm bone bed. This **Rhaetic episode** is paralleled by deposits thousands of metres thick in the Alpine area, but in Britain there are usually less than 16 m of varied shales and limestones, including the Cotham 'marble' with its darker, tree-like markings probably caused by the action of algae. Rhaetic rocks are best exposed on the shores of the Severn estuary.

Summary

The study of stratigraphy has resulted in a well-established series of events in the British Isles – a small part of the earth's surface, but one which contains a great variety of geological situations. This is not just an academic study, in which one acts as a detective, sifting the evidence and piecing together the events of the distant past. It has a most important practical and economic value in locating, and tracing the extent of, deposits of coal, oil, gas, limestone, ore minerals and other valuable earth resources. New ideas related to the theory of plate tectonics, and a wide acceptance of the mobility of continental masses over the earth's surface, have made us see the established sequence of geological events in the worldwide context, and this has given us a fuller understanding of their significance.

Mount Hadley and astronaut Irwin with the lunar Rover in late July, 1971 on the Apollo 15 mission. Note the relief of the Moon's surface, the nature of the regiolith and the support systems required by man. (NASA)

I3
Evidence from other worlds

In addition to the new view of earth history as seen through the interpretations of plate theory, a further addition to our knowledge has resulted from the exploration of the planets of the solar system since the 1960s. This has included landings by the Apollo astronauts on the moon (1969–72) and the unmanned Viking craft on Mars in 1976, as well as probes to the surface of Venus, and close photographic examination of Mercury and Jupiter. The information returned from these expeditions has broadened our horizons concerning the significance of the processes studied on the earth, and has provided an insight into conditions during periods of time for which there is no record on the earth.

The moon, Mars and Mercury have thin or no atmospheres and surfaces which have been intensively cratered by the impact of meteorites. Some sectors show rougher features than others: on the moon the 'highland' areas contrast with the flat, sea-like 'maria'. These features are the result of the older areas having been subjected to longer periods of impacting, whilst the smoother areas have been covered up by more recent floods of basalt lava. Dating the rocks brought back from the moon has resulted in rather surprising conclusions. The oldest rocks are well over 4000 million years old (cf. the oldest earth rocks are 3850 million years old), and it seems that a series of massive impacts at about that time caused the crustal rocks to crack and let out the molten lava to fill the maria. This process continued until 3000 million years ago, and even the numbers of impact craters have decreased over time: the moon has a 'fossil' landscape – preserved from the distant past. Mercury has a very similar landscape.

Mars also contains a record of such cratering – large craters filled with lava and a series of smaller, fresher crater forms – but in some areas this has been overlaid by other features. There are volcanic craters, some of which have grown into vast land-

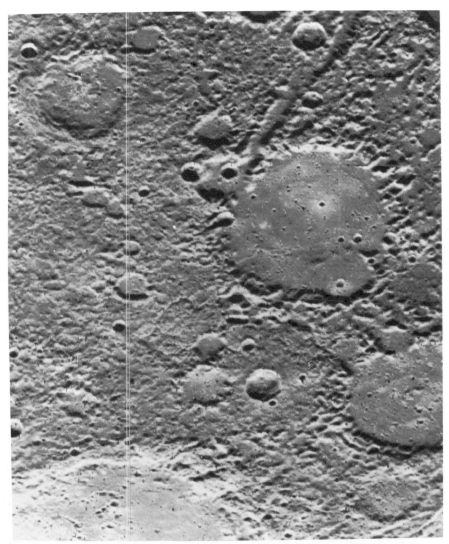

The surface of Mercury as seen from an altitude of 35 000 km (the picture is 180 km from top to bottom). The surface is dominated by impact craters—larger ones filled with lava, and a series of smaller ones with sharper, fresher edges. Work out the history of landscape formation. (NASA)

forms: Olympus Mons is 25 km high (more than twice the height of Mount Everest) and covers an area the size of England and Wales. It is the largest of several craters along what must have been a line of crustal weakness (this is suggested by the Capretes canyon features). Some interpret this as a crustal fracture related to the early stages of plate formation. Like the moon, however, the Martian surface must have become 'fossilised' at this stage. The lack of plate movements has meant that volcanic peaks have continued to build up in the same spot to form such massive edifices (cf. the Hawaiian islands). The canyon system near the Martian equator is also associated with stream-like channels, although all are dry now. This may also be a case of fossilised landforms resulting from a more humid period, perhaps as much as 3000 million years ago. It seems that any water which did exist on Mars has become frozen, and has stayed that way.

Venus is surrounded by a very dense atmosphere, which includes a high content of acids like sulphuric or hydrofluoric, both of which would weather surface rocks rapidly. The build-up of heat in the Venusian atmosphere has resulted in surface temperatures of around 500°C: just as any surface water on Mars has frozen, it has been evaporated on Venus.

Jupiter, the largest of the solar system planets, has also been photographed by passing spacecraft, and the latest interpretation of its significance is that it may be another sun in the making. It is quite different from the inner four planets, Mercury, Venus, earth and Mars, since it is radiating more energy than it receives from the sun.

The significance of these studies for geologists on the earth (the parallel study of the planets has been termed 'planetary geology') includes the following points:

(1) Application of geological principles evolved in relation to earth rocks can be made to the rocks of

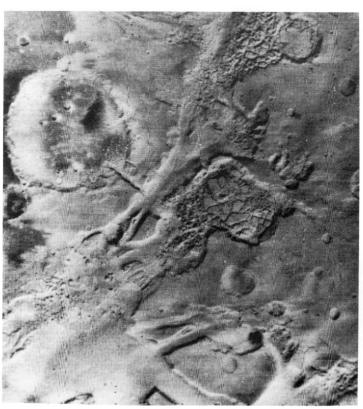

Mars: surface details of some craters and channels: the channels are up to 40 km wide and 150 km long, and were formed by sudden discharges, possibly of melted buried dry ice. (USGS)

A global mosaic of Mars, showing the North Pole ice cap (top) and the great volcano, Olympus Mons—600 km across and 25 km high (centre) (NASA)

other planets, which have been formed in a similarly orderly manner over long periods of time.

(2) Any record of the phases of massive impacting and lava eruption approximately 4000 million years ago have been lost on the earth, where the continual re-circulation of surface materials has preserved rocks only as old as 3850 million years. It is possible, however, that the earth went through such a phase, and that a study of the other inner planets therefore provides a record for the missing early stages in earth evolution.

(3) The lack of atmosphere, or the presence of a very thin atmosphere on the moon, Mercury and Mars, tends to preserve surface landforms with little alteration. There is little evidence from Venus, but many forms sensed by radar appear to be of low relief. The earth is unique in its combination of features and processes.

Implications for man and the earth environment

The visits to other planets also have broader lessons for man. None of those planets has an environment in which man could live without considerable support systems taken with him from the earth, and the raw materials available on them are extremely limited by comparison with those on the earth. It is only on earth, so far as we know, that the combined variety of geological processes has resulted in the storage, recycling and availability of so many materials vital for man's continued existence. The earth is bountiful and beautiful, but ultimately limited in its provision, and must be treated with respect. Soil erosion, desert extension, careless use of energy resources and pollution of air and water, could make the future less bright for our descendants.

A view of Venus from 720 000 km, taken by Mariner 10 in ultraviolet light to emphasise the swirls of cloud covering the planet. (NASA)

A magnified section of Moonrock: the banded crystals are of a common Earth rock-forming mineral, plagioclase feldspar. This is probably the oldest rock on the Moon (over 4000 million years old) and is known as 'Genesis' rock. (USGS)

Further Reading

M. J. Bradshaw (1977) *Earth, the living planet.* Hodder & Stoughton.

M. J. Bradshaw, A. J. Abbott, A. P. Gelsthorpe (1978) *The Earth's Changing Surface.* Hodder & Stoughton.

M. J. Bradshaw, E. A. Jarman (1969) *Reading geological maps.* Hodder & Stoughton.

M. J. Bradshaw, E. A. Jarman (1969) *Geological map exercises.* Hodder & Stoughton.

British Museum (Natural History). *British fossils* (3 volumes) H.M.S.O.

D. Brunsden, J. Doornkamp, Eds. (1974) *The unquiet landscape.* David & Charles.

N. Calder (1972) *Restless Earth.* BBC.

G. L. Herries Davies, N. Stephens (1978) *The geomorphology of Ireland.* Methuen.

P. Francis (1976) *Volcanoes.* Penguin.

I. G. Gass, P. J. Smith, R. C. Wilson, eds. (1972) *Understanding the Earth.* Open University.

A. Hallam (1973) *A revolution in the Earth Sciences.* Oxford.

W. R. Hamilton, A. R. Woolley, A. C. Bishop (1974) *The Hamlyn Guide to minerals, rocks and fossils.* Hamlyn.

C. A. M. King (1976) *The geomorphology of northern England.* Methuen.

F. W. Lane (1968) *The elements rage.* Sphere.

G. B. Oakeshott (1976) *Volcanoes and earthquakes.* McGraw-Hill.

F. Press, R. Siever (1974) *Earth.* Freeman.

F. Press, R. Siever (1974) *Planet Earth.* Freeman.

H. H. Head, J. Watson (1968) *Introduction to Geology, Volume 1 Principles.* Macmillan.

H. H. Read, J. Watson (1975) *Introduction to Geology, Volume 2 Earth history.* Macmillan.

J. B. Sissons (1976) *The geomorphology of Scotland.* Methuen.

P. J. Smith, Ed. (1975) *The politics of physical resources.* Penguin.

In addition, the following Open University courses will take the student further:

S100	Science (units 22–27)
S23	Geology
S2–2	Environment
S2–3	Geochemistry
S2–4	Geophysics
S26	The Earth's physical resources.

The United States Geological Survey publishes a number of Circulars, which are available free from: 1200 South Eads Street, Arlington, Virginia 22202, U.S.A. These cover topics from hydrology to glaciology and earthquakes.

Index